GERMINAL LIFE

'*Germinal Life* carries forward Deleuze and Guattari's project of a philosophy of vitality, in creative dialogue with the latest developments in the sciences of life and theories of technology. An innovative mediation at the boundary between philosophy, science, and cultural studies.'

Brian Massumi, *Australian National University*

'*Germinal Life* is a brilliant – and critical – reworking of Deleuze, mapping a new tradition of biophilosophy. With this follow-up to *Viroid Life*, Keith Ansell Pearson has established himself as one of Britain's most exciting young philosophers.'

David Wood, *Vanderbilt University*

'This is a very important book on a topic that is going to be of increasing importance over the next several years. It addresses the overlapping audiences of continental philosophy, cultural studies, critical theory and science studies to provoke a rethinking of the boundaries dividing life from non-life, the human and the post-human (super and sub-human).'

Elizabeth Grosz, *SUNY at Buffalo, USA*

'Keith Ansell Pearson approaches the work of Gilles Deleuze with the precision of a diamond cutter, exposing brilliant facets of thought "beyond the human condition".'

Stanley Shostak, *University of Pittsburgh*

Germinal Life: The Difference and Repetition of Deleuze is the highly successful sequel to *Viroid Life: Perspectives on Nietzsche and the Transhuman Condition*. Where *Viroid Life* provided a compelling reading of Nietzsche's philosophy of the human, *Germinal Life* is a highly original and incisive study of the biophilosophical aspects of Deleuze's thought. In particular, Keith Ansell Pearson provides fresh and insightful readings of Deleuze's work on Bergson and skillfully shows how Bergsonism is the crucial factor in any encounter with Deleuze's philosophy of life.

Germinal Life also provides new insights into Deleuze's most famous texts, *Difference and Repetition* and *A Thousand Plateaus*. Ansell Pearson investigates Deleuze's relation to some of the most original thinkers of modernity, from Darwin to Freud and Nietzsche, and explores the connections between Deleuze and more recent thinkers such as Adorno and Merleau-Ponty. Ansell Pearson also offers imaginative readings of works of literature such as Hardy's *Tess*. Concluding with reflections on the figuration of the 'fold' and 'superfold' in Deleuze, *Germinal Life* confronts what ultimately it might mean to think 'beyond' the human.

Keith Ansell Pearson is Professor of Philosophy at the University of Warwick. He is the author of *Viroid Life* and editor of *Deleuze and Philosophy*, also published by Routledge.

GERMINAL LIFE

The difference and repetition of Deleuze

Keith Ansell Pearson

Routledge
Taylor & Francis Group

LONDON AND NEW YORK

First published 1999
by Routledge
2 Park Square, Milton Park, Abingdon, Oxon, OX14 4RN

Simultaneously published in the USA and Canada
by Routledge
605 Third Avenue, New York, NY 10017

Routledge is an imprint of the Taylor & Francis Group, an informa business

Typeset in Times by Routledge

British Library Cataloguing in Publication Data
A catalogue record for this book is available from the British Library

Library of Congress Cataloging in Publication Data
Ansell Pearson, Keith, 1960–
Germinal Life: The difference and repetition of Deleuze / Keith Ansell Pearson.
p. cm.
Includes bibliographical references and index.
1. Deleuze, Gilles. 2. Bergson, Henri, 1859–1941 – Influence.
I. Title.
B2430.D454A57 2000   98–41197
194–dc21   CIP

ISBN 13: 978-0-415-18351-2 (pbk)
ISBN 13: 978-0-415-18350-5 (hbk)

FOR JASMINE AND RICHARD

Immanence can be said to be the burning issue of all philosophy because it takes on all the dangers that philosophy must confront, all the condemnations, persecutions, and repudiations that it undergoes.
(G. Deleuze and F. Guattari, *What is Philosophy?*, 1991)

This is an ancient and eternal story: what formerly happened with the Stoics still happens today, as soon as any philosophy begins to believe in itself. It always creates the world in its own image; it cannot do otherwise. Philosophy is the tyrannical drive itself, the most spiritual will to power, to the 'creation of the world', to the *causa prima*.
(F. Nietzsche, *Beyond Good and Evil*, 1886)

CONTENTS

CONTENTS

ACKNOWLEDGEMENTS

For helping this book to move along and germinate, I am indebted to the following people: Gregory Adamson, Justine Baillie, Richard Beardsworth, Catherine Dale, Robyn Ferrell, Elisabeth Grosz, Brian Massumi, John Mullarkey, John Protevi, Robert Spaven, James Williams, and Tony Bruce at Routledge. For their general support and encouragement over the past few years, as well as the inspiration provided by their own work, I extend my thanks to colleagues and graduates at Warwick and to the following: Eric Alliez, Gary Banham, Sebastian Barker, Constantin Boundas, Howard Caygill, Daniel W. Conway, David E. Cooper, Simon Critchley, Michael Dillon, Diane Morgan, David Musselwhite, David Owen, Liam O'Sullivan, Noel O'Sullivan, and David Wood.

A NOTE ON TRANSLATIONS

Readers should note that in many instances translations from the French and German have been modified (on most occasions only slightly) and this is indicated in the text only now and again. Full details of my references and sources can be found in the extensive bibliography provided at the end of the book. I am grateful to Melissa McMahon for allowing me to freely use her unpublished translation of Deleuze's 1956 essay on Bergson and difference, a new version of which is to be published in John Mullarkey's collection of essays *The New Bergson* (Manchester University Press, forthcoming).

INTRODUCTION

Repeating the difference of Deleuze

> In no small measure, the intellectual lineage is straight from
> Weismann to today.
>
> (Kauffman 1995: 274)

> Put bluntly, closed systems are bound to be finished.
>
> (Adorno 1966: 35; 1973: 27)

I

The aim of this book is to illuminate the character of Deleuze's philosophy
by situating it in the context of a neglected modern tradition, that of
modern biophilosophy, which runs from Darwin and Weismann through to
Bergson and Freud, and which also encompasses the work of a diverse and
little-known group of thinkers such as Raymond Ruyer, Gilbert Simondon,
and Jacob von Uexküll. The fact that this trajectory of thought going back
to the neo-Darwinism of Weismann, and exerting a decisive influence on the
thought of Deleuze, has been so neglected might explain why to date there
has been so little in the way of an incisive philosophical encounter with
Deleuze's work. Deleuze is difficult to place in the philosophical discourse of
modernity largely, I suspect, because of the peculiar character of his
philosophical thought, with its investments in biology and ethology.

Deleuze's turn to a conception of difference in the 1950s entailed a highly
distinctive and novel Bergsonism since to write about Bergson at this time,
and from the perspective of a concern with 'difference', was not a task that
would have been either fashionable or predictable. Bergson has been an
unduly neglected figure within recent continental philosophy. It is part of the
brilliance of Deleuze's readings to show the vital importance and continuing
relevance of his great texts on time, creative evolution, and memory, for the
staging of philosophical problems. I believe that the character of Deleuze's
'Bergsonism' has been little understood, and yet I want to show that it plays
the crucial role in the unfolding of his philosophy as a philosophy of
'germinal life'. In this study the focus is on Bergson's conception of 'creative

1

evolution' and on the way in which an encounter with it can be shown to be of crucial importance for any attempt to comprehend and work through some of the central problems of philosophic modernity. It is through Bergsonism that Deleuze seeks to re-invent this modernity and to articulate a radical project for philosophy. Philosophy is a highly autonomous practice for Deleuze and the distinctive task he prescribes for it, from his early work on Bergson to his last major work, co-authored with Guattari, *What is Philosophy?*(1991), amounts to the complex and paradoxical one of thinking 'beyond' the human condition. In the course of this inquiry we shall have the chance to track the differing stresses Deleuze gives to the meaning of this 'beyond' as his work unfolds and develops. The critical question to ask, and which I simply pose here, is this: does thinking beyond the human condition serve to expand the horizons by which we think that condition and so deepen its possible experience, or is the 'change of concept', in regard to the overhuman, so dramatic that it requires the dissolution of the human form and the end of 'the human condition'? Such a question takes us, I believe, to the heart of Deleuze's project and brings us into a confrontation with its peculiar challenge, as well as its most innovative and demanding aspects.

Each one of the three chapters in this book offers a treatment of what it means to think beyond the human condition. They are united by the attempt to examine, and to subject to critical but informed scrutiny, the character of Deleuze's 'Bergsonism'. In the first chapter I examine Deleuze's early engagement with Bergson in the major study he wrote in 1966. I also utilize the neglected and largely unknown 1956 essay on Bergson. The focus is on Bergson's notion of duration and how this notion informs his conception of a creative evolution, especially how philosophy is able, in spite of the natural bent of the human intellect, which produces a mechanistic and spatial account of the real, to think the character of this evolution in both speculative and vital terms. In seeking to think 'beyond' the human condition the task of this philosophy of creative evolution can be shown to be an 'ethical' one, concerned with opening up the human experience to a field of alterity. In the second chapter my attention shifts to Deleuze's first major attempt at 'independent' philosophy in *Difference and Repetition* (1968), in which he no longer relied on the history of philosophy but sought to articulate, still utilizing the resources of the tradition from Duns Scotus through to Heidegger and Nietzsche, a specifically modern project of thinking difference 'and' repetition. He produced in the process a unique 'schizo-scholasticism'. My attention is focused once again on the biophilosophical aspects of the work, which obviously means that key aspects of this immensely fertile and complex text are neglected. Nevertheless, in adopting such a concentrated focus I hope to yield novel and incisive insights into Deleuze's thinking of difference and repetition, especially in relation to a number of the key theoretical figures of the modern period such as Darwin, Freud, and Nietzsche. The reasons informing this particular selection of thinkers will become clearer later in the Introduction. The reading of *Difference and Repetition* is supplemented, in terms of the task of bringing out the ethics of the project, by a reading of the notion of the 'event' that is more explicitly articulated in its sister work *The Logic of Sense* (1969).

In Chapter 3 I move forward some considerable distance in the trajectory of Deleuze's work to a treatment of the essential biophilosophical dimensions of the text he co-produced with Guattari in 1980, *A Thousand Plateaus*. I examine key aspects of this work in terms of its engagement with modern evolutionary theory and modern ethology, aiming to demonstrate the innovations made in the text with regard to a 'machinic' approach to questions of 'evolution' and to an ethology which focuses not on behaviour but on assemblages. I also pay special attention to the configuration of 'Bergsonism' in this work, in particular the 'meaning' of its 'Memories of a Bergsonian'. Deleuze now approaches 'creative evolution' in terms of 'becomings' that are held to be peculiar to modes of creative *involution* (a move anticipated, in part, in *Difference and Repetition*: 'only the involuted evolves'). It is in the context of these anomalous becomings that one can, I believe, best demonstrate the experimental character, and illuminate the tensions, of an ethological ethics in respect of 'nonhuman' becomings of the human. This chapter concludes with some speculations on Deleuze's last work, co-authored with Guattari, *What is Philosophy?* (1991), especially in regard to its construction of philosophy as a form of 'absolute deterritorialization' and its invocation of a new earth and a new people.

In the conclusion I take a look at Deleuze's return to the question of the fold as it is articulated in two texts of the second half of the 1980s, the studies of Leibniz (1988) and Foucault (1986), and relate the movements of thought taking place in them to the fundamental aim of Deleuze's philosophical project to think, with the aid of a Spinozism and a Bergsonism, beyond the human condition.

Each chapter concludes with an appraisal of the moves made by Deleuze with regard to thinking 'beyond' the human condition. In order to open out Deleuze's texts to a philosophical encounter I have found it helpful to bring his thinking into confrontation with other major contemporary figures, notably Merleau-Ponty in Chapter 1 and Adorno in the denouement to Chapter 3. What is at stake in this demanding and difficult attempt to develop an *ethics* of germinal life and to produce a philosophy of the Event – whether it rests, as some have contended, on a disavowal of the human condition, or whether it succeeds in showing that a radical philosophy must necessarily think trans- or overhumanly – will, I hope, unfold dramatically in all its implication and complication in the attentive explication that follows. Although the reading of Deleuze that is cultivated here is *peculiar* to myself, and does not pretend to either define or exhaust the 'meaning' of *his* event once and for all, it is a reading I hope that will serve to provoke and challenge, in all sorts of unpredictable and incalculable ways, those who have a concern with his work and its legacy. This book amounts to an affirmation of Deleuze's event, but it is not the affirmation of a braying-ass.

II

Let me now say something at once more general and more specific about this book. There is a real danger in the recent upsurge of interest in his texts that studies will proliferate which address and construe Deleuze as a philosopher of life – as a vitalist thinker – in an all too casual and cavalier manner without any serious comprehension of, or probing insight into, the biophilosophical dimensions of his project. It is the aim of this study to counter this tendency. The biophilosophical aspects of Deleuze's thought have to be taken seriously, contextualized, and the stakes of his working out of a philosophy of germinal life need to be carefully unfolded. It would be inadequate, however, to restrict Deleuze's project to the merely biological and to claim it solely or exclusively for a novel 'philosophical biology'. To understand why this is so will require navigating intricate and interweaving lines of thought.

Deleuze conceived a thinking of difference and repetition as historically specific to capitalist modernity. The philosophy of difference emerges at that 'moment' in history when the most stereotypical and mechanical repetitions appear to have taken over the forces of life completely and subjected it to a law of entropy. It is this, which motivates his engagement with biology, with ethology, with ethics, and with literature, as he seeks to articulate a critical modernity that exposes a series of the transcendental illusions encompassing both scientific and philosophical thought. These illusions concern the nature of time, consciousness, death, subjectivity, and so on, and are manifest in our models of capital and of entropy, to give two of the most important examples. The critical questions I pose of Deleuze's philosophy concern his attempt to think of Being as immanence and in terms of the 'event'. We need to determine how Deleuze envisages the 'overcoming' of nihilism through the praxis of a critical modernity. Before returning to this question I want to provide some insight into the theoretical context which will, I believe, enable us to stage an instructive and novel encounter with Deleuze.

There is, I believe, a quite specific intellectual context within which to illuminate Deleuze's work and its engagement with biophilosophy, and this is the tradition of neo-Darwinism that stems from the revolutionary work of August Weismann carried out at the end of the nineteenth century. The idea of 'germinal life' pursued in this book resonates with a number of sources and thinkers. In addition to Deleuze these include thinkers such as Bergson and Freud, and novelists such as D. H. Lawrence, Thomas Hardy, and Emile Zola. The key figure in this lineage is Weismann (1834–1914), the founding figure in the emergence of modern neo-Darwinism. An engagement with

Weismann runs throughout Deleuze's writings, an appreciation of which provides, I want to demonstrate, valuable insights into the character of his philosophy of germinal life.

Weismann is an immensely complicated figure whose work combines elements of nineteenth-century biology that were to be discredited in the twentieth century, such as the recapitulation thesis (to be encountered in Chapter 2), as well as aspects of the new science of genetics that were to prove so seminal in terms of the consolidation of neo-Darwinism with the modern synthesis of the 1930s and the discoveries of molecular biology in the 1950s. He wanted a scientifically accurate account of heredity and began by questioning Darwin's confused theory of pangenesis (the idea that every cell of the body contributes minute particles, the 'gemmules', to the germ cells and so participates in the transmission of acquired characteristics). He then made an assault on Lamarck's account of evolution, which relied on the doctrine of the inheritance of acquired characteristics and speculation regarding the use and disuse of organs. The effect of his revolutionary work in biology was to sever Darwinism from its entanglement in Lamarckian dogma, so making way for the establishment of a strictly mechanistic and nonvitalist theory of evolution by placing all the emphasis on natural selection as the blind machine that guarantees the reproduction of life from generation to generation in terms of an unbroken descent. For Weismann life is able to replicate and reproduce itself owing to the powers of a special hereditary substance, the germ plasm (what today is called DNA), which controls and programmes in advance, and without the intervention of external factors, the development of the parts of an organism and which gets transmitted from one generation to the next in a continuous passage of descent. The germ cells differ in their function and structure from the somatic cells, and making this distinction between the two led Weismann to introducing his famous 'barrier' by which changes in the phenotype can have no effect on the genotype. Weismann is insistent that the hereditary substance 'can never be formed anew', but 'can only grow, multiply, and be transmitted from one generation to another' (Weismann 1893: xiii). His thesis was updated in the 1950s by work in molecular biology that sought to demonstrate that no information in the properties of somatic proteins can be transferred to the nucleic acids of DNA.

Weismann's work grew out of developments in cytology in the mid-nineteenth century. The idea that organisms are made of cells dates from the 1830s, while in the 1870s the cell nucleus was revealed for the first time by enhanced microscopic power and, a short while after, the chromosomes in that nucleus were demonstrated to be the birthplace of new cells. In the early 1880s the crucial discovery was made that the sex cells divide differently from the cells that make up the rest of the body. Weismann interpreted this 'difference' in terms of a division of labour between germ and soma plasms. The germ-cells are restricted to just the one task and function, that of making new organisms through the intermingling of inherited information from the parent cells. Weismann obviously did not know that *all* cells, including somatic ones, contain the total complement of inherited information. However, as Depew and Weber point out, ignorance of this fact was not crucial since Weismann was able to argue that the sequestering of the germ line of egg and sperm occurs so early in ontogeny (the sequence of cellular divisions) that any developments in the somatic cells can have no effect on the process (rats deprived of tails do not go on to breed tailless rats) (Depew and Weber 1996: 189).

Weismann did not break entirely with the developmentalist tradition and still subscribed to its central thesis on recapitulation regarding the relation between ontogeny and phylogeny. Where the break begins to emerge is in the stress placed on the 'immortality' of the germ line (Weismann 1893: 183–92), which serves to sever the Darwinian link with the tradition of epigenesis that since Aristotle had treated reproduction and growth as phases of one single process of development. Depew and Weber see Weismann's germ line, in fact, as presaging the revival of the old idea of preformationism that takes place with Mendel's genetics, and, indeed, in the preface to the English translation of his work, Weismann admits that having sought an epigenetic, and not an evolutionary, approach, he eventually became convinced of the impossibility of an epigenetic account of the organism (xiii–xiv). Weismann anticipates twentieth-century Darwinism in conceiving adaptedness on the level of changing proportions of the cells that inform a population, while evolutionary novelty is the result of internal change in the germ cells. In today's language all mutations are seen to arise from changes in the sequences of DNA (Depew and Weber 1996: 189–90). In addition, his later work on 'germinal selection', which aimed to show the possibility of conflict between *levels* of selection, anticipates contemporary accounts of genic selectionism, such as the doctrine of the selfish gene. The field of competition is seen to take place on the cellular level with germinal cells battling it out with each other 'for nourishment in an intracellular Malthusian world', so that what is good for the egg and sperm might not be good for the organism (191).

Weismann's biology of the germ-plasm is a biophilosophy of the species not of individuals. The germ line represents the skeleton of the species on which individuals get attached as excrescences. Any changes that are the result of outside influences are merely temporary and disappear from evolution when the individual reaches its end. External events are no more than transient episodes that affect particular life-forms but not the species, which goes on regardless of changes to individuals. The germ line is thus, it is alleged in Weismann's account, guaranteed the reproduction of identical cells. Although germ cells vary from species to species, new structures required by evolution are produced not by individuals, but by the hereditary arrangements contained in the germ cells. So while natural selection *appears* to be operating on the aspects of an adult organism, in actuality it is working only on the predispositions lying concealed in the germ cell. The germ line, therefore, is outside the reach of any variation that takes place in individuals of the species (for further insight see Weismann 1882, Volume II, 634ff.; Jacob 1974: 216–17; see also Ruyer 1946: 138ff.).

Weismann's intention was to account for the transformation of forms of life with the 'sole aid of Darwinian principles', denying both the existence of an 'internal developmental power' and the assumption of a 'phyletic vital force' (Weismann 1882, Volume II, 634–6). But this adherence to mechanistic explanation does not rule out for Weismann the admission of teleology: 'Mechanism and teleology do not exclude one another....Without teleology there would be no mechanism, but only a confusion of crude forces; and without mechanism there would be no teleology, for how could the latter otherwise effect its purpose?' (716). However, Weismann insists that only natural selection is able to account for the production of the purposive:

> The principle of selection solved the riddle as to how what was purposive could conceivably be brought about without the intervention of a directing power, the riddle which animate nature presents to our intelligence at every turn, and in the face of which the mind of a Kant could find no way out....The selection theory...enables us to understand that there is a continual production of what is non-purposive as well as of what is purposive, but the purposive alone survives, while the non-purposive perishes in the very act of arising.
>
> (Weismann 1909: 21)

In other words, the survival of the purposive is to be explained through the existence of an 'intrinsic connection' between the conditions of life and the structural adaptations of an organism, with the adaptations not determined by the organism itself but rather 'called forth by the conditions'. Weismann's reworking of mechanism and teleology is interesting since it shows that Darwinism does not so much jettison a notion of teleology, as commonly supposed, but rather secularizes it.

Georges Canguilhem argued that the exclusion of teleology, as classically conceived, from Darwinian theory does not mean that its conception of life excludes all value-laden terms. 'Success' in life is configured in terms of 'survival' (especially of the fittest). Conceived in this context, he notes, it is difficult not to think that some 'vital' meaning is being attached to the stress within Darwinism on adaptation, 'a meaning determined by comparison of the living with the dead' (Canguilhem 1994: 211). The crucial point of the theory is to show that variations in nature – for Darwin these are deviations in structure or instinct – remain without significance or effect without the mechanism of selection. This introduces into biology, Canguilhem contends, a new criterion of 'normality' based on the living creature's relation to life and death. It is, therefore, not surprising that from the start Darwinism was taken up normatively as both lending support to a moral theory and to a social theory (giving rise to both an ethical Darwinism and a social Darwinism). However, although Darwin did sever the notion of adaptation from any idea of preordained purpose, he did not completely divorce it from a notion of normality. Rather, the norm is no longer tied to a fixed rule but only to a transitive capacity, namely survival and the passing of successful descendants through the struggle for existence. Normality, therefore, is not a property of the living thing in terms of some given or fixed essence, but 'an aspect of the all-encompassing relation between life and death as it affects the individual life form at a given point in time' (212).

I am sure that the work of his former teacher on the normal and the pathological exerted a powerful and lasting influence on Deleuze's thinking. Canguilhem's revaluation of our categories of the normal and the pathological or deviant, of the healthy and the diseased, finds all sorts of echoes in Deleuze's work, from the emphasis in *The Logic of Sense* on the 'crack', which plays a crucial role in Deleuze's reworking of Nietzsche's notion of great health (the health that has incorporated sickness), to the emphasis in *A Thousand Plateaus* on symbiotic complexes, including monstrous couplings and unnatural participations, as the source of real innovation in evolution. It is these kinds of emphases, which mediate Deleuze's reception of Weismann. Weismann's neo-Darwinism is always in danger, whether in its initial formulation as the germ-plasm or the more recent account of DNA, of treating the matter of evolution as a closed genetic system. Although the various biophilosophical engagements which characterize Deleuze's work, such as the concern with creative evolution, the thinking of difference and repetition, and the move towards creative ethology, are highly different in their scope and focus, what motivates all these thought-experiments is a concern with the character of open systems. Indeed, the plane of immanence, which is used to explain the transversal movements of material forces and affects, is presented by Deleuze as the open system *par excellence*.

It is interesting to note that in each case where we can identify an encounter with Weismann in Deleuze we find a favourable reception and a productive reworking. This reworking takes place in three notable places: *Difference and Repetition*, where he reads him in terms of Darwin's revolution and construes his addition to Darwin's doctrine as a further contribution to a biophilosophy of difference on the level of sexed reproduction; in one of the appendices to *The Logic of Sense* where Deleuze provides an 'epic' reading of Zola's novel *La Bête Humaine* in terms of a reworking of Freud's death-drive (which, in turn and in part, is a reworking of Weismann); and, perhaps most crucial of all, in *A Thousand Plateaus*, where Weismann's germ-plasm is transformed into the 'body without organs' which becomes the site of 'intense germen'. In fact, this latter reading of 1980 is strikingly similar to the one we find in the encounter with Zola and

Freud in *The Logic of Sense* of 1969. In both cases Deleuze's aim is to show that the question of heredity is not simply one that is *given*, either by the species or by the continuity of the germ-plasm. Rather, heredity becomes transfigured, and is made vital, through the becoming of the new individual and through a 'law of life' (Nietzsche 1994: Essay III, Section 27) that goes beyond laws of genealogy and filiation. The egg, Deleuze will argue in *A Thousand Plateaus*, is an egg of germinal intensity that does not simply denote a fixed moment of birth or a determinate place of origin. Natality is always inseparable from processes of decoding and deterritorialization (this is the very 'meaning' of germinal life, whether that life be 'difference and repetition', the 'event' of the crack, or anomalous 'animal-becomings'). Once this has been appreciated the way is now opened for a conception of 'creative evolution' that has to do with 'involutions' and with communications that cut across distinct lineages, and so allows for the possibility of an 'ethological' ethics.

There are crucial moves made in Deleuze's biophilosophical thinking, therefore, that are neither consistent nor consonant with the Weismannian tradition. Deleuze is keen to avoid, for example, a purely geneticist account of evolution as well as a DNA mythology. All of this serves to make the character of his biophilosophy very different. Deleuze, of course, never systematically works through the character of his biophilosophy in any of his texts. He never addresses, for example, the tensions that might generate from so freely drawing on different strands of biophilosophical thought. For this reason this study of Deleuze is necessarily an *invention* of his biophilosophy, laying out the centrality of the moment of Weismann, but equally seeking to show that all the different resources provided by modern biology are utilized by Deleuze in terms of his principal philosophical references, notably Bergson, Nietzsche, and Spinoza. As we shall see, Deleuze's conception of biophilosophy is, ultimately, and first and foremost, 'ethical' in the quite specific sense that he reads the likes of Spinoza and Bergson. The crucial reworking of Weismann which Deleuze and Guattari carry out in *A Thousand Plateaus*, in terms of seeking to address the question of how the organism can make itself into a body without organs, will make very little sense unless this point is appreciated.

The fact that the index of the English translation of *Mille Plateaux* contains no reference to Weismann, in spite of the fact that it is the doctrine of the germ-plasm which provides the inspiration for the notion of the body without organs (Weismann is explicitly named in the text), indicates, I think, the extent to which this crucial thinker has been neglected and overlooked both in terms of his importance for understanding Deleuze's work and for understanding large chunks of intellectual modernity. Weismann's neo-Darwinian revolution was not without cultural impact at the time of his own writing. In positing the germ-plasm theory of heredity in terms of an unbroken descent and a fixed channel of communication Weismann laid down a challenge to a whole generation of writers and thinkers, including the likes of Hardy, Lawrence, Bergson, and Freud. This challenge amounted to nothing less than the challenge of a biological nihilism. The genealogy of life as set up by Weismann means that the substance of life is immortal, not subject to the influences or effects of individual lives and bodies, and so it assumes the appearance of a tremendous inhuman force. Such a view is still expressed today in biology, with its most articulate exponent being Richard Dawkins and his well-known theory of the selfish gene. The nihilism of the message of this theory is quite explicit: 'The universe we observe has precisely the properties we should expect if there is, at bottom, no design, no purpose, no evil and no good, nothing but blind, pitiless indifference...DNA neither knows nor cares. DNA just is. And we dance to its music' (Dawkins 1995: 133).

Deleuze works and reworks a conception of 'evolution' in important ways. In the material on Bergson of 1956 and 1966 attention is focused on the 'creative' and 'virtual' dimensions of evolution. In the work of the late 1960s, notably, *Difference and Repetition*, his attention shifts to 'complex systems' that 'evolve' in terms of an interiorization of their components and constitutive differences (at this point in his writings Deleuze places the word evolution in scare quotes). In his mature work with Guattari, notably *A Thousand Plateaus* (1980), a 'rhizomatics' comes to the fore that fundamentally breaks with genealogical and filiative models of evolution, to the extent that Deleuze is no longer dealing with 'evolution' as a problem of heredity. This move is, in fact, already prefigured in the earlier work with Guattari, *Anti-Oedipus* (1972). From the point of view of 'community' (*la communauté*), they argue in this work, evolution is always disjunctive simply because the cycle is marked by disjunctions. This means that generation is secondary in relation to the cycle and also that the 'transmission' – of genes, for example – is secondary in relation to information and communication. Deleuze and Guattari argue that the 'genetic revolution' of modern biology consists in the discovery that, strictly speaking, there is not a transmission of flows but rather the 'communication of a code or an axiomatic' which informs the flows. To give primacy to the phenomenon of communication in this way is, ultimately, to push into the background the problem of hereditary transmission. The boldness of Deleuze's move consists in extending this insight into the phenomenon of communication to the social field of desire (see Deleuze and Guattari 1972: 328; 1984: 276).

Deleuze's significance as a thinker of creative evolution lies in the fact that he responds to the challenge presented by Weismann (and by other thinkers of life and death, such as Nietzsche, Freud, and Bergson) by aiming to demonstrate the immanent movement beyond nihilism. Deleuze is a philosopher of the *crack*, of the cracks of life and of modes of communication that allow for novel becomings and transformations, so escaping the grim law of life implicit in Weismann's theory of the germ-plasm that would condemn 'individual' life to the eternal return of a nihilistic fate and that would dissipate the forces of the outside and minimize their influence. The aim in this study of the germinal life is to work through and navigate a way beyond two nihilisms of modernity, the potential nihilism of Weismann's germ-plasmic finality and the perceived nihilism of Freud's death-drive.

III

I now want to say something in advance about the *ethics* of this biophilosophical project, which is a concern in each of the chapters which make up the present study. A notion of ethics has to be seen not as an incidental element of Deleuze's project but as one of its most fundamental and essential elements. Deleuze is, in fact, compelled by the very adventure of thought to think ethically and even to think an ethics of matter itself. In his work we find a number of conceptions of this ethics, including an ethics of the eternal return (his book of 1962 on Nietzsche and *Difference and Repetition*, 1968), an ethics of the event (*The Logic of Sense, What is Philosophy?*), an ethics of affective bodies and an ethological ethics (the two books on Spinoza of 1968 and 1981 (first published in 1970 in a shorter version), and *A Thousand Plateaus*, 1980), and so on. These can all be shown to be differing articulations, however, of one and the same ethics of Being as Deleuze conceives it (univocal, the plane of immanence, germinal life, etc.). In this part of the Introduction I wish to speak in general terms about the figuration of ethics in Deleuze by discussing its treatment in his big book on Spinoza and expressionism of 1968 since, although this text, with its novel reading of Spinoza, does not figure in any substantial way in this study, it demonstrates in clear and powerful ways how for Deleuze we should approach the question of 'ethics'. What Deleuze has to say in this work on the question of ethics does resonate in crucial ways with his Bergsonism. We should perhaps note that Deleuze's reading of Spinoza is often inspired by his Bergsonism, so that, for example, we find in the 1968 text that the task of life is defined as one of learning to exist 'in duration' (Deleuze 1968: 289; 1992: 310). And indeed, for Bergson, such a task was supremely ethical in putting us back in contact with the 'eternity of life' that is neither the eternity of immutability nor the eternity of immortality (Bergson 1965: 156–7; compare Deleuze 1968: 292; 1992: 314). In addition, and as Merleau-Ponty astutely noted, Bergson's philosophy is a philosophy of pure 'expression' (Merleau-Ponty 1988: 28).

This is *not* to say that Spinoza and Bergson's thinking on duration are one and the same since clearly they are not. Duration (*duratio* not *tempus*) belongs for Spinoza to a quite specific realm of existence, namely, the domain of finite modes where it refers to the individuality of distinct things (Deleuze takes the line that there is no 'instantaneity of essence' in Spinoza and that the 'continual variations of existence' that characterize a mode's power of acting and its constant passages to greater and lesser perfections are only comprehensible in terms of duration; see Deleuze 1981: 57; 1988: 38–9). *Duratio* is to be understood modally rather than temporally; unlike infinite substance, a thing's existence does not follow from its essence but is dependent on external causalities for its endurance (see Yovel 1992: 110).[1] Bergson's conception of duration is radically different in that it refers not to the realm of distinct entities and things but rather to the virtual realm of creative processes and becomings. Some commentators, including Deleuze at one point in his 'Bergsonism' of 1966, will often casually read Bergson's duration as giving expression to a kind of *natura naturans*. Bergson, however, separates his thinking from Spinoza's conception of substance on this very issue, arguing that Spinoza's causalism and determinism are unable to allow a genuinely inventive character to be given to duration. I shall have something more to say on this in Chapter 1.

In his readings of both Bergson and Spinoza, Deleuze focuses the ethical question on bodies conceived as existing immanently on a plane of nature and constituted by an originary technics and artifice. This means for him that what a body can do is never something fixed and determined but is always implicated in a 'creative evolution'. His 'Bergsonism' follows Bergson's conception of evolution in conceiving life in terms of the play between two creative dimensions, that of nonorganic life and that of the organism. One of the major insights of Bergson's *Creative Evolution* is that the unity of nature consists of a complicated unfolding of an originary impulsion in which the creative energies of life are canalized in specific bodies (organisms and species). However, species are never the *telos* of evolution in Bergson's view, since the process of a creative evolution is 'without end' and its creativity implies ceaseless invention and re-invention. The ethical question addressed to bodies is one of gaining self-knowledge concerning their dynamic and 'evolutionary' conditions of existence in order to cultivate both joyful passions and enhanced relations with other bodies. This ethological dimension of ethics, which concerns the relations between affective bodies, operates both within the order of nature and also informs the ethical becoming of human bodies, to the extent that the later Deleuze explores the possibility of a becoming-animal and becoming-molecular of the human. As we shall see, this move is not unproblematic and without tremendous difficulties. In this Introduction, however, I want to restrict my attention to the earlier text on Spinoza of 1968 since in this work it is clear for Deleuze that the task of philosophy is not one of constructing a philosophy of nature, but rather one of showing how the acquisition of a 'superior' human nature is possible from a comprehension of the nature of bodies.

In *Expressionism in Philosophy* Deleuze follows Spinoza in conceiving bodies as 'finite modes' that are expressions of an infinite substance (in later work Deleuze will insist that both substance and modes presuppose a plane of immanence). The 'great ethical question' concerns whether it is possible for these bodies to attain 'active affections', and, if so, how (Deleuze 1968: 199; 1992: 219). Later in the text another dimension gets added to the ethical question which concerns how a maximum of joyful passions is to be achieved (ibid.: 225; 246).[2] One of the most important moves Deleuze locates in Spinoza's *Ethics* is that of reconciling a physical view of bodies with an ethical one. However, this requires a philosophy that is able to demonstrate that bodies are not fixed (they have no 'essence' other than that of a becoming) in terms of their exercise of active and passive affections. This is not to claim that there are no limits in nature to what bodies can achieve; rather, the claim is that these limits are 'general' ones always subject to a creative evolution and involution. Hence the 'ethical' character of Spinoza's thinking derives from his emphasis on the fact that we do not know what a body is capable of and what affections it can attain (what bodies can do always necessarily exceeds our knowledge at any given time, just as the capacities of thought always exceed the nature of consciousness).[3] The implications of this insight for thinking are twofold: that of acquiring a higher human nature through an adequate comprehension of nature, and that of raising a physics of nature to a higher plane (a *meta*-physics) by showing that bodies are capable of a potentially infinite becoming and modulation within finite limits. For Spinoza no one can tell what the body is capable of from simply observing the laws of nature (Spinoza 1955: Part III, Proposition II, Note). If the question of what bodies can become is an open question, there can only be empirical grounds, not logical ones, for closing it (Hampshire 1981: 133). Deleuze's reading of Spinoza on this issue is novel in that he does not simply restrict it to a matter of our ignorance, a deficiency in our knowledge, but also views it in terms of the experimental and open-ended character of a future affective 'evolution'.

In all the major modern thinkers who most inspire him, such as Spinoza, Nietzsche, and Bergson, Deleuze finds a similar stress and value placed on *experimentation*. Contra Leibniz's criticism, he seeks to show in the case of Spinoza that an ethics does not result in the impotence of the creatures who found themselves in nature conceived as an immanent and infinite substance. Deleuze is insistent that through the modes Spinoza shows how individuals participate in 'God's power' as singular parts (intensive quantities and irreducible degrees) of a divine power. In fact, he shows how both Leibniz and Spinoza produce a new naturalism that resists the mechanistic view of the universe promulgated by the great modern Descartes which only succeeds in devaluing the autonomy of nature by depriving it of any virtuality or potentiality as an immanent power. Both Leibniz and Spinoza manage to do this, Deleuze contends, without falling back into a pagan vision of the world which would simply produce a new and blind idolatry of nature (1968: 207–8; 1992: 227–8). It is important to appreciate, Deleuze notes, that nature is not constructed for our convenience, it is full of cruelty and lack of sympathy for our peculiar being and evolves without regard for our particular habitat in it. There are few deaths in nature that are not brutal, violent, and fortuitous. The task, however, is to comprehend this and to bestow upon such deaths a new meaning, which is the 'meaning' of a praxis that can only arise out of creating, and experimenting with, new possibilities of existence. In addition, Deleuze will claim, in his readings of Spinoza, Nietzsche, and Bergson, that there is never any finality in nature or evolution, but only a musical expression of nature involving explication, implication, and complication. Evolution is thus to be thought as a great 'fold' and we are to become those that we are: musicians of nature and artists of our own cultivation.

Fundamental paradoxes necessarily inform Deleuze's attempt to derive an 'ethics' from evolution and ethology. Whether these paradoxes are such that they imperil the coherence and sustainability of his project, we shall have chance to examine and is perhaps, ultimately, a judgement to be deferred to the practices, wise or unwise, of the reader. First it is necessary to discover and invent what work can be done with it, and attempt to determine what insights into a philosophy of life can be yielded from it in terms of its ethical and political dimensions. It is never for Deleuze a question of sustaining a parochial perspective on life, or of limiting the forces of creative evolution to the concerns of the human, narrowly defined and understood. It is necessary to combat two things: the blindness of science which would give us matter without ethics, and the blindness of faith which would give us an ethics without matter. Deleuze finds the moral view of the world especially pernicious since it is a view that enchains the body to an unknown soul and prohibits the becoming of the body (it is anti-pedagogic in this respect if we take culture and discipline to involve a true *paideia*). The moral view of the world condemns us to infinite sadness in this life. Deleuze stresses that nobody is born ethical, just as nobody is born a citizen, religious, sinful, or free and reasonable. We are neither emancipated nor condemned from the start. It is for this reason that the obligation of philosophy consists, above all perhaps, in unfolding an experimental and ethical pedagogy, one that requires a 'slow, empirical education' (1968: 244; 1992: 265). Furthermore, 'The state of reason is one with the formation of a higher kind of body and a higher kind of soul....A reasonable being may...in its way, reproduce and express the effort of Nature as a whole' (ibid.: 243, 264–5). The practical task of philosophy is a quite simple one: namely, that of freeing individuals and freeing knowledge from the claims of superstition. It is superstitious beliefs, which are by no means peculiar to religion but also characterize science and philosophy itself, which prevent our gaining access to a wholly 'positive' nature and which threaten all human becoming.

'Ethics', therefore, is an intrinsic part of Deleuze's philosophical project and it plays a role in each one of his attempts to articulate a philosophy of nature and to think 'beyond' the human condition. Ethical life arises for

Deleuze out of the context of a naturalism, a particular philosophical conception of the world that requires the cultivation, as we shall see, of a 'superior empiricism'. This conception has recourse to transcendental and critical philosophy to the extent that it is able more imaginatively and sublimely to open up the territory of this naturalism and empiricism. Deleuze himself stresses that the *biological* significance of this new conception is not unimportant. However, when 'taken as a model' its chief significance is declared to be *'juridical and ethical'* (ibid.: 236; 257) (my emphasis). The difficult challenge that this ethics provides, which will be unfolded in this study gradually and carefully, is to think 'ethics' both transhumanly and germinally, that is, in terms of the 'living beyond' and the 'living on'.

IV

The question of Being as univocal cannot be avoided in any appreciation of, or encounter with, Deleuze, and I shall draw this Introduction to a close by making some pre-emptive remarks about it. The notion of univocal Being informs the final climax of *Difference and Repetition* and it continues to inform the crucial matters at stake in his joint work with Guattari, notably *A Thousand Plateaus* (the construal of nature as a plane of consistency presupposes Being as univocal). The role played by univocal Being has now assumed a crucial importance in the interpretation of his work. Alain Badiou's *La clameur de l'Etre* – the title of which is taken from a section in *Difference and Repetition* that speaks of the 'single voice' on Being running from Parmenides to Heidegger and is also the note on which the book ends with Deleuze writing of a 'single clamour of Being for all being' – provides a great deal of insight into Deleuze in relation to Heidegger on the question of Being, and in relation to the issue of Platonism, though his book suffers from neglecting a crucial work like *A Thousand Plateaus* (Badiou 1997: 31–49, especially 34–8, 42–7). Badiou does show, however, that the thought of univocity in Deleuze is not a tautological one (the One is one) and that it is entirely compatible with multiple and infinite forms of Being (39). But if the univocity of Being does not refer to a One, to a being as such, what does it speak of?

Deleuze's engagement with the history of philosophy revolves around his adherence to this conception of univocal Being (Being can be said in a single sense 'of' all its individuating differences and intrinsic modalities). What is its precise character? This can best be approached in terms of his readings of Spinoza and Nietzsche, both of whom are drawn upon in *Difference and Repetition* to supplement the doctrine of Duns Scotus (1987: 4ff.). On Deleuze's reading, Spinoza's substance cannot be thought independent of its expression in attributes which are always dynamic. The immanent expression of substance is neither an emanation nor a creation. In the entry on

'Attribute' in *Spinoza: Practical Philosophy* (1970, revised 1981/trans. 1988), Deleuze writes: 'And *immanence* signifies first of all the *univocity of the attributes*: the same attributes are affirmed of the substance they compose and of the modes they contain'. In the entry on 'Mode' this is clarified as follows: 'there is a univocity of Being (attributes), although that which is (of which Being is affirmed) is not at all the same (substance or modes)'. In the entry on substance Deleuze makes it clear that the question of the 'one' substance is, if properly thought through, inadequate for comprehending what is in play in Spinoza. The problem with the 'One' is that it refers to a numerical distinction that is never 'real', while a 'real' distinction is never numerical. Deleuze prefers to speak, therefore, of formally distinct attributes that are affirmed of a 'singular substance' (as Michael Hardt notes: because number involves limitation and requires an external cause it can never have the character of substance, 1993: 60). In Spinoza's thinking substance is presented as though it were independent of the modes and the modes as fully dependent on substance. In order to overcome any possible dichotomy here, Deleuze suggests that susbtance must be said only of the modes. In short, Deleuze's radical and complex move is to argue that being can only be attributed to becoming, identity can only be said of difference, and the one can only speak of the multiple. This is what he means when he declares that Being can be expressed in a single and same sense when it said 'of' all its individuating differences (a colour, for example, enjoys various intensities but it remains the same colour).

The individuating factors at play in the becoming of Being are not for Deleuze simply individuals constituted in experience, but are that which acts in them as a transcendental principle, such as intrinsic modalities of being that pass from one individual to another and which circulate and communicate beneath matters and forms. The factors of individuation, conceived in Kantian and transcendental terms, are what constitute individuals as beings of time and also what dissolve and destroy them in time. Individuation not only differs in kind from the individuals it constitutes and dissolves, it is also *presupposed* by matters, forms, and extensions, and *precedes* differences that are generic, specific, and individual. It is this 'Copernican Revolution', consisting of giving difference its own concept, that Deleuze identifies with Nietzsche's doctrine of eternal return.

A classical source for the treatment of the problem of the 'copy' is Plato's *Timaeus*, where the concern is with how 'the beginning of everything' in its copied or mimetic form has to be grasped as being in accordance with nature grasped as that which is lasting and unalterable (Plato 1961: 1162ff.). Plato's discussion of the eternal, and how it can be faithfully copied, takes place, of course, in the context of his treatment of time as the 'moving image of eternity'. Both *Difference and Repetition* and *The Logic of Sense* are devoted, in part, to effecting an 'overturning' or reversal of Platonism on these points, showing how there is a founding 'philosophical decision' in Plato against the play of primary difference and in favour of the powers of the Same and the Similar. In the crucial distinctions that Plato makes between the original and the image, and the model and the copy, the first term refers to the superior, founding identity in which the Idea is nothing other than what it is (courage is courageous, piety is pious, etc.), and the copy is to be 'judged' in terms of a derived internal resemblance. Difference is only conceivable in relation to these prior founding terms. For Deleuze, however, the 'true' Platonic distinction is of another nature and lies not between the original and the image but between two kinds of images or idols, namely, copies and simulacra. In other words, Deleuze claims that the distinction between model and copy exists only in order to found and apply that between copy and simulacra. This is crucial since the function of the model is not to oppose the world of images but to provide criteria for selecting the 'good images' (those that resemble) and eliminating the 'bad' ones (the simulacra).

It is, therefore, contends Deleuze, a specifically *moral* view of the world that informs the logic of representation within Plato's thought: simulacra are 'demonic images' that have become stripped of resemblance and must be exorcized. Deleuze's inversion consists in granting to the simulacra its own model, the model of difference in itself. The 'game' of difference and repetition is designed, therefore, to replace that of the Same and representation. This is a world without prior identity and without internal resemblance, a world full of monsters of energy, demonic visitations, and dark precursors (Deleuze: 'It is the monster which combines all the demons', and see Ansell Pearson 1999, forthcoming, for further insight). It is Nietzsche's world of will to power and eternal return (Nietzsche 1968: 1067). Deleuze also argues, however, that a reversal of Platonism, involving the triumph of the simulacra, is already prefigured in Plato's dialogues themselves, notably at the end of the *Sophist* (Deleuze 1968: 165–8; 1994: 126–8; and see the first appendix to *The Logic of Sense*, 1969/1990 'The Simulacrum and Ancient Philosophy'). Paul Patton has made the important point that the contrast drawn by Plato in the *Sophist* between 'likenesses' (copies) and 'semblances' (simulacra), in which only the former is held to resemble the original, is for Deleuze a difference in kind. In other words, they are not two species of the same genus 'but two completely different kinds of being' (Patton 1994: 149; on Deleuze and Platonism see also Foucault 1977: 165–96, and the critical treatment in Badiou 1997: 42–6, 68ff., 85ff., 105ff., 118, 148–50, who makes the point: '*Le "platonisme" est la grande construction fallacieuse de la modernité comme de la postmodernité*', 149).

In the repetition that belongs to Nietzsche's eternal return Deleuze locates the affirmation of originary difference and the play of simulacra. 'The simulacrum', Deleuze writes, 'is not a degraded copy. It harbors a positive power (*puissance positive*) which denies the *original and the copy, the model and the reproduction*' (1969: 302; 1990: 262). Furthermore, it 'is not even enough to invoke a model of the Other, for no model can resist the vertigo of the simulacrum' (1969: 303; 1990: 262). It is the simulacra which enables both the notion of the copy and the model to be challenged. It is the eternal return which best articulates for Deleuze the nature of univocal Being. The affirmation of the eternal return is the affirmation of all chance in a single moment and of Being as primary excess or difference, 'one Being and only for all forms and all times' but always as 'extra-Being' (Deleuze 1969: 211; 1990: 180). On Deleuze's model the eternal return provides nothing less than a 'new image of thought', of the birth of thought and the thought of birth. It is the 'phantasm' that constitutes the site of the eternal return, endlessly mimicking the birth of a thought and the thought of a birth, beginning 'a new desexualization, sublimation, and symbolization, caught in the act of bringing about this birth' (256; 220). The eternal return is the 'force of affirmation', the force which takes everything to its highest level of intensity, affirming 'everything of the multiple, everything of the different, everything of chance', selecting only against that which would subordinate them to a principle of identity (the Same, the One, and necessary) (1968: 152; 1994: 115). The eternal return is the hidden 'other face' of the death instinct manifested in the compulsion to repeat, affecting a world that has 'rid itself of the default of the condition and the equality of the agent in order to affirm only the excessive and the unequal, the interminable and the incessant, the formless as the product of the most extreme formality' (ibid.).

What Deleuze wrote of Leibniz in his book on the fold of 1988 could be equally applied to his own project as its fundamental leitmotif: 'No philosophy has ever pushed to such an extreme the affirmation of a one and the same world, and of an infinite difference or variety in this world'. It is necessary to ask a number of critical questions concerning this ontology and onto-ethology of univocal Being, and these questions should be kept in mind as the study proceeds and unfolds. We note that for Deleuze there is no contradiction between declaring that 'all is one' and saying that 'all is multiple'. This ontology is univocal because it names, above all, a philosophy of *expression*, in which all that becomes is an expression of Being (as becoming, as difference). In his study of Leibniz Deleuze identifies the two poles at play in the creation of philosophical principles as (a) everything is always the same because there is only one and the same 'Basis' (*Fond*), and (b) everything is distinguished by degree and differs in manner. However, we can ask: if all is difference, even in resemblance and identity, what is it that *makes* a/the difference? How do we identify a difference and, moreover, a difference that makes a difference to the order of things? In *Difference and Repetition* Deleuze appreciates this point and appeals to various processes of differentiation to account for the production of difference. But the question persists regarding both the 'value' that is being placed on difference and the role it is playing in a 'critical modernity'. Is repetition the force of a (superior) law that is in thrall to novelty, change, destruction, producing a critical and clinical modernity for the sake of them? Is this to go beyond modernity and its cult of the new or simply to perpetually outdo it?

The question of philosophy and the new, of both philosophy's relation to the new and its own production of the new, is intimately connected to the matter of capitalism. The above questions remain abstract so long as they are posed outside of the context of this relation. Throughout Deleuze's work we find an attempt to think this relation between philosophy and capital, in which the question of the 'and' is perhaps the most crucial: difference 'and' repetition, capitalism 'and' schizophrenia, absolute deterritorialization 'and' relative deterritorialization. In his book on Leibniz *'and* the baroque' Deleuze argues that if the Baroque is to be associated with capitalism it is in terms of an emergent crisis of *property*, in which questions of the proper, of ownership and possession, undergo the transmutation of a certain nihilism, since the crisis is connected with the emergence of new machines in the social field and the discovery of new and myriad living beings in the organism. The question of philosophy in this context concerns the *critical* and *clinical* relation of philosophy to the nihilistic movements of capital, movements articulated by both Marx ('all that is solid melts into air, all that is holy is profaned') and Nietzsche ('nothing is true, everything is permitted'). For Deleuze it is always a question of *two* nihilisms, namely, the one kind which destroys old values in order to conserve the established order and which never produces anything new, and the other kind which extracts from nihilism something that 'belongs' to the untimely, to the monstrous future, remaining faithful to the promise of a 'time to come'. In other words, it is a question of being modern, but not at any cost, argues Deleuze. When the world loses its established principles the only radical ('im-proper') response is to invent and proliferate new ones.

We shall have the chance to critically examine the nature and force of this distinction between the two nihilisms at key moments in the drama of this study. Let me draw this Introduction to a close by expressing the principal concern I have over Deleuze's project. A philosophy of immanence always encounters the greatest dangers not only because it freely takes on and undergoes these dangers, as Deleuze readily appreciates, but also because it runs the risk of producing a purely narcissistic method and mode of critique. The conception of Being as univocal and the experiment of the plane of immanence are clearly the source of the most important innovations of Deleuzianism, but they are also the site of its principal predicaments. Both need to be attended to and the kind of attention we give them is, ultimately, an ethical matter. It is also the occasion for a speculative affirmation of the event of thought.

1

THE DIFFERENCE OF BERGSON

Duration and creative evolution

Deleuze is a magical reader of Bergson, who in my opinion is his real *maitre*, even more than Spinoza, and perhaps even more than Nietzsche.

(Badiou 1997: 62)

If the thought really yielded to the object, if its attention were on the object, not on its category, the objects would start talking under the lingering eye.

(Adorno 1966: 38; 1973: 27–8)

Introduction

It is Bergson who defines the task of philosophy as one of learning to think beyond the human condition. What exactly is the meaning of this 'beyond'? Deleuze configures the task as follows:

To open us up to the inhuman and the superhuman (*durations* which are inferior or superior to our own), to go beyond the human condition: This is the meaning of philosophy, in so far as our condition condemns us to live among badly analyzed composites, and to be badly analyzed composites ourselves.

(Deleuze 1966: 19; 1988: 28)

Why are we caught up in the investigation of badly analysed composites and why are we those ourselves? What does it mean to be open to other durations? Addressing these questions through a careful reading of Deleuze's 1956 essay on Bergson and his 1966 text, *Bergsonism*, will take us to the heart of the project of philosophy as Deleuze conceives it, and also show that going 'beyond' the human condition does not entail leaving the 'human' behind, but rather aims to broaden the horizon of its experience. Throughout the contours and detours of his intellectual life Deleuze never ceased to grant a fundamental and vital autonomy to philosophy. It was from an essentially Bergsonian influence that Deleuze cultivated the idea that philosophy could go beyond human experience in a way that would deepen and enhance it. This requires a major reorientation of philosophy, a new conception of critique, and a new logic which proceeds not via contradictions but through nuances.

Deleuze's first sustained and published encounter with Bergson has several notable features. First, there is the fact that it took place in the 1950s, less than a decade after the end of the war, when the climate would have been unfavourable to a novel reworking of Bergson's philosophy with its underlying optimism (although the essay on difference was published in 1956 it had been composed a few years earlier). Vitalism was largely a dead tradition in 1950s and Bergson's significance reduced to that of being a philosopher of insects. Second, and perhaps most significant, is that Deleuze should have returned to Bergson in order to articulate on the basis of his philosophy a thinking of difference as a thinking of *internal* difference. This conception of difference is extended to the notion that many have taken to be the most problematic aspect of Bergson's so-called vitalism, that of *élan vital*, which is conceived by Deleuze as an internal explosive force that can account for the 'time' of evolution as a virtual and self-differentiated movement of invention.

The 1956 essay begins very precisely, and seemingly uncontentiously, by opening out a problematic that the rest of the essay will then pursue in all its ramifications: 'The notion of difference must throw a certain light on Bergson's philosophy, but inversely Bergsonism must bring the greatest contribution to a philosophy of difference' (1956: 79; 1997: 1). It is a quasi-phenomenological venture since the aim, Deleuze declares, is to 'return' to things themselves. The promise, if this is got right, is nothing less than one of difference delivering Being to us. A careful consideration of the differences of nature will lead us to the nature of difference. Hitherto, thinking has confused two kinds of differences and covered one over with the other: differences of degree over differences of kind or nature. At other times, it has effected the reverse and focused only on differences of kind (such as between the perception of matter and matter itself). The task of philosophy is to grasp the thing itself in its positivity, and this requires a notion of internal difference. Deleuze fully appreciates that a certain strand of modern philosophy finds such a notion of difference to be absurd. In the Hegelian schema of difference a thing differs from itself only because it differs in the first place from all that it is not. Difference is, therefore, said to be constituted at the point of contradiction and negation. The novel modernity of Bergsonism lies, for Deleuze, in its critique of metaphysics and of a science that has forgotten the durational character of life and imposed on it an abstract mechanics. It rests on a schema that homogenizes difference by selecting only differences of degree through a spatialized representation of the real. General ideas simply present for our reflection completely different givens that get collected in utilitarian groupings. The task for Deleuze is one of breaking out of a merely 'external state of reflection', so that philosophy no longer has a merely negative and generic relation with things in which it remains entirely in the element of generality (1956: 80).

Deleuze's thinking of difference 'and' repetition is already fairly well worked out in the 1956 essay in terms of its essential components. In the following reading I shall commence with the 1966 essay since it begins in instructive fashion by seeking to expose the particular character of the Bergsonian method (intuition). I will then show how this links up to the question of duration and to the *élan vital*. Throughout I will supplement the analysis of the 1966 text by drawing on material outlined in the 1956 essay. There is a remarkable continuity between the two pieces of writing in spite of the ten-year gap in Deleuze's public engagement with Bergson. We should note, however, that some of the earlier material gets dropped or reconfigured, such as, for example, an examination of the character of historical consciousness ('If difference itself is biological, consciousness of difference is historical', 1956: 94).[1] My aim is not to assess the correctness of Deleuze's reading of Bergson (by describing it as 'Bergsonism' Deleuze is surely drawing our attention to the inventive character of his reading), which is often accused of making him sound too much like Nietzsche. Rather, I have set myself the task of providing instructive insight into its most salient features as it pertains to the matter of 'creative evolution' in terms of a thinking of difference.

The method of intuition

> The Absolute is difference, but difference has two facets, differences in degree and differences in kind.
>
> (Deleuze 1966: 27; 1988: 35)

The ambition of philosophy, Deleuze writes in the 1956 essay, is to articulate for the object a concept appropriate to it, even though it may no longer remain from that point on a concept as such given its singularity. In order to articulate this philosophy it will be necessary to have recourse not simply to a method but to a method capable of disclosing 'the *jouissance* of difference' (1956: 81). Deleuze stresses in his 1966 text that intuition is the method peculiar to 'Bergsonism' and denotes neither a vague feeling or incommunicable experience nor a disordered sympathy. Rather, it is a fully developed method that aims at precision in philosophy (see Bergson 1965: 11). Where duration and memory denote lived realities and concrete experiences, intuition is the only means we have at our disposal for crafting a knowledge of experience and reality. 'We may say, strangely enough', Deleuze notes, 'that duration would remain purely intuitive, in the ordinary sense of the word, if intuition – in the properly Bergsonian sense – were not there as method' (Deleuze 1966: 2; 1988: 14). However, intuition is a complex method that cannot be contained in a single act. Instead, it has to be seen as involving a plurality of determinations and a variety of mediations. The first task is to stage and create problems; the second is to locate differences in kind; and the third is to comprehend 'real time', that is, duration as a heterogeneous and continuous multiplicity. Indeed, Bergson himself acknowledges that other philosophers before him, notably Schelling and Schopenhauer, had drawn heavily on intuition as a method. Their task, however, he says, was not to discover duration but to search for the eternal (Bergson 1965: 30).

Concerning the first problem, Deleuze argues that we go wrong when we hold that notions of true and false can only be brought to bear on problems in terms of ready-made solutions. This is a far too pre-emptive strategy that does not take us beyond experience but locks us in it. This negative freedom is the result of manufactured social prejudices where, through social institutions such as education and language, we become enslaved by 'order-words' that identify for us ready-made problems which we are forced to solve. This is not 'life', and it is not the way life itself has 'creatively' evolved. Therefore, true freedom, which can only be a positive freedom, lies in the power to decide through hesitation and indeterminacy and to constitute problems themselves. This might involve the freedom to uncover certain truths for oneself, but the excessive freedom is not so much one of discovery as more one of invention. Discovery is too much involved in uncovering what already exists, an act of discovery that was bound to happen sooner or later and contingent upon circumstances. Invention, however, gives Being to what did not exist and might never have happened since it was not destined to happen, there was no pre-existing programme by which it could be actualized. In mathematics and in metaphysics the effort of invention consists in raising the problem and in creating the terms through which it might be solved but never as something ready-made. As Merleau-Ponty notes in a reading of Bergson that is consonant with Deleuze's on some of these points, when it is said that well-posed problems are close to being solved, 'this does not mean that we have already *found* what we are looking for, but that we have already invented it' (Merleau-Ponty 1988: 14).

Deleuze offers a Bergsonian reading of Marx on this point. When Marx says that humanity only sets itself problems it is capable of solving this is not the empiricist, or rather, positivist, trap we might think, since the problems take us beyond what we think we are and are capable of. Marx's thought, therefore, is a vital empiricism. For Deleuze, the history of humanity, considered from both theoretical and practical points of view, is a history of the construction of problems (it is a history of overhumanity, one might say). It is in this excessive sense that we can say humankind makes its own history, and becoming conscious of this praxis amounts to a drama of freedom as the 'meaning' of human life and of its germinal existence (the fact that it lives on and survives only by living beyond itself). In a deeper sense, however, the historical character of human existence is an expression of the *élan vital* which marks life as creative: 'Life itself is essentially determined in the act of avoiding obstacles, stating and solving a problem. The construction of the organism is both the stating of a problem and a solution' (Deleuze 1966: 5; 1988: 16).

We will shortly examine in greater detail this point on the organism as an invention of life and as the staging of a problem. Does it suggest that the organism is constructed for a purpose or an end? Is it an end-in-itself or merely a vehicle through which life limits itself only in order to enhance and revitalize its powers of invention? Can one speak of an end of evolution, granting it a final purpose, or is constant invention and re-invention the only end? Let us note at this stage that Deleuze locates the novelty of Bergson's approach in its attempt to think 'creative' evolution beyond both mechanism and finalism and, in this way, to grant an important role to invention in evolution, an invention that is only possible to the extent that evolution involves duration. It is with Bergson's approach in mind that Deleuze argues that in the cases of both living matter and organized matter, such as organismic life, what one finds in evolution is the stating of problems in which the 'negative' articulations of opposition and contradiction, derived from notions of obstacle and need, are secondary phenomena and of secondary importance (for a clear statement of this view see Deleuze 1968: 272; 1994: 211). We may note how remarkably close this is to Nietzsche's 'contra Darwin' position articulated in his *Genealogy of Morality*, in which the critique takes the form of challenging the emphasis placed in Darwinian theory on 'adaptation'. To construe evolution in terms of adaptation is to overlook the prime importance of the form-shaping forces or powers that provide life with new directions and from which adaptation derives only once these powers have had their effect. Speaking of Herbert Spencer, Nietzsche argues that the dominant role these powers play in 'the organism itself', in which the life-will is active and reveals itself, is denied (Nietzsche 1994: 56; for more detailed insight into Nietzsche's relation to Darwinism see my *Viroid Life*, 1997, Essay 4). This suggests that there may be a more objective basis to Deleuze's 'Nietzschean' reading of Bergson than commonly supposed. Indeed, it is astonishing how close to Nietzsche Bergson is in the way he approaches questions of philosophy. Equally surprising is the fact that Nietzsche nowhere figures in the text *Creative Evolution*, since it is Nietzsche who had already endeavoured to think evolution beyond the terms of both mechanism and finalism.

Let me return, however, to Deleuze's reading of Bergson. The second rule of intuition is to do away with false problems, which are said to be of two kinds: first, those which are caught up in terms that contain a confusion of the 'more' and the 'less'; and, second, questions which are stated badly in the specific sense that their terms represent only badly analysed composites. In the first case the error consists in positing an origin of being and of order from which nonbeing and disorder are then made to appear as primordial. On this schema order can only appear as the negation of disorder and being as the negation of nonbeing (see Bergson 1962: 223; 1983: 222). Such a way of thinking introduces lack into the heart of Being. It more or less errs in not seeing that there are *kinds* of order and forgetting the fact that Being is not homogeneous but fundamentally heterogeneous. Badly analysed composites result from an arbitrary grouping of things that are constituted as differences in kind. For example, in *Creative Evolution* Bergson contends that the cardinal error that has vitiated the philosophy of nature from Aristotle onwards is identifying in forms of life, such as the vegetative, instinctive, and rational, '*three successive degrees of the development of one and the same tendency, whereas they are divergent directions of an activity that has split up as it grew*'. He insists that the difference between them is neither one of intensity nor of degree but of kind (1962: 136; 1983: 135). Bergson wants to know how it is that we deem certain life-forms to be 'superior' to others, even though they are not of the same order, and neither can they be posited in terms of a simple unilinear evolutionism with one life-form succeeding another in terms of a progress towards perfection in self-consciousness (175; 174). Life proceeds neither via lack nor the power of the negative but through internal self-differentiation along lines of divergence. Indeed, Bergson goes so far as to claim that the root cause of the difficulties and errors we are confronted with in thinking creative evolution resides in the power we ascribe to negation, to the point where we represent it as symmetrical with affirmation (286; 287). When Deleuze says that resemblance or identity bears on difference *qua* difference, he is being faithful to Bergson's critical insight into the character of negation, chiefly, that it is implicated in a more global power of affirmation.

It is through a focus on badly analysed composites that we are led, in fact, to positing things in terms of the more and the less, so that the idea of disorder only arises from a general idea of order as a badly analysed composite. This amounts to claiming, as Deleuze cognizes, that we are the victims of illusions which have their source in aspects of our intelligence. However, although these illusions refer to Kant's analysis in the *Critique of Pure Reason*, where Reason is shown to generate for itself in exceeding the boundaries of the Understanding inevitable illusions and not simple mistakes, they are not of the same order. There is a natural tendency of the intellect to see only differences in degree and to neglect differences in kind. This is because the fundamental motivation of the intellect is to implement and orientate action in the world. For the purposes of social praxis and communication the intellect needs to order reality in a certain way, making it something calculable, regular, and necessary. Nietzsche applies this insight not only to our action on the world but also to the constitution of our 'inner world' (the 'soul'). Human consciousness and memory have been cultivated by making the individual self itself something regular and calculable and this has been done by constituting the self as a site of discipline and culture. Bergson is incredibly close to Nietzsche on some of these points. In terms reminiscent of Nietzsche he attacks the notion of a timeless substratum lying behind all reality, adding that an ego which does not change does not endure either: 'If our existence were composed of separate states with an impassive ego to unite them, for us there would be no duration' (Bergson 1962: 4; 1983: 4). What is obtained by construing the flow of life in this way is an 'artificial limitation' of the internal life, a stasis which lends itself to the requirements of logic and language. For Bergson it is possible – indeed necessary – to think of movement without the support of an invariable substance that is posited as underlying all becoming, while Nietzsche is insistent that logical-metaphysical postulates, such as the belief in substance, are the result of our habit of treating all our deeds as consequences of our will in which the ego is protected from vanishing in the 'multiplicity of change' (compare Nietzsche 1968: Section 488; see also Sections 484–5, 515–16; and 1974: Section 354). The intellect schematizes in order to impose upon the chaos it encounters the degree of regularity and uniformity that its practical needs require. Therefore, the 'meaning' of our knowledge, Nietzsche insists, is to be approached in 'a strict and narrow anthropocentric and biological sense'. In order for a certain species to maintain itself and increase its feeling of power over the world it is necessary that it develop a conception of calculable and constant reality in order to establish a schema of behaviour on it (Nietzsche 1968: Section 480). Informing the development of our 'organs of knowledge', therefore – organs that have evolved in terms of their formation through a culture of external technics – is the 'utility of preservation' (ibid.).

To bring into play a different kind of intelligence is for Deleuze to introduce the 'critical' element into philosophy that will enable us to go beyond the human condition and to widen the canvas of its experience. It is intuition which allows this critical tendency to express itself through two procedures: the discovery of differences in kind and the formulation of criteria for differentiating between true and false problems. But at this point things get even more complex. This is for several reasons. Thought can only move beyond dualisms in the activity of their construction (extensity and intensity, space and time, matter and memory, instinct and intelligence, etc.). Second, if intuition is to be conceived as a method which proceeds via division – the division of a composite into differences of kind – is this not to deny that reality is, in fact, made up of composites and mixtures of all kinds? For Bergson, Deleuze argues, the crucial factor is to recognize that it is not things that differ in kind but rather tendencies: 'a thing in itself and in its true nature is the expression of a tendency before being the effect of a cause' (Deleuze 1956: 83; 1997: 4). In other words, what differs in nature are not things (their states or traits) but the tendency things possess for change and development. A simple difference of degree would denote the correct status of things *if* they could be separated from their tendencies. For Bergson the tendency is primary not simply in relation to its product but rather in relation to the causes of productions in time, 'causes always being retroactively obtained starting from the product itself' (ibid.). Any composite, therefore, needs to be divided according to qualitative tendencies. Again, this brings Bergsonism close to Kant's transcendental analysis, going beyond experience as given and constituting its conditions of possibility. However, these are not conditions of all possible experience but of 'real' experience (the experience of inferior and superior durations). Thought moves beyond experience, including the experience of consciousness, but only ever in a direction that is drawn from experience itself. We must appreciate the indissoluble character of Deleuze's 'transcendental empiricism' which aims to articulate this double movement, progressing thought to the point where it is, uncannily, able to think a 'pure perception identical to the whole of matter' and a 'pure memory identical to the totality of the past' (Deleuze 1966: 17–18; 1988: 27).

Deleuze speaks of Bergsonism as a method that seeks to reveal the 'peculiarities' of real experience (the experience of duration). In order to broaden the field of this experience, indeed to open up this experience, we cannot restrict philosophy to the unfolding of concepts since concepts, Deleuze argues in his 1966 work, only define the condition of possible experience in general. It is thus necessary to go beyond concepts and to determine the conditions of experience in 'pure percepts'. Of course, such percepts must be united in a concept but such a concept will be modelled on the thing itself. Ultimately, the task, perhaps the most difficult of all Deleuze as readily acknowledges, is to identify the sufficient reason of the composite. This point can only be understood by appreciating the extent to which once the 'turn' has been made – 'the turn in experience' – the lines that diverge and the tendencies which differ in kind link together again and give rise to the thing as we come to know it. Beyond the turn, therefore, the lines intersect once again but not at the 'point' from which they started but rather at a 'virtual point': 'Turn and return. Dualism is therefore only a moment, which must lead to the re-formation of a monism' (ibid.: 20; 29). This, in short, is one of the principal elements of Bergson's 'superior empiricism' (a term Deleuze borrows from Schelling and applies to Bergson; on the relation between Bergson and Schelling see Merleau-Ponty 1988: 145–6; 1994: 78–81; Lovejoy 1961). If the conditions of experience can, in fact, be grasped in terms of an intuition it is because they are no broader than the conditions of 'real' experience.

Central to this vital empiricism is an attack on mechanism, for it is mechanism that repeatedly invokes a spatialized model of time which prohibits it from gaining insight into reality of duration and invention. On the mechanistic model beings are represented solely in terms of differences of degree (position, dimension, and proportion). It is mechanism that informs, contends Deleuze, modern evolutionism that posits a unilinear model that moves from one living system to another via simple intermediaries, transitions, and variations of degree. If everything is implicated in tendencies then movement can be said to be primary. However, reality is always double, involving a difference between two orders of difference. The real is both that which is cut out (*se découpe*) according to natural articulations and differences of kind *and*, at the same time, that which intersects again (*se recoupe*) along paths that converge toward a virtual point. It is the virtual which allows for divergence and convergence. The virtual stipulates the dynamic and inventive conditions of possibility of evolution as a creative process, as the open system *par excellence*. We shall attempt to determine the character of this virtual in some detail shortly. Before doing so it is necessary to say something *en route* about the relation between intuition and duration in Bergson.

Intuition and duration

Intuition is not duration, but rather the movement by which thought emerges from its own duration and gains insight into the difference of other durations within and outside itself. It both presupposes duration, as the reality in which it dwells, but it also seeks to think it: 'to think intuitively is to think in duration' (Bergson 1965: 34). Without intuition as a method duration would remain for us a merely psychological experience and we would remain prisoners of the given, of what is given to us. This is why not to think differences of kind – differences that only duration can access for us – is for Deleuze to betray philosophy. Informing Deleuze's Bergsonism from the start, therefore, is a philosophical critique of the order of need, action, and society that predetermine us to retain a relationship with things only to the extent that they satisfy our interest, and of the order of general ideas that prevent us from acquiring a higher human nature. Deleuze insists that no differences in kind or nature can be located except in duration. Duration is both the 'totality' and the 'multiplicity' of such differences. The difference of kind is not between duration and space, but solely within duration, simply because space is held to be a quantitative homogeneity. To illuminate the difference of kind involved in duration, and to show how there comes into being specific kinds of duration, including superior ones, we can give the banal example, chosen by Bergson and also treated by Deleuze, of a lump of sugar. If such a lump is grasped solely in terms of its spatial organization then only differences of degree between it and any other thing will be revealed. However, it can be approached as an object that enjoys a specific rhythm of duration, in which its 'way of being time' is revealed, in part, in the process of dissolving. Here the sugar shows itself to be both a difference of kind and to generate an internal difference (to differ from itself). The 'substance' of a subject, such as the cube of sugar, consists in its participation in duration. The act of witnessing the sugar dissolving also reveals different kinds of durations that have their own rhythms, including my own and those in things around me. Duration can, therefore, be conceived as the 'environment' of differences in kind.

Merleau-Ponty's reading of Bergson is, again, also helpful on these points. He notes that for Bergson many traditional questions of philosophy, such as 'Why have I been born?', 'Why is there something rather than nothing?', and 'How can I know anything?', can be held to be 'pathological' in the sense that they *presuppose* a subject already installed in being, that is, they are the questions of a doubter who no longer knows whether he has closed the window (Merleau-Ponty 1988: 12). A strictly 'positive' philosophy, therefore, notes Merleau-Ponty, will not aim to 'resolve' classical problems, but rather 'dissolve' them. However, Merleau-Ponty is right to stress that Bergson's new approach does not simply aim to pit intuition against the intellect and logic, or spirit against matter, and life against mechanism. Here his adversaries 'have missed the point', since in Bergson's 'positive philosophy' the 'negative begins to reappear' and, as it does so, is 'progressively affirmed' (13). It is not, moreover, a question of absorption into being, of losing oneself in it, but of being 'transcended' by it. Merleau-Ponty expresses it in helpful terms as follows: 'It is not necessary for him [the philosopher] to go outside himself in order to reach the things themselves; he is solicited or haunted by them from within' (14–15). An ego that finds itself 'in *durée*' can only grasp another being in the form of 'another *durée*', in which there is 'an infinity of possible *durées*' (15). The 'singular nature' of the *durée* that makes up one's manner of being is also implicated in a 'universal dimension' in which what is considered to be 'inferior' and 'superior' to us 'still remains "in a certain sense, interior to us" ' (ibid.) (for example, as in my relation to the dissolving sugar as I wait for a glass of sugar water). Merleau-Ponty is able to declare, therefore, that when one is at the 'source' of one's *durée* one is also 'at the heart of things', simply because they are 'the adversity which makes us wait'. Thus, the relation of the philosopher to being is not that of spectator and spectacle but involves a special kind of complicity, 'an oblique and clandestine relationship' (ibid.). The relationship between our *durée* and that of things is not one of coincidence but of 'co-existence', it is a kind of 'nonidentity' in Adorno's sense, as when Merleau-Ponty writes, in regard to Bergson, of the pebble:

> We are not this pebble, but when we look at it, it awakens resonances in our perceptive apparatus; our perception appears to come from it. That is to say our perception of the pebble is a kind of promotion to (conscious) existence for itself; it is our recovery of this mute thing, which from the time it enters our life, begins to unfold its implicit being, which is revealed to itself through us.
>
> (1988: 17)

Intuition, therefore, calls forth a development that contains a double reference to the mute being to which it is related and to the 'tractable meaning' that is derived from it. This is why for Merleau-Ponty the experience of a 'concordance' between durations amounts to a *reading*, that is, 'the art of grasping a meaning in a style before it has been put into concepts...*the thing itself* is the virtual focal point of these convergent formulations' (19–20). This is also why, in a truly astute insight, Merleau-Ponty characterizes the 'internal movement of Bergsonism' in terms of a development from a philosophy of '*im*pression' to a philosophy of '*ex*pression'. Philosophy thus enjoys a peculiar strangeness, never entirely in the world and yet never completely outside it.

The 'transhuman' element of duration consists in the power it has to encompass itself. What is meant by this? The point refers to those 'inferior' and 'superior' durations that Bergson's philosophy seeks to open us up to. Deleuze insists – in the context of a discussion of Bergson and Einstein on time – that it is not a question of asking whether time is one or many, but rather of asking after the kind of multiplicity that is peculiar to the single, universal, impersonal 'Time'. Let us take the example of a triplicity of fluxes contained in sitting on the bank of a river: the flowing of the water, the flight of a bird, and the murmur of my life passing by deep within. Here my duration is an element of the fluxes taking place and contains the others in a co-existence. Two of the fluxes can only be said to be simultaneous or coexistent to the extent that they are involved in a third. It is, however, my duration that has the power to disclose other durations, encompassing them as well as itself ad infinitum. Although access is provided to other durations, both inferior and superior, Deleuze does acknowledge that these durations 'remain internal to my own' (1966: 80–1; 1988: 80–1).

The point being made contra Einstein and Relativity is that duration is a 'virtual multiplicity' that has a special mode of division and a special kind of co-existence: elements that differ in kind are divided and these elements exist only to the extent that the division is carried out in an actualization (which takes place when the virtual stops). The divisions of a virtual duration which differ in kind – such as, for example, the race of Achilles and the tortoise in the famous paradox of Zeno – can only be 'lived' in the perspective of a single time. The differences here concern the differences of relaxation and contraction that communicate in this single time. These differences belong to the virtuality that encompasses the differences in kind and which is actualized in them. The recognition of an infinity of actual fluxes does not rule out there being a monism of time but actually presupposes it. The converse praxis can only be to spatialize time and so lose insight into the kind of knowledge that Bergson is trying to produce, namely, of the virtual multiplicities that imply a single time in which communication of the heterogeneous is possible. The calculations of the physicist actually presuppose a single universal time. For example, when it is said that the time of Peter is expanded or contracted at the point where Paul is located, this does not express what is lived by Paul who sees everything from his particular point of view and has no reason to experience the time that elapses within him and around him any differently from Peter. The physicist absolutizes the view of Paul by attributing to him the image Peter forms of his (Paul's) time, assuming that he is the spectator of the entire world. Merleau-Ponty contends that the physicist is guilty here of doing what the philosopher is so often accused of, making a myth of time (Merleau-Ponty 1964c: 196; and compare Deleuze's treatment of Peter and Paul in 1966: 84–5; 1988: 83–4). It is on this point – what is the time of Relativity? – that Bergson argues we need to be more Einsteinian than Einstein.[2]

Although the order of space, conceived independently of durational movement and qualitative change, is an illusion, it is not grounded simply in our nature but rather in the nature of things. Matter is, in actuality, says Deleuze, that 'aspect' by which things present themselves to each other and to us in terms of differences in degree. To this extent, we can say that it is experience which gives us composites. Matter itself presents a point of view or perspective on states of things that prevents the experience of differences in kind. The illusion, therefore, is not merely the expression of a deficiency in our nature but arises from the actual world that we inhabit. This explains why for Deleuze the task of Bergsonism is not simply one of coming up with a new psychology of time but must, instead, be one of articulating a complex ontology or an ontology of the complex. We are not simply caught in subject–object relations that are the result of the limitations of our knowledge; rather, reifications arise from out of the tendency of the real itself, in the very terms of its evolution, to proceed via differences of degree. For Bergson philosophy, as the vocation of the overhuman, must assume a violent form, doing violence to both scientific practice and to the habits of the mind. Science is not, however, to be demoted to the status of a merely relative knowledge, becoming little more than a symbolic discipline that proceeds via a quasi-Darwinian principle of success through tests of fitness; rather, science is part of the ontology of the complex. To allow the science of thought to make contact with the science of the experience of duration is to re-enter the absolute and to go 'beyond' the human condition as the peculiar and singular practice of philosophy.

Deleuze always stresses the importance for Bergson of the *empirical* character of the *élan vital*, denoting the fact that is both real and concrete. This can only be demonstrated, however, to the extent that we go beyond a geometrical definition of intuition. This presupposes a critique of 'intelligence', the matter of which we shall explore in some detail below. Intelligence commences from the immobile and seeks to reconstruct movement by juxtaposing immobilities. This is because it is concerned, above all, with static things, in which change is viewed as accidental and as a supplementary dimension to be merely added to them. Intelligence, therefore, cuts out of the (virtual) whole – out of a becoming – a thing and makes of this thing a substitute for the whole. On this model the new is always treated as simply a new arrangement of pre-existing elements. Now Bergson is fully aware of the deeply paradoxical character of his attempt to think the real 'in duration', namely that thought must always utilize language and, as such, it will find itself lodged in concepts that have a tendency to harden and congeal, such as, he notes, 'duration, qualitative or heterogeneous multiplicity, unconsciousness – even differentiation' (Bergson 1965: 35). His response is to appeal to another encounter with the clarity of thinking. He contends that a concept of intellectual origin is always 'immediately clear', while that which springs from intuition begins by being obscure and it remains obscure to the extent that it *compels* us to think beyond our ordinary conception. The problem with the intellect is that on account of its constitution it always desires to see in the new something of the old, and in this way it feels itself secure since it is on ground that is familiar to it. However, he proposes that there is another kind of experience of clarity and to which we freely submit when we undergo the experience of time as it *imposes itself* upon us. This is the clarity of the 'radically new' that is more than a mere reassembling of existing elements. Initially, our response to this radically new might be to declare that it is incomprehensible. But although it strikes us initially as obscure we discover, by undergoing a new kind of thinking, that it also 'dissipates' obscurities and problems which seem to us insuperable dissolve either by disappearing or by presenting themselves to us in a different manner.

Bergson insists that his method of intuition contains no devaluation of intelligence but rather a determination of its specific facility. If intuition transcends intelligence this is only on account of the fact that it is intelligence which in the first instance gives it the 'push' to rise beyond, so that without it intuition would remain wedded entirely to instinct and thus riveted to the particular objects of its practical interests. It is clear, perhaps, that contained within the method of intuition is an 'ethics' of thought. This is an ethics that is both human and super- or overhuman. Why is this? Bergson seeks to show this point by comparing the natural perception of the eye, which looks upon the features of the living being as though they were merely assembled as opposed to being mutually organized, with the eye of the artist which is able to expose the 'intention of life', namely that there is a movement through the divergent lines of life that binds them together and gives them significance. By an effort of intuition and through sympathy with things themselves the artist is able to break down the barrier that space creates between the intellect and life in duration. However, the limitation of modern art is that it concerns only an individual. For this reason it is necessary to conceive of an activity that is turned in the same direction as art but which is able to take life 'in general' as its peculiar object. This is the specific task of philosophy and is entirely different from – though not opposed to – that of science, which takes up the utilitarian vantage point of external perception and prolongs individual facts into general laws. In cultivating the method of intuition philosophy may serve to show that life does not fit into the categories of the one and the many, that it is the movement of difference in its self-differentiation (this does not denote any originary plenitude, though it needs to be linked up with the notion of the virtual to demonstrate this, which is what we are about to do), and that neither mechanical causality nor finality are able to provide an adequate interpretation of the vital movements of a virtual and actual process of creative evolution.

The 'ethical' character of this method of philosophy resides, therefore, in the cultivation of a 'sympathetic communication' that it seeks to establish between the human and the rest of living matter (Merleau-Ponty is one of the few commentators to note that an 'ethics' informs the entirety of Bergson's thinking, 1988: 31–2). By expanding consciousness 'it introduces us into life's own domain, which is reciprocal interpenetration, endlessly continued creation' (Bergson 1962: 179; 1983: 178).

What is duration?

The time of the virtual

For Deleuze the Bergsonian revolution consists not in the banal claim that duration is subjective, but rather in the more demanding and compelling claim that subjectivity belongs to time. This non-chronological time that constitutes our interiority is necessarily a time of excess in relation to the subject, an insight that partly informs Deleuze's claim that we cannot own our subjectivity in terms of some fixed and secure property: 'Subjectivity is never ours, it is time, that is, the soul or the spirit, the virtual' (Deleuze 1985: 110–11; 1989: 82–3). As a modern philosopher Bergson is novel not because he does not accept the restrictions placed on the philosophy of nature or life by Kant. His originality resides in the manner in which he resists Kant. The conceptions of homogeneous space and time that characterize modern thought are neither properties of things nor essential conditions of our knowledge of them. Rather, they articulate what Bergson calls the 'double work of solidification and division' that we effect on the world – 'the moving continuity of the real' – as a means of obtaining a fulcrum for our action: 'They are the diagrammatic design of our eventual action upon matter' (Bergson 1959: 237; 1991: 211). Like Hegel, Bergson makes the charge that Kant's Copernican revolution has the effect of making matter and spirit unknowable. Navigating a way through and beyond the poles of metaphysical dogmatism (whether mechanism or dynamism) and critical philosophy becomes necessary in order to demonstrate that the 'interest' of space and time is not 'speculative' but *vital* (238; 212; see also the discussion of Kant at 259ff.; 230ff.). This is why Deleuze insists that it was important to Bergson to demonstrate the entirely empirical character of the *élan vital*, that is, as something that is lived. It will then become possible to gain an insight into the germinal character of life in which the separation between things, objects, and environments is neither absolutely definite nor clear-cut, for 'the close solidarity which binds all the objects of the material universe, the perpetuality of their reciprocal actions and reactions, is sufficient to prove that they have not the precise limits which we attribute to them' (235; 209). Kantian realism cannot be taken as a solution to the problem of metaphysics and as constituting a decisive post-critical break simply because it has raised homogeneous space as a barrier between the intellect and things without real philosophical foundation.

Throughout his work Bergson performs an important critique of natural perception, which is a model of perception that reduces the activity and becoming of life (movement) to a centred subject of perception. Hence he writes: 'to perceive consists in condensing enormous periods of an infinitely diluted existence into a few more differentiated moments of an intenser life, and in this summing up a very long history. To perceive means to immobilize' (233; 208). Not only is the border between an organism and its environment never clear-cut, being always porous and sympathetic, but so are the boundaries which separate and divide bodies: 'the close solidarity which binds all the objects of the material universe, the perpetuality of their reciprocal actions and reactions, is sufficient to prove that they have not the precise limits which we attribute to them' (209). 'True evolutionism', therefore, must assume the form of a study of 'becoming' (Bergson 1962: 369; 1983: 370). But this requires that we do not follow the path of perception which would reduce an 'infinite multiplicity of becomings' to the single representation of a 'becoming *in general*' (303–4; 304).

In Bergson's thinking of time as duration the emphasis is on its virtual character, in particular on time's past which always 'grows without ceasing' and which possesses an infinite capacity for novel re-invention. Duration is essentially the 'continuation of what no longer exists into what does exist' (Bergson 1965b: 49). It is in this context that he outlines his conception of memory, which is to be understood neither as a drawer for storing things away nor simply as a psychological faculty of recollection. Whereas a faculty works only intermittently, switching on and off as it pleases, the reality of the past is a deep and productive unconscious that evolves automatically. Memory (*le souvenir*) is not involved in representation, it cannot be said to represent anything (if we must speak of it as a representation Deleuze suggests that we conceive it as a memory of the present and the future) (Deleuze 1956: 100; 1997: 17). We thus arrive at the definition that duration 'is the continuous progress of the past which gnaws into the future and which swells as it advances' (Bergson 1962: 5; 1983: 5). Duration involves a process of repetition and difference (Bergson 1959: 122; 1991: 111ff.). It is irreversible since 'consciousness cannot go through the same state twice. The circumstances may still be the same, but they will act no longer on the same person, since they find him at a new moment of his history' (Bergson 1962: 5; 1983: 5). Even if states can be repeated and assume the character of being identical, this is merely an appearance, so we cannot live over and over again a single moment. The future, which is implicated in the past's becoming, is both irreversible and unforeseeable. Bergson speaks here of agents but his analysis can readily be extended to systems. Indeed, as Deleuze insists in his essay on Bergson, movement cannot be reduced to a question of merely psychological duration since it belongs to 'things' as much as to consciousness (Deleuze 1966: 44, 76; 1988: 48, 76). Moreover, duration is not for Bergson, Deleuze rightly claims, reducible to psychological experience, but constitutes the 'variable essence of things' and provides the theme for a complex ontology (27; 34) (on memory itself as a system property that is not reducible to the synaptic mechanisms necessary for its establishment, see Edelman 1994: 238).[3]

Indeed, this emphasis by Deleuze on a complex ontology captures a crucial aspect of Bergson's theory of duration, one that Bergson is keen to explore in his text on Einstein and relativity, *Duration and Simultaneity*. The key problem that is in need of an account and a demonstration, he states, concerns how we are to pass from the 'inner time' of a perceived duration to the 'time of things' (1965b: 45). How is it possible, furthermore, and by extension, to comprehend the 'duration of the universe', so as to be able to access that 'impersonal consciousness' that links all individual consciousnesses and the rest of nature? Can one advance the hypothesis of a physical time that is one and universal? We cannot examine the complex details of Bergson's argument here, but let us take cognizance of one point he makes that sheds a certain light on his usage of the word 'consciousness'. Bergson appreciates that using the word 'consciousness' to account for the endurance of things is likely to induce an aversion in many readers simply because of its anthropomorphic attachments. His response is to argue that in order to conceive of duration it is not necessary to adopt one's own memory and transport it into the interior of the thing. In thinking the impersonal character of the consciousness that belongs to duration we are not to preserve the personal 'human' character of memory, but rather pursue the opposite course (48). We are able to link consciousness with duration once we appreciate that 'perceived' time is also 'conceived' time.

We may also note one further aspect of Bergson's conception of duration, which is that a 'becoming' which flows out of previous forms while always adding something new to them does not entail a commitment to Spinozism, or what Bergson takes to be Spinoza's conception of the 'one complete Being' which manifests forms. For Bergson this would be to deny 'effective action' (*action efficace*) to duration (Bergson 1962: 361; 1983: 362). *Creative*

Evolution reiterates a point Bergson initially made in *Time and Free Will*, in which he argued that both Cartesian physics and Spinozist metaphysics seek to establish a relation of logical necessity between cause and effect and in so doing 'do away with active duration' (Bergson 1960: 208–9).

Bergson's relation to Spinoza, and the extent of the affinity between Bergsonism and Spinozism, is an immensely complicated issue which, unfortunately, cannot be explored here in any concerted way. I shall simply note a few points. In his instructive study of the '1960s' Deleuze, Michael Hardt argues that Deleuze approaches the opening of Spinoza's *Ethics* as a way of rereading Bergson, presenting the proofs of the existence of God and the 'singularity' of substance in terms of a meditation on the positive nature of difference (Hardt 1993: 60). He goes further and claims that a common ontological grounding of difference can be identified in the two thinkers. This has two major aspects: first, internal (efficient) causality, in which the cause is internal to its effect, and second 'immersion in the absolute' (61–2). However, he never contends with those moments in Bergson's text when an ontological affinity with Spinoza is explicitly repudiated. It is clear that for Bergson duration is not to be modelled along the lines of a Spinozism (the precise role of efficient causality in Spinoza's conception of God or nature is itself a highly complicated issue, see Spinoza 1956: Part 1, Proposition 16, Corollary 1 and Proposition 18). A principal reason informing his attempt to distance his conception of creative evolution as involving duration from Spinozism must surely reside in the radically inventive character he wishes to give it. In Deleuze's 1956 essay on the conception of difference in Bergson, we find the emphasis being placed not on an efficient causality, as Hardt maintains, but rather on the manner in which Bergson transforms duration into the only conceivable notion of substance, however strange it might seem at first to treat substance in terms of processes and movements rather than things and objects. As the tendency of self-differentiation what differs from itself is, contends Deleuze, '*immediately* the unity of substance and subject' (1956: 89; 1997: 8). If difference becomes what it is (self-differentiation) as and through duration, then this can only mean that 'movement is no longer the character of something, but has itself taken on a substantial character' (ibid.).

Here Deleuze is simply remaining faithful to Bergson's position that 'change' is primary in relation to things and has no need of a support: '*There are movements, but there is no inert or invariable object which moves: movement does not imply a mobile*' (Bergson 1965: 147). Hence his seemingly paradoxical argument that there is change although there are not *things* which change. Indeed, in the 1956 essay Deleuze argues that elemental causes essentially belong to the domain of quantity and, as we have already noted, that the only 'things' which can be said to differ in nature are not states or traits of things but tendencies. The tendency is said to be primary in relation to both its product and causes in time, which are retroactively obtained from the product itself (Deleuze 1956: 83; 1997: 4). Deleuze gives the example of human and animal brains. If considered in terms of a product the human brain will show only a difference of degree in relation to the animal brain. If it is viewed in terms of its tendency, however, it will reveal a difference of nature (this stands in marked contrast to Darwin's construal of the issue in the *Descent of Man*, where the difference is taken to be one only of degree, Darwin 1993: 329–30, 347ff.; the same appears to be true of Spinoza if we consider the difference between the finite mode of the human mind-body and that of an animal, in which the difference is one of degree of complexity – the finite mode of the human suffers a greater array of affects in relation to the environment, see Hampshire 1981: 76). To neglect the move being made from causes to tendencies is, in effect, to reduce differences of nature to differences of degree and to lose sight of the inventive character of duration which actualizes itself only as 'novelty': 'Indetermination, unpredictability, contingency, liberty always signify an independence in relation to causes' (Deleuze 1956: 111; 1997: 25; and see Nietzsche 'contra Spinoza and his causalism', 1968 Section 627). Of course, one could generously read Spinoza's substance in such terms, but then one would be reading him in terms of a Bergsonism (for further insight into Bergson on substance see Lacey 1993: 104–8). At the end of *Difference and Repetition* Deleuze argues that 'all' that is required for Spinozism to turn the univocal into an object of pure affirmation is 'to make substance turn around modes' (Deleuze 1968: 388; 1994: 304) (on this elevation of the modes compare Hegel 1995: 289; and for Deleuze's reading of duration and eternity in Spinoza as a 'difference of nature', see Deleuze 1968: 289ff.; 1992: 310ff.).[4]

Hardt makes the turn to efficient causality as he wants to understand how it is that Deleuze can posit a difference that is able to sustain its being, that is, as positive and productive being difference cannot be merely accidental, which it would be if it was subject to merely external determinations or causalities. His claim that Deleuze does not criticize causality *tout court*, but only rejects external conceptions in favour of an internal, efficient notion, may be legitimate and incisive. But if this is the case then the crucial move to be made resides with the conception of difference being advanced in terms of this radical notion of duration, which is never noted in Hardt's study (indeed, duration does not figure in his book; for Deleuze on efficient causality in Spinoza see 1968: 147–52; 1992: 162–7, where he argues that Spinoza's immanence 'expresses' a double univocity of cause and attribute that rests in turn on the 'unity of efficient and formal cause').

Let us return to the character of a creative evolution for Bergson. The process of evolution is not only one of change but of invention, since the forms do not exist in advance. The process involves not a realization of the possible but an actualization of the virtual, in which the virtual enjoys its own 'consistency' (*consistance*) as a productive power of differentiation (see

Deleuze 1956: 98). Here what is actualized does not come to resemble the virtual since evolution does not merely follow rules of resemblance and limitation.[5] At the same time it is important to appreciate that the virtual is only real in so far as it is actualized. There is no fixed metaphysical opposition between the virtual and the actual; rather, the virtual 'is inseparable from the movement of actualization' (Deleuze 1966: 36; 1988: 42–3). This is a movement that takes place through the creation of lines of differentiation. Being is internal difference, an explosive force, and can only be approached in terms of a 'logic of nuances', not a logic of contradiction. A combination of two contradictory concepts is unable to present either a diversity of degrees or a variety of forms. Against the dialectic, therefore, Bergson insists that difference is not determination but differentiation (Deleuze 1956: 92; 1997: 14). And against Darwinism it is maintained that creative evolution cannot be a mere accidental property of life since this would be to reduce being as primary difference to a mechanism governed by a purely external causality (Deleuze 1966: 102; 1988: 99). A purely mechanistic biology conflates the passive adaptation of inert matter, submitting to environmental influence, with the active adaptation of an organism. Bergson gives the example of the eye, noting that to say that the eye 'makes use' of light is not simply to declare that the eye is capable of seeing. Rather, it is to refer to the exact relations between an organ and the apparatus of vision. Moreover, it would be absurd, he insists, to say that the influence of light has 'physically caused' the formation of those systems that are continuous with the apparatus of vision, such as a nervous system and an osseous system. Given the limitations of such a reductive view it becomes necessary to attribute to 'organized matter a certain capacity *sui generis*...of building up very complicated machines to utilize the simple excitation that it undergoes' (1962: 72–3; 1983: 71–2).

The emphasis on the virtual rules out any notion of a simple unilinear evolutionism in which evolution's task would be reduced to one of simply bringing into realized being something that already existed in a nascent state. Although this may involve a process of invention such a process is no more than the execution of a plan or programme (the notion of the virtual is treated at length in Bergson 1959/1991: Chapter II; and the contrast between the processes of the virtual/actual and the possible/real is discussed in the Introduction to Ansell Pearson (ed.), *Deleuze and Philosophy*, 1997). Bergson's emphasis throughout is on divergent lines of evolution in which evolution does not entail a simple movement from one actual term to another in a homogeneous unilinear series, but rather partakes of a movement from the virtual to the heterogeneous that actualize it along a ramified series (Deleuze 1966: 103; 1988: 100). Life is not like geometry in which things are given 'once and for all'. If time is a virtual reality then it makes no sense to break the movement into distinct instants since there cannot be an instant which immediately precedes another instant any more than we can say that one mathematical point touches another (Bergson 1962: 21; 1983: 21). It is only the confusion of space and time, and the assimilation of time into space, which generates the illusion that the whole is given in advance of its becoming (Deleuze 1966: 108; 1988: 104). The whole is neither given nor giveable 'because it is the Open' whose nature it is to constantly change and to give rise to the new (Deleuze 1983: 20; 1986: 9).[6]

Bergson insists that there is no universal biological law that can be automatically applied to every living thing. Rather, there are only directions from which are thrown out relative finalities. Bergson holds that there is an unbroken continuity between the evolution of the embryo and the complete organism: 'The impetus which causes a living being to grow larger, to develop and to age, is the same that has caused it to pass through the phases of embryonic life' (Bergson 1962: 18; 1983: 18). Development is nothing other than a perpetual change of form. In the case of the organism, for example, do we say that it is growing old or that it is still the life of an embryo continuing to evolve? In the case of the human organism the crises of puberty and menopause are 'part and parcel of the process' of ageing, comparable to changes which occur in larval or embryonic life. This is to say that such events do not occur *ex abrupto* from without. Puberty is 'in course of preparation at every instant from birth' (19; 19). Thus, Bergson is able to claim that what is 'properly vital' in growth and ageing is nothing other than the 'insensible, infinitely graduated, continuance of the change of form' (ibid.). The phenomena of organic destruction is an integral feature of this evolution of the living being which, like that of the embryo, implies 'a continual recording of duration', that is the persistence of the past in the present:

> At a certain moment, in certain points of space, a visible current has taken rise; this current of life, traversing the bodies it has organized one after another, passing from generation to generation, has become divided amongst species and distributed amongst individuals without losing any of its force, rather intensifying in proportion to its advance.
>
> (1962: 26; 1983: 26)

One source of intensity for Bergson is germinal life. He refers to, and takes cognizance of, Weismann's distinction between soma- and germ-plasms. Although he does not fully endorse Weismann's conception of the continuity of the germ-plasm contained in the sperm and the egg, Bergson does want to advance the idea of there being a continuity of 'genetic energy' which involves the expenditure of energy at certain instants, just enough to give to embryonic life the requisite impulsion, which then is 'recouped' again in order to assure the creation of new sexual elements in which it 'bides its time'. When treated from this perspective, life can be construed as '*like a current passing from germ to germ through the medium of a developed organism*' (Bergson 1962: 27; 1983: 27). Life appears in the form of an organism as if the organism were only an 'excrescence, a bud caused to sprout by the former germ endeavouring to continue itself in a new germ' (ibid.). However, this genetic energy does not guarantee the production of the same but is a source of novelty in which life perpetually reinvents the character of its own evolution or becoming. Although Bergson is close to Weismann he does not wish to locate the vital principle in something as specific and self-contained as a germ-plasm, since this would reduce the scope within 'creative evolution' for invention and innovation. For Bergson what is transmitted is not simply the physico-chemical elements of the germ-plasm but also the vital energies and capacities of an embryogenesis and morphogenesis that allow for perpetual invention in evolution.

Bergson's *Creative Evolution*

Introduction

Until well into the 1960s there was no well-established tradition of Darwinian theory in France. Indeed, it was not until 1965 that a Darwinian was appointed to the Chair of Evolutionary Biology at France's premier Faculty of Science in Paris. Bergson, with his alleged 'vitalism', is traditionally portrayed as being outside the Darwinian tradition. However, caricatures of Bergson's position prevail which mean that we often overlook the extent to which there is a serious and informed engagement with Darwinian theory in *Creative Evolution*. Moreover, a great deal of Bergson's position can be seen to resonate with contemporary developments in biophilosophy, such as complexity theory, where the focus is on an understanding of living systems as dynamical systems, in which organisms do not simply passively adapt to changes in the environment but rather are seen to develop internal structures which serve to mediate the environment, including the 'meaning' an environment has for an organism as a complex living system. There is also in current work in evolution and ethology a questioning of how we configure and delimit the boundaries of an organism that is highly consonant with Bergson's approach. This work will be examined in some detail in Chapter 3. In this chapter I want to remain within the parameters of Bergson's argument, showing that, although it has problematic aspects, such as a questionable, albeit highly complex, hylomorphism (see Merleau-Ponty 1988: 66), it contains much that remains important for an appreciation of Deleuze's thinking. In his two major pieces of writing on Bergson, of 1956 and 1966, Deleuze's analysis remains within the ambit of the context in which Bergson developed his conception of creative evolution. This context is that of Weismann's neo-Darwinism and its break with the Lamarckian account of evolution and its principal doctrines of the inheritance of acquired characteristics through the effort of the animal or organism itself and the use and disuse of organs. This means that Deleuze's early readings focus on the challenge Bergson presents to evolutionist thinking independent of any cognizance of the revolutions that have characterized neo-Darwinism in the twentieth century. But it is cognizance of these revolutions, notably, the modern synthesis of the 1930s and 1940s (population thinking, for example) and the discovery of DNA and the rise of molecular biology in the 1950s, which mediate his reading and utilization of Bergson in the 1980 work *A Thousand Plateaus*. By this time Deleuze has taken on board neo-Darwinism as we conceive it today, as opposed to how it would have been understood in Bergson's time, and the work of key *French* Darwinian thinkers, such as Jacques Monod and François Jacob, has come to occupy a prominent place in Deleuze's biophilosophical thinking. But, as we shall see in Chapter 3, this is not at the cost of Deleuze's attachment to a certain 'Bergsonism'. The character of this Bergsonism does undoubtedly undergo significant modifications in Deleuze's later work, but the attempt to formulate a 'science of multiplicities' can be shown to rest on a novel synthesis of neo-Darwinism and the Bergsonism propounded in the 1966 text. In what follows I aim to provide a fairly close and incisive reading of Bergson's text before turning to examine how its chief elements are configured by Deleuze in his study of 1966. I shall concentrate my attention on the first three major parts or books of *Creative Evolution*, ignoring its final part on the Nothing and the history of philosophy from Plato through to Spencer, which repeats principal insights that are articulated to the similar effect in the preceding parts. The question 'what is creative evolution?' will be addressed and its meaning unfolded through this close analysis of the text's three major parts.

Beyond mechanism and finalism (Darwin and Lamarck)

I begin with some general remarks about the conception of 'evolution' Bergson advances in *Creative Evolution*, first published in 1907. Bergson claims that time (heterogeneous and continuous duration) has not been taken seriously in previous science, which remains one of his most enduring contributions to a creative mapping of evolution. It is his contention that both common sense and science deal with isolated systems, which are systems that realize themselves 'in the course of time' (1962: 9; 1983: 9). Time is reduced to a process of realization on account of the fact that mechanical explanation treats both the past and the future as calculable functions of the present (37; 37). As a result time is deprived of efficacy and, in effect, reduced to nothing, having just as much reality for a living being as an hour-glass. This is true of both mechanism and finalism for Bergson; indeed, he contends that finalism is merely an inverted mechanism, substituting the attraction of the future for the compulsion of the past and conceiving the order of nature on the model of a realization of a plan (39; 39). Our intellect is not designed to perceive creative evolution at work. For all sorts of practical needs and utilitarian calculations it needs to divide time into instants and segments. In *Creative Evolution* Bergson endeavours, therefore, to steer a course beyond the opposition of mechanism (neo-Darwinism) and finalism (neo-Lamarckism) by developing a conception of evolution which placed the emphasis on the creation of forms and the continual elaboration of the 'absolutely new' through 'invention' (Bergson 1962: 11; 1983: 11). He is careful to distinguish his own work from illegitimate vitalisms and not to appeal to anything mysterious. After having noted that the inventive character of vitality resides in its tangential relationship to physical and chemical forces at any and every point, he acknowledges that while an appeal to a 'vital principle' may not explain much it does at least serve to remind us of our ignorance, when mechanism simply invites us to ignore the ignorance. The error of both mechanism and finalism is that both treat the past and future as calculable functions of the present and so both are unable to think how evolution involves an inventive duration.

In *Creative Evolution* Bergson will oscillate between identifying in matter itself a tendency to constitute isolable systems that can be treated geometrically, and locating such stratification solely and squarely within the consciousness of the intellect which is compelled to effect a spatialization of time in order to procure the satisfaction of its practical needs. However, while the subdivision of matter into distinct bodies is relative to our perception, is it also true to claim for Bergson that *nature itself* has closed off and made separable the living body (Bergson 1962: 12; 1983: 12). What is the difference between the two procedures? Bergson notes that there is an 'order of nature' in the sense, as the Greeks recognized, that the idea of genus corresponds to an objective reality, namely the fact of heredity. There can only be genera, therefore, where there are individual objects. But while an organized being is 'cut out from the general mass of matter' by this organization and individuation, which is to say 'naturally' Bergson notes, it remains the case that the schema of life and evolution we confer on this organized individual body is necessarily decided by our perception that always cuts into matter distinct bodies, in which, and on account of the interests of action, 'genera and individuals determine one another by a semi-artificial operation entirely relative to our future action on things' (228; 227). Bergson insists, therefore, that although it is the case that nature itself has made separable the living body of life, creative evolution is only possible to the extent that it involves 'sympathetic communication' and the activity of 'reciprocal implication and interpenetration' between the parts (Bergson 1962: 179, 190; 1983: 178, 189).[7]

This problem will come to inform the argument in *A Thousand Plateaus*. Deleuze and Guattari remain Bergsonian by maintaining, for example, that the organism is merely an expedient invention by which life opposes itself in order to limit and reinvent itself. The truly intense and powerful life *remains* anorganic (Deleuze and Guattari 1980: 628; 1988: 503). Bergson stages the play between the nonorganic and the organismic as one of co-implication (the play between the organism and nonorganic life is also staged in terms that remind one of Bergson in Vernadsky's now widely recognized classic study of the biosphere. Although not explicitly referenced Bergson is one of the most important influences on Vernadsky, 1988: 53, 97–8, 111). Evolution is creative for Bergson precisely because it is characterized by an unending conflict between the cessation of the flow of becoming – in the creation of forms such as organisms and species – and the tendency of the flow of becoming to break out of any fixed or stable determination. Deleuze, as we shall see, stresses that for Bergson evolution is not accidental but creative precisely because it proceeds via indetermination. For Bergson matter never goes 'to the end', with the result that the isolation of life 'is never complete' (this is one of the reasons why Bergson is such an interesting and complex finalist). Whatever remains excessive to an isolable system is untouched by science since it regards the existence of this excess as negligible to its own monitoring of a system's activities. But it is this excess which is important to Bergson: individuality is a characteristic property of life even though it is difficult to determine precisely its boundaries and borders. As Bergson acknowledges, individuality admits of any number of degrees and is never fully realized anywhere, 'even in man' (ibid.: 12; 12). He gives the obvious example of plants, the subject of a more recent treatment by the biologist Richard Dawkins (Dawkins 1983: 253–4). The reason why individuality can never be fixed once and for all is due precisely to the vital character of life. A perfect definition could only be intelligible in relation to a completed reality, but 'vital properties' are never such realities since they exist only as 'tendencies' never as 'states'. Contemporary biologists who work within a certain Kantian tradition of thought, and place the emphasis on the self-organizing aspects of life, similarly locate the vital properties of life in the nonequilibrium conditions and specific kinetic pathways which inform the organization of living systems such as cells and organisms (compare Wicken 1987: 30–1).

Bergson admits that his thinking of creative evolution necessarily partakes of a certain finalism. But he has little time for the idea to be found in Leibniz, and best expressed in his well-known conception of monads as *windowless*, meaning that nothing can enter into or depart from them, that organismic life is characterized by an 'internal finality' (Leibniz also describes monads as autarkic entelechies, 1991: 58–9, 87–8; compare Kant 1974/1982: Section 64–5). First, he points out that each of the elements which are said to make up the organism may be organisms themselves. Second, he contends that the notion of a finality that is always internal is a self-destructive notion since it is incapable of accounting for a whole range of phenomena, from phagocytes which attack the organism that feeds them, to germinal cells which, he holds, following Weismann, have a life of their own alongside the somatic cells (Bergson 1962: 41; 1983: 42). It is precisely because there exists neither internal finality nor absolutely distinct individuality that the position of vitalism is so intangible. For Bergson if there is finality in the world of life it can only refer to the 'whole of life in a single indivisible embrace'.

It should now be apparent that Bergson means something quite specific with 'evolution'. He wishes to privilege not the thing produced but the activity of evolution itself: that which is responsible for the generation of new forms. This is important since it allows him to distance himself from accounts which posit evolution as a simple or straightforward passage from the homogeneous to the heterogeneous (1962: 49; 1983: 49). Bergson's point is that this is to assume a vantage point by which one can appraise the complex pathways that have characterized evolution, turning evolution in the process into a linear descent and imposing on it a logic of perfection and progress. This betrays an anthropocentric conceit, namely, the view that that which is more complex is higher in the scale of nature, with man being posited as is the ultimate object and goal of evolution. Bergson's position is more complex than this, though even informed readers end up attributing such a position to him. Like finalism Bergson's model of creative evolution is committed to a principle of harmony. This harmony, however, proceeds by way of discord and can never be pinpointed as such (its 'vitality' is virtual). The 'identity' of life is due to an 'identity of impulsion' rather than a 'common aspiration' (Bergson 1962: 51; 1983: 51). It is in this context that he clearly states it is illegitimate to assign to life an 'end', since this is human, all too human. On the model of creative evolution the positing of an 'end' of evolution makes no sense since this would be to reduce the creative process to the realization of a pre-existing model. Only once the road has been travelled is the intellect able to mark its direction and judge that where it has got to is where it was going all along. But this is no more than a deception since 'the road has been created *pari passu* with the act of travelling over it, being nothing but the direction of this act itself' (ibid.).

The movement of evolution is a complicated one simply because the evolution of life has not been characterized by a single direction. Rather its movement can be compared to that of an exploding shell bursting into fragments: shells which in turn continue to burst into other fragments. Continuing this analogy further Bergson speaks of evolution in terms of the breaking of a shell that involves both an explosive force (the powder it contains) and the resistance it encounters (in the metal). Thus, the way life itself evolves into individuals and species depends on two similar causes, namely, the resistance of inert matter and the explosive force that life holds within itself owing to an unstable balance of tendencies. Life enters into the 'habits of inert matter' and from this learns how, little by little, to draw from it animate forms and vital properties. The 'complex and quasi-discontinuous organism' clearly arises from smaller, more elemental, prototypes but in advancing in complexity such an organism introduces into life new components and evolves via new habits (the contrast with Freud's account in 'Beyond the Pleasure Principle', to be examined at length in the next chapter, is striking). Life for Bergson is essentially a 'tendency' that creates, through its growth and becoming, divergent directions among which its 'vital impetus' gets divided (Bergson will later admit in the text that the notion of 'impetus' offers only an 'image' for thought to think creative evolution). Unlike an individual life which must choose the interwoven personalities that characterize it, nature preserves the different tendencies that bifurcate.

I have hinted at the claim that Bergson is in some sense a finalist, albeit a complex one. Indeed, he acknowledges a link with finalism in the course of his argument. It is particularly interesting to note the way that he both identifies with, and radically divorces his argument from, neo-Lamarckism. This is in the context of a discussion of the now widely discredited thesis of orthogenesis or directed evolution (which also exerted an influence on Nietzsche). In accounting for the evolution of a complex organ such as the eye, neo-Lamarckians, Bergson notes, resort to the intervention of a psychological cause. Bergson suggests that this is the strength of its position. Where he departs from it is with the idea that conscious effort is a property of the individual. The role of conscious effort in evolution is for Bergson limited to a small number of cases. He correctly notes that where it does act it is unable to account for a change so profound as an increase in complexity. He notes further that this would require not simply the transmission of acquired characters but also their transmission in a cumulative form. Such a hereditary change in a definite direction, the direction of a more and more 'complex machine', Bergson holds not to be possible via Lamarckian means. Instead he sides with Weismann when he argues that the 'effort' of evolution, the work of its energy in the direction of more complex self-organization, must be of a much greater depth than individual effort and more independent of circumstances; in short, it must be an effort common to most of the representatives of a species 'inherent in the germs they bear rather than in their substance alone' (1962: 88; 1983: 87). To what extent this is consonant with Darwinian theory we shall not speculate upon until we have examined more closely the notion of the *élan vital* in a later section and in connection with Deleuze's interpretation of the significance of Bergsonism in relation to Darwin.

Bergson claims in *Creative Evolution* that both mechanism and finalism are, ultimately, attempts to conceive evolution along the lines of the workings of the human mind. He thus claims, somewhat radically, that the reproach mechanism makes against finalism of anthropomorphism can also be applied to it. Both conceive of nature working like a human being in bringing parts together and proceeding via association and the addition of elements. Bergson suggests that a glance at the development of an embryo will readily show that life works in a very different manner, namely, via dissociation and division (self-differentiation). Bergson never denies that selection plays a crucial role in evolution and that it is a necessary condition of it. Species would simply disappear, he notes, should they not bend to the conditions of existence in which they aim to survive. But he goes on to make two critical points. The first is that evolution is equally impossible without an 'original impetus'. It is as if Bergson is here positing a kind of 'life-drive' as essential to any understanding of the process of creative evolution. When we glance at fossil species, he says, we are exposed to the fact that life need not have evolved at all and might have evolved only in very restricted limits. The second critical point he makes is that if we accept that life is inventive even in its adaptations then these inventions serve to make life itself more complex in the very matter and mechanism of its evolution. In other words, life does not only involve adaptations dictated by external circumstances. Again, he is insistent that to entertain this idea is not at all to posit evolution in terms of some plan or programme. Neither is it perfectionist or progressivist in any straightforward sense since, he notes, evolution is not only a movement forward but equally a deviation and a turning back:

> Before the evolution of life...the portals of the future remain wide open. It is a creation that goes on for ever in virtue of an initial movement. This movement constitutes the unity of the organized world – a fecund unity, of an infinite richness, superior to any that the intellect could dream of, for the intellect is only one of its aspects or products.
>
> (Bergson 1962: 106; 1983: 104–5)

In short, his argument with Darwinism is that it lacks the notion of 'activity' that has to be seen as essential to any understanding of the evolution of life (the closeness to Nietzsche is, let us note again, striking). He accepts the fundamental principles of neo-Darwinism that what is passed on in heredity are accidental modifications of the germ and that natural selection may eliminate forms of instinct that are not 'fit to survive'. He also concurs with Darwinism in the view that evolution involves and takes time. His critical point, however, is to stress that the instincts of life could not have evolved in complexification by a process of simple accretion since each new element or piece requires a 'recasting of the whole', a recasting which, he contends, mere chance could not effect. In other words, complexity in evolution cannot itself be simply the result of an exogenous mechanism (such as natural selection) (1962: 170; 1983: 169).

The significance of this conception of creative evolution for a philosophy of difference will be brought out in the section on Deleuze and the *élan vital* (p. 65). Bergson's quarrel with mechanism and finalism can be summarized as follows: the former errs in making evolution a purely accidental process, while the latter errs in reducing it to a phenomenon of individual effort.

Plant and animal

Bergson begins this part of his second book by noting that there is no definite or certain characteristic that separates these two forms or kingdoms of life. Properties of vegetable life are found in degrees in animal life, while characteristic features of the animal have been actualized in the vegetable world at certain moments in time. He goes on to claim that there is no expression of life that does not contain, whether in a rudimentary state or whether as latent, essential characters associated with diverse expressions. Thus, the difference between forms of life could be said to lie in the proportions. For Bergson the only truly valid way to demarcate zones of life is in terms of their virtual tendencies and not their actual states, that is, the more a form of life develops the more it tends to emphasize particular characteristics. Between plant and animal, therefore, we can locate divergent tendencies within creative evolution, such as alimentation where the vegetable derives the elements necessary to maintain to life directly from the air, water, and soil, while in the case of the animal the activity of assimilation is always mediated.

Another tendency marking a divergence between plant and animal is the balance in favour of fixity in the one and that of mobility in the case of the other, with the latter clearly playing a crucial role in consciousness and the development of the nervous system. However, Bergson refuses to reify the role played by the nervous system, noting that neither mobility nor consciousness require such a system as a necessary condition (a position we may simply note that is in line with current research in neurophysiology and neurobiology). The nervous system has merely brought to a higher degree of intensity a quite rudimentary and vague activity that can be shown to be diffused throughout the mass of the organized substance: 'it would be as absurd to refuse consciousness to an animal because it has no brain as to declare it incapable of nourishing itself because it has no stomach' (1962: 111; 1983: 110). This demystification of the brain informs Deleuze and Guattari's stress on 'microbrains', as an expression of 'the inorganic life in things' in the final chapter of *What is Philosophy?*. Here they seek to go beyond the subject-centredness of phenomenology that relies on a subject of natural perception in order to posit a world by claiming in opposition to the proposition that 'consciousness is always consciousness *of* something' that 'consciousness *is* something' (1991: 200; 1994: 213). Similarly, in his book of the late 1980s entitled *The Fold*, Deleuze claims that *proteins* already 'attest' to the activity of a certain perception and discrimination, and he remarks that here are revealed the existence of those 'primary forces' that physical impulses and chemical affinities are unable to explain (Deleuze 1988: 122;

1993: 92). For Bergson the most humble organism is conscious in proportion to its capacity to move freely. The nervous system, in other words, does not create the function but brings it to a higher degree of intensity and precision through a division of labour that modifies reflex movement and voluntary activity (on the relationship between the nervous system and memory see Bergson 1959: 279–80; 1991: 248–9). So even where there has not taken place a canalization of nervous elements and their concentration into a system it is still possible for a living organism to be conscious. In the case of plant life Bergson does not want to quickly jump to the conclusion that, owing to its fixity, it is solely unconscious simply because even in its fixity there is a kind of mobility as well as aspects of its life which reveal the 'liberty of movement'. The difference between animal and plant is, therefore, proportional and not absolute. Bergson argues that both tendencies of life were co-present in earlier forms of life and that they arise from a common ancestor. In explaining the divergence of life into certain tendencies he insists further that it is not necessary to appeal to any 'mysterious force', but simply a matter of appreciating that vegetable life and animal life have selected two different kinds of convenience in the way of procuring the chemicals they need to sustain life.

Does this emphasis on proportions not contradict Deleuze's reading of Bergsonism with its emphasis on differences of kind? I would argue that it does not; rather, what it shows is that Bergson's thinking is informed, above all, by a logic of nuances. A thinking of difference as the difference of kinds is always implicated in such a logic. Equally there is a logic of the trace in Bergson's thinking, since although he is committed to the idea of evolution being characterized by elementary tendencies, he also holds that these tendencies bear the 'marks' or traces of an original tendency prior to its dissociation (this would be to think the existence of these traces not spatially but durationally) (Bergson 1962: 119; 1983: 118).

The crucial element that Bergson wishes to grant to life is not a mysterious force but rather a principle of 'indetermination'. It is this indetermination, and with it the capacity for novel adaptation, that he sees as being 'engrafted' onto the necessity of physical forces, so making possible a 'creative', as opposed to a purely mechanistic or deterministic, evolution. This 'effort' of life towards indetermination cannot result in any creation of energy; rather, what it does is to creatively utilize the energy that pre-exists, accumulating from matter and from the sun potential energy that is then stored in reservoirs to be utilized at a later time (a 'desired moment' and a 'desired direction'), so suspending on the surface of the earth a continual outpour of usable energy. In this argument it is as if Bergson is introducing into his conception of creative evolution the idea of there being within this evolution a 'time' of energy, a time that is created by the conditions of life on earth and without which this life would not even be possible. Life is informed by the ability of its forms and expressions to hold chemical energy in a potential state and which serve as little explosives that need only a spark to set free the energy stored within them. If Bergson is a hylomorphic thinker then he is one of a peculiarly interesting sort. He is an 'ideal materialist', as in his account, to be treated below (p. 59), of the *ideal* genesis of matter, holding that without the indetermination introduced into matter by life there would be no evolution (time enjoys an entirely 'positive' reality, Bergson 1965b: 63). Although he appeals throughout his writings to the spiritual character of this 'vital impulse' in the form of a consciousness that traverses matter, it is still based on an attention to the peculiarities of matter. It grants a high degree of importance to the role of contingency in evolution (although life is marked by tendencies it did not have to evolve in the way it has, a point to which we shall return). Bergson severely critiques radical finalism for overestimating the powers of a directed evolution since for him the emphasis needs to be placed on the limited character of the forces, the obstacles they encounter, and the desire to transcend these limits while living within them. The evolution of the living world cannot be modelled on that of a construction analogous to a human work: 'Even in its most perfect works', he writes, 'though it seems to have triumphed over external resistances and also over its own, it is at the mercy of the materiality which it has had to assume' (Bergson 1962: 128; 1983: 127). He then makes an analogy with the experience of human life and human freedom:

> Our freedom, in the very movements by which it is affirmed, creates the growing habits that will stifle it if it fails to renew itself by a constant effort: it is dogged by automatism. The most living thought becomes frigid in the formula that expresses it. The word turns against the idea. The letter kills the spirit.
>
> (Bergson 1962: 128; 1983: 127)

The 'animality' of a higher organism consists in the capacities it enjoys for self-regulation. The sensori-motor system operates in relation to systems of digestion, respiration, circulation, secretion, and so on, the task of which is to create an 'internal environment' for the organism by which it is then able to monitor and regulate itself. However, from the perspective of the *virtual* whole, life can be conceived as a ceaseless play between the limited inventions of complex living systems, such as organisms and species, and the desire of the impulse of life for ever renewed vitality. Life requires organisms in its very inventiveness; that is, it requires their limits, since left to its devices it would fain to do all at once and proceed in a straight line. It is in 'maternal love', says Bergson, that we may be exposed to life's secret since it shows us that the living being is, above all, a 'thoroughfare' and 'that the essence of life is in the movement by which life is transmitted' (129; 128). This is an insight into life's germinal character, emphasizing the dual importance of both the life that is living and the life that is lived on and gets passed on. The best reading of Deleuze, I would contend, would be the one that emphasized the indissoluble character of the organismic life and the nonorganic life, and which did not blindly and naively affirm the powers of the nonorganic independent of the vital contribution made to life in the virtual time of its actual creative evolution by the life of the organism (for Vernadsky the autonomy of organisms stems from the distinctive character of their thermodynamic fields, 1998: 97). It is one of the great achievements of Bergson's text, I would argue, to reveal the logic of nuances through which 'life' is to be grasped independently of a logic of contradiction with its attachment to principles of negation and supersession.

I wish to treat one more important point before moving on to Bergson's treatment of intelligence and instinct, which is this: how well does his argument stand up when placed alongside the theory of natural selection? I appreciate this is to raise an enormous question, so let me here restrict myself to making a single comment. Bergson, throughout his text, fully concedes the importance played within creative evolution by adaptation and by selection. Mobility among animals would not have evolved and developed unless there were quite specific advantages to be gained from it in particular conditions and particular environments. The transformation of species can thus always be explained on such a basis. However, while conceding this point and recognizing the importance of natural selection, Bergson insists that this is to give us only the 'immediate cause' of variation and not the most 'profound' one. For him this can only lie in the 'impulse' that introduces vital-durational life into the world, dividing it into vegetables and animals, granting to the animal a suppleness of form that made possible the evolution of greater mobility as the changed conditions necessitated such a direction and adaptation. Again, this is to reiterate Bergson's major contention against Darwinism, namely, that while it can very ably explain the 'sinuosities' of the movement of evolution, it is much less able to explain the movement itself. The Darwinian would reply to this, however, by arguing that Bergson has simply set up a spurious problem, namely, that of 'movement'. In other words, for the Darwinian 'evolution' is nothing other than its 'sinuosities', and any appeal to an independent movement of evolution is in danger of falling back on a mystical principle to account for transformism. Bergson's great challenge, however, one which the theory of evolution has perhaps yet to really confront or even comprehend, is to argue the need for a notion of duration, which for him is not merely applicable to psychological phenomena but also to organic phenomena too. The nature and significance of this argument may not yet be sufficiently clear to be graspable. It may become so when I look at Bergson's engagement with thermodynamics later in the chapter.

Merleau-Ponty claimed to have identified the contradictory character of the *élan vital* in Bergson's argument. He notes that the *élan* resumes the problem of organic nature as it was initially articulated by Kant and

Schelling, namely, as the attempt to describe a 'natural production' that proceeds from the whole to the parts without owing anything to the premeditation of the concept and without admitting any simple-minded teleological explanation. However, Bergson contradicts himself, Merleau-Ponty, argues, when he speaks of evolution as a 'simple act' and assigns a reality to the *élan* 'in advance of its effects as a cause which contains them pre-eminently' (Merleau-Ponty 1988: 146). When placed in the context of this kind of critical, but also cursory, reading of Bergson's so-called vitalism it is possible to appreciate just how bold and original Deleuze's 1956 essay on Bergson was in laying the emphasis on the radically inventive character of duration. In order to be inventive creative evolution must have the features of indetermination and unpredictability. It cannot be a 'simple act' in which effects are contained pre-emptively or prefiguratively in the *élan* conceived as a causal power.

Intelligence and instinct: the technics of the human and nonorganic life

Intelligence and instinct are said to be both opposite and complementary, with neither existing in a pure state. As with plant and animal, the two tendencies are originally interpenetrating and even in their severed states they continue to remain haunted by each other. Bergson refuses to equate technical intelligence with human endeavour but argues that many kinds of animals are capable of either fashioning instruments or utilizing objects created by man. 'Intelligence' is to be defined as the 'faculty of manufacturing artificial objects, especially tools to make tools' (1962: 140; 1983: 139). The crucial claim he makes, however, in an effort to comprehend the specific character of human intelligence, is that although an animal can be said to possess and manipulate tools and machines these remain instruments that form an essential part of the animal's body. This means that tools are used by animals on the order of instinct. By contrast, the specific character of human intelligence resides in the fact that in its case technics enjoy a functional indeterminacy. From the first, human praxis has been mediated by the invention of technics, says Bergson, as its 'essential feature'. We fail to recognize the extent to which the ontogenesis of the human is dependent on this technogenesis, and the extent to which human culture is wholly due to the utilization of artificial instruments, simply because our individual and social habits survive beyond the circumstances in which they were formed. The effects of technical inventions are thus not noticed until their novelty is well out of sight. Bergson thus argues that '*Homo sapiens*' can be defined as '*Homo faber*' to the extent that it has been formed prehistorically and historically by the invention of technics.

Bergson, it should be noted, is not positing an opposition between technics and 'life', but arguing that the activity of instinct at work in technics and its formation of intelligence is also at work in nature itself. The key concept at play here is that of 'organization', so that, for example, in the metamorphoses of the larva into the nymph and then into the perfect insect one witnesses no sharp distinction between the instinct of the animal (which requires an initiative) and the organizing work of living matter itself. The difference between instinct and intelligence resides in the utilization of instruments specific to each: in the former activity is concentrated on instruments that are already organized, while in the latter it consists in fabricating and utilizing *unorganized* instruments. The instrument that is fashioned by intelligence is always an 'imperfect' or indeterminate one – being made of unorganized matter it can assume diverse forms and be used for different purposes. Such an instrument is inferior to the natural instruments deployed by the instinct since it does not satisfy an immediate need. Its advantage is that it allows for the expansion of need and opens up an unlimited field of activity. The decisive point Bergson makes is that the construction and cultivation of imperfect instruments results in a fundamental transformation of 'the nature of the being' that utilizes them, to the extent that the natural organism becomes extended by the artificial organ. In the human we see perfected a tendency that is immanent within animal life as a whole and which, in turn, is itself consistent with the movement of living matter. The technics of the human amount to a transformation of nonorganic life into an organ itself. The paradox that currently unsettles much contemporary thinking about the bios of technics, of whether technics is an invention of man or man is to be viewed as an invention of technics (as an invention of life itself), is encountered in Bergson's text:

> Life, not content with producing organisms, would fain give them as an appendage inorganic matter itself, converted into an immense organ by the industry of the living being. Such is the initial task it assigns to intelligence. That is why the intellect always behaves as if it were fascinated by the contemplation of inert matter. It is life looking outward, putting itself outside itself, adopting the ways of unorganized nature in principle, in order to direct them in fact. Hence its bewilderment when it turns to the living and is confronted with organization. It does what it can, it resolves the organized into the unorganized, for it cannot, without reversing its natural direction and twisting about on itself, think true continuity, real mobility, reciprocal penetration – in a word, that creative evolution which is life.
>
> (1962: 162–3; 1983: 161–2)

The paradox that Bergson is trying to unfold here is that while the human is entirely a production of evolution conceived as the invention of time – a time that ultimately belongs to evolution and not to man – it itself, as a product and not as the process, is unable to think evolution 'in the proper sense of the word – that is to say, the continuity of a change that is pure mobility' (164; 163). Because the intellect has evolved as a means to organize matter through the mediation of technics it has a natural tendency to conceive only what is useful to it. An integral part of the formation of such an intellect is the tendency to conceive of everything spatially through modes of representation. This is what informs Bergson's claim that the intellect always represents 'becoming' as a series of homogeneous states that do not themselves change. The irony is that intelligence is constantly engaged in the creation of the new but is itself unable to recognize the novelty of its creative evolution. On account of its need for stable forms and controllable objects the intellect resolves the unforeseeable and the new into old and known elements, so that the finality it thinks is always the finality of industry: that is, working on a model that is given in advance and composed of elements already known and familiar. Bergson then notes another paradox of the intellect, namely, that while it is so skilful in dealing with the inert it is incredibly awkward when it comes to touch the living. The intellect finds itself much more at home, he contends, with the immobile and the dead. It is, he says, giving the history of hygiene and pedagogy as two examples of human experimentation, too blunt an instrument when it comes to dealing with what is fluid and continuous. For Bergson this limitedness of the human condition – to think the life and the duration it is the result of – actually reveals a paradoxical feature of the process of creative evolution itself as he sees it. This is the fact that through the creation of life through bounded species and distinct kingdoms it becomes 'cut off from the rest of its own work' (168; 167). In the case of human life we find that we trail behind ourselves all the time, unaware of the whole and virtual past, indeed unaware of the creative character of our own individual evolution or becoming.

Matter and consciousness: 'man' as the 'end' of evolution?

The question 'what is consciousness?' is one that continues to occupy and haunt the attention of all philosophers whatever their theoretical allegiances. In *Creative Evolution* Bergson evinces a unique conception of consciousness that places it firmly within the order of nature as he has so far unfolded it. He has reached the conclusion that instinct and intelligence represent two different and divergent, and equally fitting, solutions to one and the same problem (a problem posed by life to particular expressions of life). He does not deny that instinct is a form of 'consciousness' or that it is devoid of intelligence. The logic of nuances that is to be deployed in order to explore the interpenetrating character of all these things and tendencies must, however, discover the differences of kind which enable us to differentiate them. A state of unconsciousness, for example, can be more than one thing: it can be a state in which consciousness is merely absent, or it can be a state in which it is nullified (a difference between a stone falling and an animal or plant instinct).

Consciousness is defined by Bergson in very specific terms as the 'arithmetical difference between potential and real activity'. This 'difference' measures the interval between representation and action. Consciousness can be conceived as a 'light' which plays around a zone of possible actions and potential activity that always encompass any real action (action carried out in real time) by a living being. In signifying 'hesitation or choice' consciousness can either be intense, for example when there is endless deliberation over many equally possible actions, or reduced to nothing, as when an action performed is taken to be the only possible action. Consciousness is only possible to the extent that there is a 'deficit' of instinct. It may, therefore, appear in intense forms only in specific habitats and under specific environmental conditions, and it may be especially acute in its intensity in the case of the human owing to the particular contingent circumstances of its evolution (such as evolving on the open savannah, to give but one factor in what is a complex evolution involving many components). However, the evolution of the human may have been well prepared for in evolution through the development of intelligence in animal life, for here, Bergson argues, deficit, understood as the distance or gap between act and idea, is nothing other than the normal state of intelligence: 'Labouring under difficulties is its very essence' (146; 145). Moreover, because intelligence is implicated in the invention of technics it is compelled to evolve in terms of a futurity, involving, for example, the satisfaction of needs which always generate the creation of new needs. The evolution of needs, therefore, has to be seen as part of a virtual history of technical evolution.

Bergson now argues that human ontogenesis through technogenesis only takes place in the context of a community of intellects, namely, society. A community, whether it be an insect society or a human one, is made possible by a language of signs. However, in the case of the insects the subdivision of labour is natural in the sense that each individual member is riveted by its structure to a specific function. Equally such a formation is based on instinct in which action and fabrication are dependent to a great extent on the specific form of the organs. In addition, the signs which characterize the language of insects are limited in number in that each one remains invariably attached to a specific object and performs a specific operation: 'the sign is adherent to the thing signified' (159; 158). In the case of human society the picture is quite different on all these fronts. Here sign and fabrication assume a variable form and there are no preordained structures for individuals to occupy. The signs of human language may be finite in number but they are also capable of being extended to an infinity of things. The tendency of the sign to transfer itself freely from one object to another is for Bergson quite specific to human culture. This is observable, he says, in the child as soon as it begins to speak: ' "Anything can designate anthing" such is the latent principle of infantine language' (ibid.). It is not, he insists, so much a faculty of generalizing that is specific to the human animal, since other animals also generalize. Rather, what is specific to the signs of human language is that whereas the instinctive sign is 'adherent', that of the intelligent sign is 'mobile'.

Bergson's argument is more innovative than simply coming up with these interesting insights into the specific complex character of human language. He goes on to contend that on account of the human invention of its life through technics it is able to create for itself a 'surplus of energy' over and above the utilitarian and, in the process, becomes 'a consciousness that has virtually reconquered itself' (160; 159). It is language which serves to *actualize* the virtual intelligence of the human. Without it intelligence would have remained riveted to the material objects that it consumes and invests its energy in. The word is not only free in being able to be passed from one thing to another; it is also transferred from a perceived thing to a recollection of that thing, and from a precise recollection to a more fleeting image, and finally to the conception of an 'idea'. It is in this way that the human acquires for itself an entire new 'inner world'. In short, Bergson arrives at the view that the human is constituted in and through the supplementarity provided by both technics and language. From these insights it follows for Bergson that it is specific to the intellect that it should trouble itself with a philosophy of life, including a theory of inanimate matter and nonorganic life.

Consciousness is not so much a phenomenological property of man but can be seen to be distributed throughout life and always along divergent lines, including the division through which matter gets organized into the two complementary expressions of plant and animal life. The question Bergson now considers at the end of book two is this: Is consciousness to be considered as simply an aid to action or might action be the instrument of a consciousness? Consciousness can be understood as the power of choice or selection that brings into play a 'zone of potentialities' surrounding an action, constituting an interval between what is performed and what might be performed. In this way we might view consciousness as indeed a simple aid to action, whether conceived as effect or cause. Viewed from another angle we might equally claim that action offers a means for consciousness to evolve and cultivate itself, for in the complication and opposition of action consciousness is set the test of constantly having to establish and liberate itself. From this Bergson will now seek to argue that another aspect of life that is peculiar to the human is the ability of its consciousness to outstrip its physical concomitant (the brain). In a dog, for example, recollection is always a captive of perception so that what is brought back to consciousness follows the recollection of an analogous perception. The dog thereby remains fixed on the present perception. By contrast, the human has the capacity to bring up any recollection at will in complete independence from any present perception.

It is on the basis of this difference in *kind* between the human and animals that Bergson reflects on the possibility that man occupies a 'privileged place' within evolution *if* evolution is conceived as the emancipation of consciousness from its containment in matter. It has to be emphasized that for Bergson this is by no means to be thought in terms of either a necessary or a linear and progressive development. If it can be claimed that consciousness has entered matter as a virtual duration and invention then such a consciousness has to be viewed as loaded with an enormous multiplicity of interwoven potentialities in which movement is always retarded and divided. In evolution we find more often than not consciousness in a torpid state. The fabrication of intelligence in man, which has taken place through the fabrication of ever more technical evolution, leads to a condition in which invention always exceeds the demands and requirements of intelligence and its need for utilitarian calculations. Indeed, Bergson argues that the disproportion between the effect and the cause of invention is so great that it becomes difficult to simply regard the cause as the 'producer' of its effect. It is not simply a question of saying that evolution belongs to 'man', but rather that the human belongs to duration conceived as the time of a 'creative evolution'. Moreover, it is not simply the case that in the human the virtual becomes identical to the actual since, as should be clear from this reading of his argument, Bergson conceives the becoming of the human as a becoming that never ceases to take place in the dimension of the virtual. Implicated as it is in a virtual evolution the human always lives 'beyond' its own condition. This living beyond is the 'meaning' of its condition. As we shall see, it is only in a highly specific sense for Bergson that it is ever legitimate to speak of the human as the 'end' of evolution.

Philosophy's task: reconciling inorganic and organic life

In the opening section of book three of *Creative Evolution* Bergson establishes the conditions for a 'method of philosophy' that may prove adequate to a comprehension of nonorganic life. He begins by noting that if the division of unorganized matter into separate bodies is relative to our senses and intellect (to its peculiar tendency to spatialize), and if matter could be viewed from the perspective of an undivided whole, then it would be possible to conceive of life or creative evolution as a flux and not a thing. In this way it might be possible to prepare the way 'for a reconciliation between the inert and the living' (187; 186).

The first task is to attempt a genesis of both the intellect and material bodies, while recognizing that intellectuality and materiality have been constituted through reciprocal adaptation. In carrying out this task Bergson rejects both the evolutionist philosophy of Spencer and the critical philosophy of Kant as viable approaches. The claim against the former is that it assumes as ready-made the intelligence it is seeking to show in its genesis. Spencer accepts, Bergson is contending, the fact that bodies appear as distinct individualities and occupy distinct positions in space. Kant's deduction of the a priori categories of thought is more subtle and self-conscious. However, in drawing out of it what has been virtually put into it, compressing the intellect and reducing it to its quintessence, it becomes impossible, Bergson claims, to demonstrate its genesis. It thus becomes necessary to find a way beyond the critical approach – conceiving the mind in its most concentrated state and then expanding it into reality (Fichte is the thinker he now mentions) – and the evolutionist approach which begins obversely with external reality and then condensing it into the intellect. The problem with these systems of philosophy for Bergson is that they assume that the unity of nature is accessible to thought through an abstract and geometrical form. In both the systems of Fichte and Spencer, he contends, one finds only the articulation of differences of degree in nature, whether they be of the order of complexity or the order of intensity. On this basis intelligence 'becomes as vast as reality; for it is unquestionable that whatever is geometrical in things is entirely accessible to human intelligence, and if the continuity between geometry and the rest is perfect, all the rest must indeed be equally intelligible' (191; 190). But if nature is assumed to be one, and the faculty of knowing taken to be coextensive with the whole of experience, then it becomes impossible to conceive of an engendering of an experience. It is indeed a generative critique, including the engendering of the very categories through which we think life, that Bergson is after. The discipline of thought is one that dissolves it into the 'whole' (the Absolute), and in this way endeavours both to 'expand' the humanity of the human condition and to go beyond it.

Bergson is well aware of the vicious circle supposed in this task, namely, that of going beyond intelligence through the use of intelligence. He replies, however, that this is to misunderstand the nature of a habit, including the habits of thought and of intelligence. It is collective and progressive action which breaks vicious circles. Bergson's most daring move in this part of his book, however, is to suggest that philosophy cannot simply occupy itself with the form of knowledge while leaving the matter of knowledge to the physico-chemical and biological sciences. What the critique of metaphysics as a critique of the faculty of knowledge underestimates is the extent to which this metaphysic and critique come to it ready-made from positive science. As a result the critical philosopher finds his role reduced to one of formulating in more simple and precise terms the unconscious and inconsistent metaphysic and critique that inform the approach science adopts in relation to reality. In short, critical philosophy rests on a highly questionable distinction between form and matter. In its conception of space, which supposes matter to be wholly developed into parts that are absolutely external to one another, Kantian philosophy has simply taken up an approach to reality established by modern science.

As a work of the pure intellect positive science feels most at home in the presence of unorganized matter since it is this matter that it is most familiar with in terms of the manipulation afforded by its own mechanical inventions. This inevitably leads science to viewing this matter in strictly mechanistic terms. Ultimately, the same approach is adopted in relation to living matter and with disastrous results. It is, therefore, Bergson contends, the 'duty' of philosophy to intervene in order to contest the applicability of the geometrism of natural logic to a comprehension of matter, and thus to examine the living and the dead without concern for their practical utility or exploitability. Philosophy is faced with the paradoxical task of freeing itself from forms and habits of thought that are strictly and exclusively intellectual. The 'will to truth' that Bergson identifies within science is not, however, wholly negative for him. On the contrary, he argues that the more it penetrates the depths of life the more it discovers the heterogeneous and the more it encounters the strange phenomena of duration which exceed its understanding. It is here that philosophy is able to supplement science in order to disclose the 'absolute' in which 'we live and move and have our being' (200; 199). This 'absolute' refers not to a linear historical evolution of consciousness but rather to the 'whole' of creative evolution conceived as a virtual duration. Bergson concedes that the 'absolute knowledge' of this whole we acquire will always be incomplete. Speaking positively of it, however, one can also say that it will be neither external nor relative. This is because it has gone beyond the factitious unity imposed on nature by the understanding from outside and the knowledge secured by science that is always relative to the contingencies of our actions and diagrammatic designs upon reality.

Bergson's critical engagement with science consists, therefore, of a number of crucial components. On the one hand, it contends that the physical laws of scientific knowledge are, in their mathematical form, artificial constructions foreign to the real movement of nature since its standards of measurement are conventional ones created by the utilitarian concerns of the human intellect. This does not prevent Bergson from appreciating the success of science; on the contrary, it is such insights into the specific character of science that enables him to appreciate the reasons for its success, namely, the fact that it is contingent and relative to the variables it has selected and to the order in which it stages problems. On the other hand, in not accepting the terms of Kant's critique of metaphysics, Bergson is not compelled to accept as a foundation of knowledge the methods of positive science and to stage questions of human freedom and the absolute in solely regulative terms of faith and not knowledge. He thus transcends the oppositional or antinomical framework of Kant's critique and presents, in the process, a highly novel conception of the task of philosophy as one of thinking a superior human nature 'beyond' the human condition.

Merleau-Ponty argues that in certain important respects Bergson 'perfectly defined' the approach of metaphysics to the world. Science not only has to do with completed formulas about the world but is also implicated in margins of indetermination that separate these formulas from the data to be explained. It thus presupposes an intimacy with data that is still-to-be-determined. Metaphysics entails an exploration of this world prior to the object of science to which science refers (Merleau-Ponty 1964b: 97). For Merleau-Ponty such an insight into the character of science and metaphysics provides us with an appreciation of their 'spontaneous convergence': 'A science without philosophy would literally not know what it was talking about. A philosophy without methodical exploration of phenomena would end up with nothing but formal truths, which is to say, errors' (ibid.; see also 1988: 152ff.). In addition, Deleuze makes the important observation that Bergson's aim was not to produce an epistemology of science but rather to invent autonomous concepts that could correspond with the new symbols of science (Deleuze 1983: 89; 1986: 60; see also the critical remarks *vis-à-vis* the danger of philosophy producing a bad caricature of science in Deleuze and Guattari 1991: 146; 1994: 155).

The ideal genesis of matter: the matter and meaning of evolution

It is not for Bergson simply a question of denying that life evolves without finality but rather of showing that this finality is always implicated in the invention of a duration. Creative evolution proceeds 'without end' in the precise sense that its inventions do not exist in advance and neither can they be simply represented spatially as distinct points of life. But this does not lead him to accepting the view that evolution is a mere blind mechanism. Evolution is marked as dissociative owing to the continual flux of duration *and* the limits set by its inventions such as the evolution of species (an actualization of a virtual evolution along lines of divergence). As specific inventions of nature we humans experience within ourselves, on a subrepresentational level, the actual individuations which inform this continual elaboration of the new, even though our intellect makes it difficult for us to comprehend their durational character. There are two ways in which we fail to recognize this durational character: one, from wanting the genesis of the universe to have been accomplished in a single stroke, so reducing evolution to a series of discrete points or instants, and, two, from wanting the whole of matter to be eternal. In both cases the 'time' of evolution is not being thought as duration.

It is at this juncture in his argument that Bergson considers the two principal general laws of modern science: the first and second laws of thermodynamics relating to the conservation and degradation of energy and in the context of relatively closed systems. He begins his examination of them by noting that the two principles or laws do not have the same metaphysical scope, in the sense that the first is a quantitative law relative to our methods of measurement, while the second law does not necessarily bear on magnitudes. Let me be more precise concerning Bergson's claim about the first law. The law of conservation concerns itself with the relation of a fragment of the solar system to another fragment and not to the whole. It is thus only intelligible as a law of energy which posits the constancy of energy in the context of this particular frame of analysis. In the case of the second law, Bergson does not deny that it arose out of the attempts by Carnot and Clausius to calculate magnitudes in the form of 'entropy'. Bergson holds it to be the most 'metaphysical' – and therefore the most interesting – of the laws of physics because without interposed symbols and the aid of artificial devices and instruments it discloses the direction in which evolution is going. This law states that all physical changes have a tendency towards degradation in the form of heat that becomes distributed among bodies in a uniform manner. What was once heterogeneous and unstable becomes homogeneous and stable.

One might suppose that such a law poses the greatest challenge to Bergson's upholding of a creative evolution since it amounts to the claim that the mutability of the universe will in the course of time be exhausted, so bringing an end to evolution's capacity for ceaseless invention. Bergson entertains the possibility of the sum of mutability in the universe being infinite owing to its passing on across different worlds, only to dismiss it as a hypothesis that is both irrefutable and indemonstrable. The idea of an infinite universe only makes sense where we admit a perfect coincidence of matter with abstract space, but this is to deprive evolution of its real 'substance', namely, duration. He then considers the theory of eternal return in which life alternates between periods of increasing mutability and diminution that succeed one another for all time, but appeals to the work of Boltzmann to shows its mathematical improbability.[8] Ultimately the problem remains insoluble, Bergson argues, as long as thought remains within the realm of physics and its attachment to the conception of energy as extended particles, that is, to a spatial conception of energy (even if these particles be thought as reservoirs of energy).

The issues raised by this latter point are stated clearly, and perhaps more helpfully, in the earlier text *Time and Free Will*. In this work Bergson is concerned to show the limitations of physical determinism by arguing that the science of energy rests on a confusion of concrete duration and abstract time. His intention throughout is to keep science and its investigations *open* by not coming up with laws or theses which make constant testing, and hence, revisability, in experience impossible. This is the case, he argues, with modern mechanisms in which the aim is to calculate with absolute certainty past, present, and future actions of a living system from a knowledge of the exact position and motion of the atomic elements in the universe capable of influencing it. It is this quest for certainty, he argues, that has informed the science which has been built up around the principle of the conservation of energy. To admit the universal character of this theorem is to make the assumption that the material points which are held to make up the universe are subject solely to forces of attraction and repulsion that arise from the points themselves and have intensities that depend only on their distances. Thus, whatever the nature of these material points at any given moment, their relative position would be determined by relation to the preceding moment (this can readily be recognized as an accurate account of Helmholtz's attempt to develop a mechanistic ontology of matter in motion by combining the new insights into energy with the laws of Newton and Kant's metaphysics of nature; see Harman 1982: 41–4). This is the point that Bergson then makes in relation to the principle of the conservation of energy:

> in order to foresee the state of a deterministic moment, it is absolutely necessary that something should persist as a constant quantity throughout a series of combinations; but it belongs to experience to decide as to the nature of this something and especially to let us know whether it is found in all possible systems, whether, in other words, all possible systems lend themselves to our calculations.
>
> (Bergson 1960: 151)

Bergson's main concern is to argue that it is illegitimate simply to extend this conception of matter to a deterministic and mechanistic understanding of psychic states (perhaps by making them reducible to cerebral states). Along the way he makes some important points that might help us to understand better his theory of duration in relation to energetics. Bergson's approach is not to deny that the principle of the conservation of energy appears to be applicable to a whole array of physico-chemical phenomena, especially the case, he notes, since the development of the mechanical theory of heat. The question he wants to pose for an open science, however, is whether there are new kinds of energy, different from kinetic and potential energy, which may 'rebel' against calculation (he is thinking of physiological phenomena in particular). His principal point is to argue that 'conservative' systems cannot be taken to be the only systems possible. For these conservative systems time does not 'bite' into them. Without duration can these systems be said to be *living* systems? Bergson is keen not only to point to the *irreversible* character of living systems but also to think of their evolution in durational terms. On the model of modern mechanism the isolable material point can only remain suspended in an 'eternal present' (153). For sure, a conservative system may have no need of a past time (duration), but for a living one that exists in a metastable state it is a prerequisite. Finally, in this section of his text on time and freedom, Bergson makes the interesting critical point that the setting up of an abstract principle of mechanics as a universal law does not, in truth, rest on desire to meet the requirements of a positive science, but rather on a 'psychological mistake' that derives from reducing the duration of a living system to the 'duration which glides over the inert atoms without penetrating and altering them' (154). For mechanism this is the only kind of duration conceivable. For Bergson this is no duration at all.

Let me return to my analysis of *Creative Evolution*. Bergson once again reconfigures the question of a creative evolution in terms of the relation between matter and consciousness, maintaining that matter is always in descent and is only prevented from achieving a complete descent into stasis (as in Freud's conception of the death-drive, as we shall see in the next chapter) owing to an inverse process. It is beyond question, he argues, that the evolution of life on earth is attached to matter. Pure consciousness would be pure creative activity lacking in invention (there would be no obstacles to overcome for a start). In being riveted to the organism evolutionary life is subjected to the general laws of inert matter. In resisting the descending tendencies of this matter it is not necessary to posit life as a tremendous neg-entropic force. Rather, we need only speak, says Bergson, of a certain retardation in which life is seen to follow from an 'initial impulsion' that has brought into being through the contingencies of evolution more and more powerful explosives. These 'explosives' can be construed as amounting to a storing-up of solar energy in which its degradation meets with a provisional suspension 'on some of the points where it was being poured forth' (1962: 247; 1983: 246). It is the organism, therefore, that represents an arrest of this dissipation of energy. It is not simply the case, therefore, that the organism negatively limits life (as Deleuze and Guattari seem to want to contend in *A Thousand Plateaus*); rather, Bergson's argument is that in imposing this limit the organism is what makes 'life' possible (life as invention and duration): 'The evolution of living species within this world represents what subsists of the primitive direction of the original jet, and of an impulsion which continues itself a direction the inverse of materiality' (248; 247). It is this insight which informs Bergson's conception of the '*ideal*' (as in virtual) genesis of matter. He wishes to stress that the life of species is often a determined life in contrast to that of the creation of a world which exceeds the limits of species and which is always a 'free act'. There is always the reality creating itself in the reality that desires to unmake itself. It is inevitable, he suggests, that each species will behave as if the general movement of life came to and end with it, instead of passing through it. The notorious 'struggle for existence' results as much from the limited perspective of each species as it does from the brutal, nasty, and short character of life, he speculates.

Bergson's so-called 'vitalism' has no desire to appeal to an external cause in order to endow life with creative principles. Rather, his vitalism emerges directly out of his critique of the spatial habits of modern scientific and philosophical thought:

> the difficulty arises from this, that we represent statically ready-made material particles juxtaposed to one another, and, also statically, an external cause which plasters upon them a skilfully contrived organization. In reality, life is a movement, materiality is the inverse movement, and each of these two movements is simple, the matter which forms a world being an undivided flux, and undivided also the life that runs through it, cutting out in it living beings all along its track.
>
> (1962: 250; 1983: 249)

This organization of life assumes for our intellect the appearance of solidity and fixity as it fails to grasp the unity of the impulse that passes through generations of life-forms, linking individuals with individuals and species with species.

Creative evolution involves a play between order and disorder, between consciousness and inert matter, between tension and extension, within which contingency plays a major role. This contingency applies both to the forms adopted and invented, and to the obstacles that are encountered at any given moment and in any given place. The only two things required for evolution to take place are, first, an accumulation of energy, and, second, an 'elastic canalization of this energy in variable and indeterminable directions' (1962: 256; 1983: 255). This is certainly how life and death have evolved on Earth, but, as Bergson correctly points out, it wasn't necessary for life to assume a carbonic form. He is prepared to go as far as to assert that it was not even necessary for life to become concentrated in organisms as such, that is, in definite bodies which provide energy flows with ready-made and elastic canals.

Bergson reaches the conclusion that there has not been a project or plan of evolution (266; 265). The animal 'man' is neither pre-figured in evolution nor can it be deemed to be the outcome of the whole of evolution. In addition, it is also the case that the rest of nature does not exist for man. It is the divergent character of evolution, the fact that it has taken place on many diverse lines, that is decisive. It is only in a special sense that man can be considered to be the 'end' of evolution. Here Bergson seems to be suggesting that although man is an entirely contingent product of an inventive process of evolution – there is no way that we can say that evolution has been searching for man – it is in the case of man that the virtual becomes equal to the actual, so opening up the possibility of a general scrambling of the various codes of life. One might locate in his thesis a metaphysics of presence in which 'man' is the privileged life-form, amounting to an apotheosis of the entire movement of matter and spirit, because in the becoming of his being the virtual finds an actual adequate and proper to it.[9] Nevertheless this criticism is unsustainable simply because on Bergson's model life is never fully present or actual to itself. This applies to the case of the human, which can only be adequately conceived in terms of a perpetual going beyond through the inventions of technics that create a history of virtuality. A creative evolution always transcends finality if by finality we mean the simple realization of something that is conceivable in advance. It is on account of his attachment to a 'creative' model of evolution that Bergson prefers a virtualist as opposed to a finalist conception of life.

The interesting question to be posed is why life has assumed the forms it has (plant, animal, etc.). Bergson answers by suggesting that life is an accumulation of energy, which then flows into flexible channels and performs various kinds of work. It is this activity of energy which the vital impulse would 'fain to do all at once' were it not for the fact that its power is limited: 'But the impetus is finite....It cannot ovecome all obstacles. The movement it starts is sometimes turned aside, sometimes divided, always opposed; and the evolution of the organized world is the unrolling of this conflict' (255; 254). Bergson insists that all that is necessary for creative evolution (the generation of 'free acts') is the accumulation of energy and the canalization of this energy in variable and indeterminable directions. In the lecture on 'Life and Consciousness' which opens *Mind-Energy* Bergson expresses it as follows: 'But life as a whole, whether we envisage it at the start or at the end of its evolution, is a double labour of slow accumulation and sudden discharge' (1975: 19). Living forms are the vehicles through which the vital impulse discharges its energy and reorganizes itself for further invention and creation: this is the life in death and death in life.

It is only in a quite specific sense that it can be contended that man constitutes the 'term' and the 'end' of evolution. In considering Bergson's speculative claim that in the case of the human being nature has created a machine that goes beyond mere mechanism we should not overlook its essential paradox: the human intellect is the ultimate mechanism. Thinking beyond the human condition necessarily, therefore, has an appeal to an *over*human. What this overhuman involves will be left to explore in the ensuing chapters and the conclusion of the study. It is clear that within this conception of creative evolution there resides deep within it an ethics of life. What this might amount to we will attempt to explore in the final section of this chapter (p. 69), where we will once again return to our encounter with what it means to think 'beyond' the human condition as the very 'meaning' of that condition.

Deleuze and the *élan vital*

In this penultimate section of the chapter I now want to examine how Bergson's presentation of a creative evolution gets configured in Deleuze's 'Bergsonism'. Deleuze's analysis is helpful in that it brings into clearer focus some of the more condensed aspects of Bergson's account, such as the crucial notion of the virtual.

There is one vitally important aspect of Deleuze's reading of Bergson as a thinker of difference we have yet to examine, and this is the precise significance of his stress on a notion of *internal* difference. It is now appropriate to examine this as it is the stress on internal difference which reveals for Deleuze the true significance of Bergson's *élan vital*. In both the 1956 and 1966 readings this is brought out in the context of a reading of Darwin's theory of natural selection. It is important to read Deleuze carefully here since he is not going to argue, and wisely so, in my opinion, that the notion of 'difference' is a biological one, but rather that its status be grasped solely and strictly as 'metaphysical'. Let us begin with the comments made in the 1956 essay, before moving on to the more detailed reading offered in 1966.

Deleuze does want to grant an importance to Bergson's emphasis on differences of nature or kind but not an exclusive importance. This is because he recognizes that a difference of kind remains an 'external difference' (1956: 87). In showing us that abstract time is a mixture of space and duration and that space is a mixture of matter and duration, Bergson presents us with mixtures that divide into two tendencies (matter is a tendency involving a slackening or relaxation, while duration is a tendency involving contraction). The important point to realize, however, is that the difference of nature does not lie between these two tendencies since it is, in fact, one of these tendencies, namely, duration conceived as that which differs from itself and not simply what differs from something else (by contrast, Deleuze suggests that matter is what merely repeats without the difference of duration). In short, difference 'is' substance in which the decomposition of a mixture does not merely provide us with two tendencies that differ in nature but rather gives us difference of nature as one of the two tendencies. Movement can now be thought of not as the character of a thing, something or nothing, but as a substantial character that presupposes no moving object. As the tendency to differ from itself in terms of a logic of nuances duration can be said to be '*immediately* the unity of substance and subject' (1956: 89; 1997: 8).This means, suggests Deleuze, that external difference has become *internal* difference since it is no longer between two tendencies but is one of the tendencies, so that '*Difference of nature has itself become a nature*' (*La différence de nature est devenue elle-même une nature*) (90; 9). In this way, Deleuze argues, internal difference must be distinguished from notions of contradiction, alterity, and negation. He mentions Plato's dialectic (a dialectic of alterity) and Hegel's (one of contradiction) as both involving the power and presence of the negative. In order to differ, however, internal difference does not have to go to a point of contradiction since it enjoys an originary power of differentiation. It is duration that includes within it all qualitative differences.

Deleuze now addresses in his 1956 essay the paradox of something differing from duration but which nevertheless still involves a duration of sorts, such as the other tendency (space decomposing into matter and duration, or organic form decomposing into matter and the *élan vital*). To appreciate this point requires modifying our understanding of duration in which it is seen to be, as an indivisible englobing power, not so much that which never gets divided but rather that which changes its nature in dividing itself. This, says Deleuze, is to introduce into duration the *virtual* dimension. It is to *Creative Evolution* of Bergson's texts that Deleuze looks to provide the necessary information here since it is biology which shows us the process of 'differentiation' (*différenciation*) at work. The difference that is neither one of degree nor intensity, neither contradiction nor alterity, is, he argues, the *vital* difference – although this concept is itself not biological (1956: 92; 1997: 11). In thinking differentiation Bergson, Deleuze argues, is thinking less of embryology and more the differentiation of species, which is, he says, to speak of 'evolution'. It is now at this juncture that he addresses Bergson's relation to Darwin. Against the mechanism of Darwinian theory Bergson wants to show that the vital difference, including the difference 'of' evolution, is internal difference. Moreover, this internal difference is badly thought if it is conceived as involved in a simple determination. Any determination can be thought as accidental in the sense that its being can be attached to a cause, an end, or the play of chance. As a result difference is reduced to a subsisting exteriority. In stressing the importance of indetermination in the becoming of vital difference, Bergson's aim is to demonstrate the unpredictable character of all living forms, so that by indeterminate he means unpredictable. Deleuze claims that for Bergson the indeterminate is not the realm of the accidental but, on the contrary, the essential element in a creative evolution. From this Deleuze draws several important insights into Bergson's conception of evolution: first, that if the tendency to change is not accidental then neither can the changes brought about be regarded as accidental. This means that the *élan vital* is, in some ultimate sense, the most profound 'cause' of variations in which difference is not a determination but a differentiation. This is not to deny that differentiation arises from the resistance matter presents to life – there is, therefore, a materiality to any process of differentiation – but rather to insist that primary difference comes from the internal explosive force of 'life'. This is not, however, to identify a life that is present to itself or which could ever achieve a state of total or full presence. This is because life evolves in the element and power of the virtual:

'Self-differentiation is the movement of a virtuality which actualizes itself'

(93; 11).

I now wish to now turn my attention to the 1966 essay in order to better understand the work the notion of the virtual is doing in Deleuze's reading of Bergson on creative evolution. In this 1966 reading Deleuze explores in more detail the role the virtual plays in Bergson's thinking. He is compelled to do this in part owing to the danger of reintroducing into thought a dualism between differences in kind and differences in degree. This applies, he says, to the analysis offered by Bergson in both *Matter and Memory* and *Creative Evolution*. Deleuze's solution to the problem is to claim that all the degrees coexist in a 'single Nature' that gets expressed in two ways, differences in kind and differences in degree. In other words, he is claiming that for Bergson there is ultimately a 'single Time' or a 'monism' of time, which is said to be nature itself (1966: 95; 1988: 93; see also Bergson 1965b: 67ff. 'Concerning the Plurality of Times'). He insists that this is not to posit a contradiction between the moments of the method since any duality is valid when it concerns only *actual* tendencies. The important point to appreciate is that it is on the virtual plane that unification is to be sought. The 'whole' is 'pure virtuality'. Moreover, differentiation is only an actualization to the extent that it presupposes a unity, which is the primordial virtual totality that dissociates itself according to lines of divergence but which still subsists in its unity and totality in each line. The virtual, therefore, *persists* in the very matter, or activity, of its actualization.

Deleuze now introduces his argument on the virtual as always involving invention beyond the realization of the possible that is governed by rules of resemblance and limitation. This is to stress that the virtual can only actualize itself by creating for itself its own lines of differentiation in a series of positive acts. It is also here that he presents Bergsonism as moving beyond both preformism and evolutionism. The advantage of the latter is that it has a notion of evolution involving the production of differences. Its weakness, according to Deleuze at this stage in his development, is that such differences can only be produced from the perspective of a purely external causal mechanism. He then reiterates the arguments he had first enunciated in his 1956 essay concerning the principal requirements of a 'philosophy of life', and which we encountered in our own independent analysis of Bergson's text on evolution: (a) that the vital difference be conceived as an internal difference in which the tendency to change is not accidental; (b) that the variations produced do not simply combine in terms of relationships of association and addition but involve dissociation and division; and (c) these variations involve an actualized virtuality that proceeds via lines of divergence always involving heterogeneous terms that are actualized not in a homogeneous unilinear series but rather along a ramified one.

Deleuze next considers a question he had not previously dealt with, namely how it is that the original 'one', which to complicate the matter of the one can only be the whole as a virtuality, has the power to be differentiated. Deleuze argues that the answer lies in the peculiar reality that belongs to the virtual, namely, a coexistence of degrees of expansion and contraction that can be extended to 'the whole universe'. This is to speak of the virtual in terms of a 'gigantic Memory' conceived as 'a universal cone in which everything coexists with itself, except for the differences of level' (1966: 103; 1988: 100). This means that whenever virtuality is actualized and differentiated it is so in accordance with divergent lines each of which corresponds to a particular degree in the virtual totality. Deleuze speaks of duration dividing itself into matter and life (perhaps not sufficiently drawing out the extent to which duration is this division), then into animal and plant, as involving such an actualization of the different levels of contraction that do not cease to coexist in so far as they remain virtual. He then makes the important point that no matter how much these lines of actualized divergence correspond to virtual degrees of expansion and contraction, they do not simply confine themselves to tracing these degrees by, for example, reproducing them through a simple resemblance. It is this fact which makes the lines of differentiation genuinely creative. It is only by excluding this plane of the virtual from the process of evolution, and by focusing on the actuals that determine each line, that we are able to locate the differences between the inventions of nature – such as plant and animal, animal and man – as mere differences of degree and locate in one the negation or inversion of the other.

Towards the conclusion of the 1966 text, which ends with this chapter on the *élan vital* and reading of *Creative Evolution*, Deleuze addresses the play within evolution between life as movement and the creation of material form in a highly interesting manner. Deleuze recognizes that in actualizing and differentiating itself life, taken as the virtual whole, runs the risk of alienating itself in the form of losing contact with the rest of itself. Thus, 'Every species is thus an arrest of movement; it could be said that the living being turns on itself and *closes itself*' (1966: 108; 1988: 104). He recognizes, however, that life cannot proceed in any other way since it involves the nonidentity of the virtual and the actual. Life cannot, therefore, reassemble its actual parts that continue to remain external to one another, so that in the actual *qua* actual there reigns an 'irreducible pluralism' constituted as 'many worlds as living beings, all "closed" on themselves' (ibid.).

It is precisely this conception of a virtual creative evolution that involves divergent lines of actualization that the Deleuze of *A Thousand Plateaus* will seek to go beyond, seeking to map life in terms of an etho-logic of living systems that involve transversal modes of becoming and communication – that is, modes which, in cutting across distinct phyletic lineages (differences in kind), introduce the possibility of a more 'machinic' model of evolution and so allowing for 'the dance of the most disparate things'.

Perhaps the most important difference between Bergsonism and Darwinism, the implications of which Deleuze never confronts in his 1956 or 1966 readings, is that the theory of natural selection does not concern itself with the question of 'Evolution'. Darwin preferred to speak of 'descent with modification' and uses the word 'evolution' only once in his *Origin of Species* at the very end of his presentation when he speaks of the 'grandeur' contained in the view of life offered by the theory of selection since it shows how, through the implacable means of a blind mechanism, there has evolved from a simple beginning endless forms of life at once wonderful and beautiful (Darwin 1985: 460). According to Stephen Toulmin this avoidance of the word 'evolution' on Darwin's part is not accidental. He argues, incisively, that his aim was to produce a general economics (my phrase) of nature that would explain a large number of distinct and separate natural processes, and 'not to reveal the essential plot in a cosmic drama' (Toulmin 1982: 121).

It may be significant that from 1968 onwards and commencing with *Difference and Repetition*, Deleuze should deploy the word 'evolution' with caution. In this work, for example, whenever the argument has recourse to a notion of 'evolution' it is always placed in scare quotes. By the time of *A Thousand Plateaus* Deleuze's attention has shifted quite dramatically and the focus is on modes of creative *involution*. Indeed, in these subsequent works we find a more serious and informed engagement with both Darwin's theory of selection and with modern neo-Darwinism. In Bergson's favour, however, it can be noted that his conception of creative evolution does not treat evolution as some grand, supra-cosmic entity that can be understood apart from its contingent and multifarious pathways. Neither does it select a main channel or pathway of evolution through which the 'meaning' of everything else is to be filtered. Any attempt to enlist Bergson's great text for the cause of some dubious postmodern vitalism will simply fail to appreciate the rich and nuanced character of his conception as well as what is at stake in his thinking of duration and the tremendous challenges it presents to thinking in biology, physics, and cosmology (a challenge, one could contend, that cosmologists are only now responding to; see, for example, the recent attempt by Smolin, 1997, to bring together Leibnizian and Bergsonian questions with Darwinian theory in an understanding of the evolution of the universe).

Conclusion: Deleuze beyond Bergson and towards an ethics of creative evolution

The Bergsonian philosophy of duration does not provide a new psychologism or a new phenomenology of time. Its significance for Deleuze resides in its formulation of rules for the development of a complex ontology. Duration, therefore, is not to be reduced to merely lived experience, something essentially poetical or mystical, but rather is 'experience enlarged or even gone beyond' in the precise philosophical sense of stipulating new conditions for the possibility of experience in its virtual reality. The virtual is the becoming specific to open systems in contrast to the realm of possibility which is always specific to matter and to closed systems (Deleuze 1966: 36–7; 1988: 43).

In the French context Bergson was taken to task for his conception of duration by the likes of Merleau-Ponty and Sartre, while in the context of biophilosophy Raymond Ruyer provided a critical reading of his models and claims (see Ruyer 1946: 29–30, 42, 53ff., who focused his attention largely on *Matter and Memory*; for insight into Ruyer's relation to Bergson see Alliez 1998: 256ff., and Deleuze himself in a footnote to *Bergsonism*, 1966: 103; 1988: 132–3). The limitations of the critique of Bergson offered by Merleau-Ponty and Sartre stem from the fact that their criticisms were advanced from the perspective of the concerns of phenomenology with its attachment to a norm of natural perception (see Merleau-Ponty 1962: especially 180–1; and Sartre 1989: esp. 135, 166–7, 170, 610). In other words, this critique failed to address seriously the challenge offered by Bergson's attempt to cultivate a method of philosophy that would take thinking 'beyond the human condition'. (For a recent account of Bergson's 'durance' and its 'going beyond' the phenomenological notion see Moore 1996: 115–22.) It is clear, for example, that Sartre was criticizing Bergson for failing to disclose duration as a principle of subjectivity – he speaks of it, for example, as a perpetual temporalization that never temporalizes *itself* (Sartre 1989: 167). In France today there are those, notably Eric Alliez, who construe the significance of Deleuze as lying in his attempt to provide for thinking a *post*-phenomenology (see Alliez 1995 and 1997: 81–9). This view is in line with the philosophical moves sought by Deleuze and Guattari in *What is Philosophy?* Moreover, it is very much in these terms that Deleuze reworks Bergson and his 'Bergsonism' in his two books on cinema composed in the 1980s (see especially Deleuze 1983: 85ff.; 1986: 57ff.). Deleuze contends that the 'opposition' between Bergson and phenomenology is a radical one in respect of the question of consciousness. Let us examine a little more closely what is involved in this 'opposition' and Deleuze's staging of it.

For Deleuze the nature of this opposition amounts to conceiving consciousness as immanent to matter, rather than bestowing upon it the privilege of a centred natural perception located in a subject. 'Our' consciousness of fact is merely the opacity without which light or luminosity gets propagated without its distributed source ever being revealed. The aim of phenomenology, Deleuze suggests, was to embark on a renewal of concepts that would serve to awaken us to the world not as babies or hominids but as beings with 'proto-opinions' who, by right, could lay claim to constituting the foundations of their world with such opinions. However, by restricting immanence to a subject, and invoking the primordially lived, phenomenology was unable to fight against the tyranny of perceptual and affective clichés but surrendered the subject to the domain of the ordinary and the everyday, imprisoning the movement of life within the realm of opinion and common sense (Deleuze and Guattari 1991: 142; 1994: 150). Although these remarks come from the much later work *What is Philosophy?*, the extent of their consistency with Deleuze's early 'Bergsonism', with its concern to define a mode of philosophy that could think beyond the human condition, should be apparent. The Bergsonian critique of phenomenology challenges the privilege accorded to 'natural perception' as a norm of perception, in which existential co-ordinates allow for a privileged subject of perception to be anchored as a 'being in the world', leading to the famous phenomenological deduction: 'all consciousness is consciousness *of* something'. Movement becomes understood as a sensible form that organizes the field of perception and as the function of a situated consciousness that is always intentional. On the Bergsonian model, however, movement is always distributed and nonlinear, 'a flowing-matter in which no point of anchorage nor centre of reference' is assignable (Deleuze 1983: 85; 1986: 57). This is how Deleuze reads the perception of cinema – as completely acentred, inventively using Bergson's own thinking on matter and movement to challenge his hasty dismissal of cinematography's misconception of movement (see Bergson 1959: 35; 1991: 38). This is to introduce us into a world of 'universal variation', 'universal undulation' and rippling, without fixed axes or centres, left or right, high or low, experiencing the universe as a kind of 'metacinema' (Deleuze 1983: 86–7; 1986: 58–9). The 'eye' is not in the 'I' as in things, diffused everywhere, so that the Bergsonian reply to phenomenology becomes: 'all consciousness *is* something'.

We need to note, however, that the phenomenology of Merleau-Ponty as it unfolds becomes more resilient to Deleuze's criticism.[10] The case of Merleau-Ponty is an especially important one in the context of a reading of Deleuze because one finds in his work an engagement not only with Bergson, but also with developments in neo- and post-Darwinian biology and ethology, such as the work of Uexküll, that connect with Deleuzian thinking (see, for example, the detailed notes in Merleau-Ponty 1994: 220–34). Let me just cite a few examples, the importance of which will become evident as we pursue Deleuze's biophilosophy in the next two chapters. First, there is the fact that Merleau-Ponty utilizes developments in biology in order to challenge a series of metaphysical oppositions and prejudices that inform philosophical thinking. In addition, he engages with this work in order to show that the endless debate concerning the internal or external direction of evolutionary trends is fruitless and unnecessary. He further seeks to show how new work in morphology challenges the Darwinian tradition by showing that 'relations of descent' are not the only ones meriting consideration. Work in embryology, on the other hand, is important in refusing to accept the option of either preformation or epigenesis, while the appearance of notions of 'gradient' and 'fields', denoting 'organo-formative territories' that impinge upon one another, 'represents a mutation in biological thought as important as anything in physics' (Merleau-Ponty 1988: 194).

Merleau-Ponty draws upon these revolutions in biology, as well as Uexküll's work on the ethology of *Umwelten* which shows the extent to which the 'environment' is structured and mediated by the specific *Umwelt* of the organism, in order to dispel the idea that life is simply an 'object' for a 'consciousness'. Such an insight, he claims, marks the point at which phenomenology breaks with idealism (1988: 196). He argues life and nature are unthinkable without reference to perceived nature. For him this perception is no simple affair of consciousness but refers to a body that is affective and libidinal much more than it is reflective. Indeed, the most important feature of the 'primacy of perception' concerns its embodied aspects: 'The body proper embraces a philosophy of the flesh as the visibility of the invisible' (197). It is not 'consciousness' that perceives nature but the human body that also inhabits it. The 'animation' and 'animality' of the body are always caught up in 'a metamorphosis of life' (196). The principal insight of Merleau-Ponty's 'philosophy of nature' can be expressed in the following terms: life involves a 'global and universal power of incorporation' in the particular sense that corporeality is the search for the external in the internal and the internal in the external. It is such an insight that informs Deleuze's later reading of Spinoza through the lens of Uexküll's ethology.

Bergson provides a certain conception of the organism in *Creative Evolution* that Deleuze will follow through in his work of the late 1960s. If the organism is marked by duration the less it can be conceived in terms of a mere mechanism. However, while articulating a certain autopoietic conception of the organism, as opposed to viewing in terms of a mere passive adaptor, Bergson still wishes to stress the extent to which the life of the organism is implicated in the continuous flow of the *élan vital*. Towards the end of book three of the text he addresses the Weismannian revolution in modern biophilosophy which shows the extent to which life is supra-individual in the specific sense that organisms and bodies can be construed as the 'mere' vehicles through which the continuity of the life of genetic energy is assured and passed on. Bergson's concern is with how this revolution affects our understanding of the life of 'spirit'. He argues that the great error of previous doctrines of spirit was to isolate its life by suspending it 'beyond the earth' in a realm of pure space and beyond time. However, no doctrine of spirit can remain uncontaminated by the revolution in modern biology and genetics that has shown that although we are right to believe in the 'absolute reality of the person', including its independence towards matter, we cannot ignore science's showing of the interdependence of conscious life and cerebral activity. Neither, in rightfully attributing to man a privileged place in nature and holding the distance between human and animal to be immense, can we ignore the fact that the history of life shows us that species evolve through gradual transformation to the extent that the human is reintegrated into animality. Bergson's response is to claim that his conception of the *élan vital* shows the extent to which the life of the spirit proceeds via the duration of a creative evolution and only when viewed in these terms does it become possible to see the life of the body 'just where it really is', namely, as implicated in a journey that leads to the life of the spirit (1962: 269; 1983: 269). It is in these terms that he attempts to reconfigure Weismann: 'The matter that it bears along with it, and in the interstices of which it inserts itself, alone can divide into distinct individualities. On flows the current, running through human generations, subdividing itself into individuals' (270; 269). This is the 'great river of life' that divides itself in individuated matter and which does not cease to flow through the body of humanity, the ever-undulating river of duration.

The ethical question of creative evolution itself can perhaps be posed in the following terms: what does it mean to live a life 'in duration'? Moreover, what does it mean for the human organism to 'become' this life that is marked by the durational flow of a germinal life? This refers not only to a life that lives on and is passed on at the point of somatic death but also to the life that is already, within the duration of a life, being lived beyond and passed on. In the text on Bergson of 1966 no ethics is ever actualized. This task awaits the supplement provided by Nietzsche with an ethics of eternal return conceived as the creative evolution of a germinal life that can be lived and affirmed by the individual.[11]

In his later work Deleuze conceives 'the nonorganic life' as the *nonpsychological* life of the spirit, a life that gets more and more progressively separated in his writings from the field of human consciousness and subjectivity. This is not to say that human life cannot 'participate' in this animated nonorganic life of things, but only to insist that this life does not exist for it and neither can it be construed as the 'meaning' and 'end' of this life. In an effort to distance himself from Bergson's residual humanism the later Deleuze will attack the mnemotechnics of 'man' as the constitution of a 'gigantic memory' which prohibits novel becomings on the plane of immanence. A number of commentators have argued that the Deleuzian move must confront the charge of resulting in a disavowal of the human being in its specific bodily individuation and technical articulation. This, however, is no easy matter to negotiate because, for Deleuze, the questions inspired by Spinoza and Nietzsche – what becomings are open to the human being? and what may still become of the human? – must always remain open as ethical ones. The matter becomes even more complicated if one takes seriously Deleuze's position that the 'human' constitutes not so much a compound form, but is the site for the transmutation of forces. In addition, one must be attentive to the precise character of the notion of corporeality that Deleuze is working with and the challenge that it is intended to present to traditions of thought that have been dominated by hylomorphic schemas of matter and which assume that there are fixed forms and a homogeneous materiality. In contrast to this tradition, and in direct opposition to it, Deleuze will privilege what he calls an 'energetic materiality' that is always in movement and that carries with it 'singularities', 'haecceities', and intensive affects. All of this will be returned to in Chapter 3 when we look at *A Thousand Plateaus*, and I also have important things to say on the life of the event in the final section of the next chapter that may have a bearing on the question of Deleuze and disavowal (p. 129). In the next chapter I want to draw out and address more explicitly the 'ethics' of Deleuze's unfolding conception of germinal life. It is also here that we shall initiate the encounter with Deleuze's strange new world of singularities, events, and deviant becomings.

Let us draw this chapter to a close by noting that an important break with Bergson is announced in *Difference and Repetition* and this is over the question of the nature of intensity. Bergson's critique of the notion of intensive magnitudes – which centres for Bergson on the misguided attempt to treat singular qualities as measurable quantities (Bergson 1960: 224–5), and which features in Deleuze's Bergsonism of 1956 and 1966 (although in the latter Deleuze notes that the critique of intensity in Bergson's *Time and Free Will* is highly ambiguous since it raises the question whether it might be intensity that 'gives' to experience the qualities that inform it) – is now held to be unconvincing since it rests on the error of assuming ready-made qualities and preconstituted extensities. It is this preoccupation with the difference of intensity and a 'continuum of intensities' that will inform Deleuze's approach to the matter of 'creative evolution' in both the work of the late 1960s and later work such as *A Thousand Plateaus*. The changes are quite dramatic in relation to what it now means for philosophy to think beyond the human condition, and for us to become less than human and more than human.

In essence, Deleuze now argues that the primary difference, which concerns those differences that carry their reason within themselves, is not the one staged between differences of degree and differences in kind, but rather resides in the differences that belong to intensity and which inform both, so constituting the real nature of difference (differences constituted by temperature, pressure, tension, potential, etc.). Where the difference of degree denotes the extensity that ex-plicates difference, the difference of kind names only the 'quality' that is im-plicated in this extensity. What is left out of the picture for Deleuze is that which lies beneath the two orders, namely, the 'intensive', and which is now understood to constitute 'the entire nature of difference' (1968: 309; 1994: 239). In itself, Deleuze maintains, difference is neither qualitative nor extensive (even qualities are involved in orders of resemblance and identity). The result of this primacy accorded to the (tautologous) principle of a difference of intensity (difference *is* intensity) is a major shift in the character of Deleuze's Bergsonism and his conception of a creative evolution. The focus is no longer on the nature of a qualitative duration, but rather attention is drawn to the 'graduated scale' (which will become in the later Deleuze the flat plane of immanence) that is characterized by 'phenomena of delay and plateau, shocks of difference, distances, a whole play of conjunctions and disjunctions' (308; 238). In large measure, Deleuze is motivated to adopt this perspective on difference on account of the need to produce a critique of thermodynamic thinking (whether in philosophy or science). If 'intensity' was not allowed as the transcendental principle (the 'being' of the sensible) which supports 'quality', then, the only kind of duration that we could attribute to quality would be a thermodynamic 'race to the grave', a thirst for the annihilation of difference in a corresponding extensity (such as the Freudian death-drive perhaps), a race that will also produce at the same time a uniformization of the qualities themselves. It should be borne in mind that informing the emergence of 'energetics' – the theory of energy – in the second half of the nineteenth century was the desire to produce a set of axioms about energy free from the uncertainty of hypotheses about the nature of matter, and so produce general principles that would refine and redefine the programme of a strictly mechanical explanation of phenomena (see Harman 1982: 59ff.). Deleuze produces the following crucial modification to Bergsonism (in this whole modification Deleuze says nothing about Bergson's own powerful critique of entropy which we looked at earlier in the chapter).

> In short, there would be no more qualitative differences or of kind (*nature*) than there would be quantitative differences or of degree, if intensity were not capable of constituting the one in qualities and the other in extensity, even at the risk of appearing to extinguish itself in both.
>
> (Deleuze 1968: 308; 1994: 239)

It becomes necessary, therefore, to posit two orders of implication and degradation: one in which intensities are always enveloped by the qualities and the extensity that explicate them, and another in which intensity *remains* implicated in itself, as both enveloping and enveloped. This is said to amount to a difference between a scientific concept of energy (entropy) and a philosophical one. It is on account of the second order that it is possible to grant to difference a 'subterranean life', which is the life that gets covered over by both philosophies of identity with their attachment to the illusions of the negative (the cancellation of difference) and modern science with its investment in laws of nature. Entropy, for example, partakes of a transcendental illusion in that, although a factor involved in extension and explication, it nevertheless remains implicated in intensity, simply because it makes possible the general movement of the implicated undergoing explication.

The illusion consists, then, in the reduction of implicated intensity to an order of extensity and is generated by the fact that empirically we perceive only localized forms of energy that are already distributed in extensity.

In the later Deleuze the thinking of creative evolution no longer relies, as in Bergson's original conception, on an anti-entropic and inventive 'consciousness' that traverses matter, but rather is built around an attention to the movement of nomadic singularities and fields of intensities that are posited as part and parcel of a *transcendental* energetics of matter. Contra the 'good' sense of both modern philosophy and science, Deleuze will think at the altar of a peculiar god, a saturnine power that devours at one end what has been created at the other, both legislating against creation and creating against legislation. It is the underworld of Nietzsche's Zarathustra – the figure who is made to destroy what he has created (a world of good and evil) – and the world of an Antichrist. It is to this world, and its secrets, we now turn.

2

DIFFERENCE AND REPETITION

The germinal life of the event

'And believe me, friend Infernal-racket! The greatest events are not
our noisiest but our stillest hours'
(Nietzsche, *Thus Spoke Zarathustra*)

True Entities are events, not concepts. It is not easy to think in terms
of the event.
(Deleuze, *Dialogues*)

Introduction

In the two texts of the late 1960s, *Difference and Repetition* and *The Logic of
Sense*, the relation between the virtual and its actualization is mediated by a
more elaborate theory of individuation than was provided by Bergson's
Creative Evolution. The introduction of the term 'individuation', derived in
large measure from the work of Gilbert Simondon, serves to ensure that
evolution is not a repetition of the same and that repetition signals only ever
difference. Indeed, Deleuze utilizes the notion of individuation to show that
the evolution of 'species' is merely a 'transcendental illusion' in relation to
more profound envelopments/developments. On this model evolution cannot
be reduced or restricted to the continuity of the germ-plasm in which
organisms are conceived as mere vehicles for genetic replication. There is also
the play of foldings and the 'evolution' (the reasons for the scare quotes will
become apparent) of various complex systems (biological, psychic, etc.), in
which individuation becomes less and less dependent on determination by an
exogenous mechanism. But Deleuze's thinking remains modern and quasi-
Weismannian in its idea that organisms and individuals are sites of
transformation through which the play of singularities and intensities
articulates itself. The 'ethical' moment of difference and repetition concerns
how the self that finds itself implicated in an individuated 'evolution' of pre-
individual singularities is able to live the germinal life that flows through it.
Deleuze will argue, for example, that the *cogito* or the 'I think' is only
established in a field of intensity that already informs the sensibility of a

thinking subject. The radical move proposed by Deleuze consists in viewing individuality not as a characteristic of the self, but rather as that which informs and sustains 'the system of the dissolved Self' (1968: 327; 1994: 254).

In *Difference and Repetition* the encounter between Bergsonism and Darwinism now takes place in much more complex terms. In part the reason for this is that Deleuze has now taken on board a much more serious appraisal of the challenge presented by Darwinian thinking, although he is concerned to remove a theory of evolution from certain misconceptions, such as the idea that evolution primarily bears on the evolution of species. In addition there is an important return to Nietzsche in these texts and which some readers, notably Constantin Boundas, have seen as providing a crucial way out of the shortcomings of Bergson in relation to future time (Boundas 1996: 100–1). Deleuze now has recourse to the 'superior' time of Nietzsche's doctrine of eternal return which, in affirming the 'pure' and 'empty form' of time, disturbs both the repetitions of habit and the repetitions of memory that are too caught up in a linear succession of lived presents (101).

In a recent attempt to rework philosophy in terms of an ontology of the event Andrew Benjamin has argued that Bergsonian duration, in which the stress is on the prolongation of the past in the present, is unable to generate a notion of 'transformation' in which the present would not simply be the site to be worked over, but is understood as the site of transformation itself (Benjamin 1993: 196). The past in Bergson's account seems only to 'endure' in the present, not to dramatically effect it or be itself radically transformed by the present. I believe it is precisely this kind of criticism that is informing Merleau-Ponty's reading of Bergson in *The Visible and the Invisible*. In this posthumous work Merleau-Ponty argues that there can be no fusion of philosophy with the Being of the past since philosophy comes after nature, life, and thought, finding them already constituted. There is a return to all things but never as a return to the immediate. The immediate is always 'at the horizon' (Merleau-Ponty 1978: 123). Bergson is correct in declaring intuition to be a method of philosophy, but he partakes of a 'supralapsarian bias' in the sense that the 'secret of Being is an integrity that is behind us' (124). Bergson lacks a notion of originary differentiation that would serve to introduce the required fracturing of time:

> The originating [*originaire*] is not of one sole type, it is not at all behind us; the restoration of the true past, of the pre-existence is not all of philosophy; the lived experience is not flat, without depth, without dimension, it is not an opaque stratum with which we would have to merge.
>
> (Merleau-Ponty 168:124)

(For further insight into Merleau-Ponty's continued wrestling with Bergson on duration and fusion see 1968: 267; 1988: 12ff., 22ff.) It is always 'in the present that we are centred' and that constitutes the measure of our spontaneity: 'We are not in some incomprehensible way an activity joined to a passivity, an automatism surrounded by a will, a perception surmounted by a judgement, but wholly active and wholly passive, because we are the upsurge of time' (Merleau-Ponty 1962: 428).

It is partly the inadequacy of Bergson's account in this regard that motivates Deleuze's reading of Nietzsche in both *Difference and Repetition* and *The Logic of Sense*. However, it is important to appreciate that for Deleuze an 'ontology' of the event concerns only the *virtual* event and in so far as it relates to the event of a germinal life. In addition, one needs to point out that the 'subject' of this ontology, the subject of the working-through and working-over, is a fractured 'I' that finds itself implicated in a field of forces, intensities, and individuations that involve pre-individual singularities. It could be argued, however, that the 'excess' of time is what prevents any adequate 'working-through' in Deleuze's model of repetition, to the extent that the charge levelled against Bergson by Benjamin might remain equally valid in the case of Deleuze. In other words, one might contend that the openness to the future as radical heterogeneity and difference is never concretely or praxially worked out in Deleuze in relation to the exigencies of any 'present'.

This matter cannot be properly appraised, however, until one has appreciated the extent to which for Deleuze the present can only be thought in relation to the event of one's time and the present becomes constituted (becomes what it is, the becoming of an event) only through the deployment of concepts and diagrams that implicate it in the fold of Being. The present does not denote a ready-made neutral space but can only be opened up with the aid of a topology of thought, one that 'frees a sense of time' for us by bringing about an encounter between the past and future, conceived as an inside and an outside, 'at the limit of the living present'. This is the task that is implied in Deleuze's conception of praxis, inspired by Foucault's conception of a 'critical ontology of the present', and which constitutes for him the 'sole continuity' between past and present as well as the way in which the past becomes for the present (Deleuze 1986: 122; 1988: 115). Admittedly, this conception of praxis is not worked out in the texts of the late 1960s, though it could be shown to be implied in Deleuze's conception of the event. Praxis is always determined by the horizon of the future. It is always the futurity of the future that is speaking to us in a hundred signs.

Published in 1968, *Difference and Repetition* can readily be interpreted as part of the philosophy of difference that emerged in France in the 1960s. The only problem with this appraisal is that it neglects the extent to which Deleuze had formulated philosophy as a thinking of difference as early as 1956. Moreover, it downplays the extent to which Deleuze's preoccupation with difference arose from the biophilosophical concerns of his Bergsonism. Although this Bergsonism is not explicitly at the centre of the book, I believe that its preoccupation with processes of individuation, which inform both the reading of Darwin and biology as well as those of Nietzsche and Freud on death and repetition, only makes sense when viewed against the backdrop of the conception of creative evolution that Deleuze derives from his reading of Bergson. There is a deep rapport between *Difference and Repetition* and its allied work *The Logic of Sense* in that both are preoccupied with a theory of individuation, an ethics of the eternal return (the event of time), and the exploration of a superior empiricism that entails a thinking of the human beyond the human condition.

In the preface he wrote for the English translation of *Difference and Repetition* Deleuze says that this was the first of his books in which he tried 'to do philosophy', as opposed to remaining within the ambit of his studies on the history of philosophy. But his preface is misleading if it suggests that this turn to doing philosophy for himself amounted to a novel attempt to introduce into philosophy a thinking of difference and repetition since, as we have demonstrated, this is already the problematic of the 1956 essay on Bergson. Its major claim that philosophy has only gone as far as introducing difference into the identity of the concept, as opposed to making the more radical move of putting difference into the concept itself, is already outlined in the essay of 1956. The 1968 text covers an extraordinary range of topics and thinkers, providing challenging readings of Kant, Freud, and Nietzsche, as well as covering the fields of mathematics, physics, and biology. Here the focus will be on Deleuze's philosophical biology. In particular, I want to examine how with a thinking of difference and repetition Deleuze reworks modern biophilosophical thought. There can be little doubt that an exacting modernity accompanies Deleuze's project. Its placing of a thinking of difference and repetition at the heart of the philosophical task only makes sense when understood in the context of the attempt to engage with biology and physics in order to work against the entropy of differences which, for Deleuze, characterizes modernity in the form (a) of the mechanism of capital that produces, in the law of exchange value, a system of equivalent values and (b) conceptions of entropy that posit extensity as the 'law' of all intensities and individuations which are seen to cancel themselves out and in which difference is construed in terms of a natural tendency to evolve into the homogeneous and self-identical (physical equilibrium, biological death). However, as Deleuze notes, entropy enjoys a fundamental paradox in that it is unlike all other extensive factors in being an extension and an explication that is always implicated in intensity and which does not exist outside implication (see Deleuze 1968: 294–5; 1994: 228–9; and 1969: 134–5; 1990: 111). In addition, Deleuze claims that a 'principle of degradation' is unable to account for the creation of the most simple systems (1968: 328; 1994: 255).

Deleuze will go so far as to posit a *transcendental* principle of energy, as opposed to limiting its purview to that of a solely scientific concept. This refers to energy 'in general', as an 'intensive quantity', that does not restrict it to the qualified factors of extensity, which for Deleuze would be to reduce its application to the tautology of a principle of identity as in 'there is always something that remains constant'. As a transcendental principle 'general energy' is not to be confused with a uniform energy at rest since it enjoys only the constancy of a superior *transformation* (only particular forms of energy that are qualified in extensity can be at rest). Deleuze appreciates that this brings the philosophy of difference and repetition into direct conflict with the 'laws of nature' and the attempts by science to establish knowledge of the world solely on the basis of empirical concepts and principles. For Deleuze, however, science's laws of nature govern only 'the surface of the world', while in the 'other dimension' of the 'transcendental or volcanic *spatium*' it is the doctrine of 'eternal return' which rules (Deleuze 1968: 310–11; 1994: 241).

Difference and Repetition is decidedly modern in the way in which it places a notion of *variation* at the centre of its unfolding and enfolding of a thinking of difference and of repetition. Variation, Deleuze notes in his 1994 preface, is not added to either difference or repetition in order to conceal their true nature (an originary being of identity through self-sameness and self-presence). Rather, it is the case that variation is repetition's constitutive element, it is its interiority which guarantees that what is repeated is not merely a return of the same but the differentiation of difference as a productive and positive power. This fidelity to differentiation as a positive power determines much of the critical character of Deleuze's encounter with modern science. The radical notion of repetition that is being put forward in the text must deal with the seemingly implacable 'laws of nature' posited by both modern philosophy and modern biology (such as the laws of natural selection). It is with the notion of repetition that Deleuze seeks to give primacy to the dissolution of form and the freeing of life from entropic containment in organisms and species. The issue he is compelled to consider is what kind of 'power' nature is and what status a 'law' of repetition as essential novelty may have. Let me stress that the 'value' that Deleuze is placing on the dissolution of forms is not an argument in favour of entropy or some originary death-drive (Deleuze, as we shall see, is keen to distance himself dramatically on the question of death from Freud's account of this drive). In other words, Deleuze configures dissolution and decomposition in terms of the vital and virtual role they play within the *creative* evolution of matter and of complex systems. In this regard, as in many others, he is close to Nietzsche (see notably Nietzsche 1994: Essay 2, Section 12, p. 56).

Deleuze acknowledges that his transcendental and superior empiricism is dependent upon certain innovations that have taken place in modern thought in both philosophy (Bergson, Heidegger) and the sciences themselves (notably biology). If there is a renaissance of, and a return to, the question of ontology within our time, this is the result of critical attention being focused on the 'question–problem complex', that is, the question of the question (keeping the question open) is no longer simply conceived as the expression of a subjective state in the representation of knowledge but is the 'intentionality of Being *par excellence*' (1968: 252; 1994: 195). In upholding the question and repeating it we are not simply dealing with a generalized method of doubt and a peculiarly modern scepticism, but rather with the discovery of the question and the problematic as involving, in terms of their complex unfolding and enfolding, a transcendental horizon, namely, the insight that a transcendental element necessarily and essentially belongs to things, beings, and events. Just as the question 'what is writing?' has become the primary concern of modern literature, and the question 'what is art?' for modern painting and the arts in general, so the question 'what is thinking?' has become the overriding concern for philosophy. Thought no longer thinks in accordance with the regulative norms of common sense and empirical dogma, but rather on the basis of an unconscious. Deleuze argues that if the 'imperatives of Being' have a relation to the 'I' it is always with the *fractured cogito*, which is the self that finds itself implicated in a complex order of time (the time of dissolution and the caesura). The only origin of the question to be found lies within 'repetition'. It is with repetition that the question begins but, as such, it is always engaged in a labour of self-overcoming. What 'returns' in the question is difference and the *monstrosity* of a difference (the new as the daimonic, for example). Where 'ordinary' repetition (a 'bare' repetition that has no clothes or disguise) rests on a prolongation of a length of time that becomes stretched into an order of duration, that of an 'extraordinary' kind (a *clothed* and *disguised* repetition) involves the 'emission of singularities' that are always accompanied by the echo and resonance that makes each the double of the other and each constellation the redistribution of another. For Deleuze it is Nietzsche's thought-experiment of eternal return which provides thought with the model of such a repetition. This is a repetition that involves neither continuation nor prolongation, nor even the discontinuous return of something that might be able to prolong itself in a partial cycle; rather, it affirms the 'reprise' of pre-individual singularities that presuppose the dissolution of all previous identities and their novel transformation. Being is thus the recommencement – the repetition – of itself (it 'is' the difference of itself, self-differentiation).

The problem of nihilism is not overcome in Deleuze's thinking but its difficulty and complexity are staged as such. The critical question, however, concerns whether the attempt to read univocal Being in terms of the primary play of difference and repetition amounts to a coherent philosophy of difference or results in a mode of thinking that exists only as in thrall to the new and the monstrous, and within which the death-drive is never adequately worked through but abstractly affirmed as the supposed 'superior' law of a creative evolution. In short, the question to be posed concerns whether Deleuze has produced a fetish of repetition. Deleuze does not appear to recognize the depth of the difficulty that faces his thinking of difference and repetition. For example, while recognizing that the law of repetition as a law of novelty might, in fact, nowhere be present in the laws of nature, he holds that the conception of *Physis*, as articulated in Nietzsche for example, discovers something 'superior to the reign of laws', namely, a 'will' that wills itself through all change, amounting to a 'power' opposed to 'law' like the interior of the earth that is opposed to its surface (1968: 13–14; 1994: 6). However, the play that is staged between conservation and destruction, between entropy and evolution, only ever reaches the stage of a pure formalism. Deleuze acknowledges that Nietzsche's eternal return amounts to a 'formalism' that overturns Kant and his categorical imperative on their own ground. It thus has little to say on the character of change, destruction, revolution and repetition, except that *there be* such phenomena.

Deleuze's commitment to thinking beyond the human condition in *Difference and Repetition* remains a pre-eminently philosophical one. It retains the Bergsonian conviction that thought must assume a monstrous form and that the break with established forms of thought must be a violent one. This time, however, the acknowledgement of the necessity of this violence takes its inspiration more from Artaud than from Bergson.[1] Nevertheless, whatever the particular inspiration that informs the idea that the exercise of thinking requires a certain violence, as well as a certain athleticism, which is not that of a fitness fanatic but more that of the 'fasting-artist' type who has seen too much in life, the cultivation of thinking necessitates a culture or *paideia* that must affect the entire individual. In *Difference and Repetition* Deleuze suggests that this does not involve a method, since method implies a collaboration of the faculties that is far too harmonious and equilibrial, as well as too determined by social values and established norms of discipline. Culture, by contrast, suggests a different kind of control and discipline, more of an involuntary adventure that finds itself caught up in the excessive repetitions of a compulsion to live beyond the old morality. If the traditional journey of thinking aimed to broaden the mind in order to achieve a consummate recognition of itself (say in the form of absolute knowledge), the journey which takes us beyond and over may go so far as to destroy this mind. Deleuze thus entertains the possibility of a new '*Meno*' in which the apprenticeship into thought would no longer be subject to the mythical forms of resemblance and identity. Instead it would freely explore the groundlessness of thought uniting difference to difference without mediation or negation (1968: 216; 1994: 166).

It is also in the text of 1968 that we see the first flowerings of the precise pedagogic conception of philosophy Deleuze will formulate with Guattari in *What is Philosophy?*, namely the pedagogy of the concept. In *Difference and Repetition* this pedagogy revolves around the demonstration of a concept of difference and is fully implicated in the attempt to construct a new image of thought. This concern of Deleuze's with the trials and procedures of thought, of how thought can construct models that take it beyond the human condition, emerges out of his earlier Bergsonism. Deleuze himself notes in his 1994 preface that it is the third chapter of the book on the image of thought that now seems to be the most concrete and apposite. It is for this reason, and because I take Deleuze's commitment to philosophy to be primarily Bergsonian in inspiration, that I shall begin this chapter by looking at Deleuze's reworking of transcendental philosophy before developing my reading of the *Difference and Repetition* and *The Logic of Sense* in terms of an ethics of the event.

Transcendental encounters

Deleuze's thinking is distinguished and marked by its attempt, inspired chiefly by Bergson, to think the intense life that is both germinal and nonorganic:

> not all Life is confined to the organic strata: rather, the organism is
> that which life sets against itself in order to limit itself, and there is a
> life all the more intense, all the more powerful for being anorganic.
> (Deleuze and Guattari 1980: 628; 1988: 503)

Although the notion of nonorganic life is not found in either *Difference and Repetition* or *The Logic of Sense*, it could be argued that these books are devoted to nothing other than a thinking with and beyond the organism (incorporating for Deleuze, as we shall see, the 'self' [*moi*] and the 'I' [*Je*]), and situating the becoming of a life within a dynamic field of intensities and singularities.

Deleuze's contribution to a new way of thinking and existing resides in the way in which he approaches questions of life and death, seeking to remove them from the restrictive economy of a personalist ethics (an ethics of the 'I' and the 'self') in order to open up the human to the *over*human. This 'beyond' of the human denotes for Deleuze nothing transcendent but is implicated fully in the infinite speed that characterizes the movement of a plane of immanence. Deleuze's avowed aim is to produce an 'art of living', in which the line upon which life and death exist in implicated and complicated movement can be mapped out in terms of the logic of the 'Outside' (Deleuze 1990: 149–51; 1995: 110–11). It is the forces of the 'outside' which impinge and impact upon us, upon what we think we are and what we think we are capable of becoming. It is the peristaltic movements of the outside which serve to destratify fixed and stable identities and produce through doubling processes new possibilities for an intenser and more creative existence. The double is never, therefore, a projection of the interior, so that the process involves not a 'One' but a redoubling of the 'Other', not a reproduction of the same but a repetition of difference, not the radiant emanation of a possessive 'I' but the immanent production of a non-self: 'It is never the other who is a double in the doubling process, it is a self that lives me as the double of the other' (Deleuze 1986: 105; 1988: 98). This means that in an encounter it is never a question of meeting oneself on the outside but only ever the other within the one that is always caught up in doubling and folding processes (Deleuze compares the process to the invagination of tissue in embryology and to the act of doubling in sewing that involves twisting, folding, stopping, repeating, and so on). The 'outside' denotes a field of immanence in which, strictly speaking, there is neither an internal self nor an external one ('I' and 'not-I'). The 'absolute Outside' is devoid of selves 'because interior and exterior are equally a part of the immanence in which they have fused' (Deleuze and Guattari 1980: 194; 1988: 156).

The aim of this new art of living is not to identify with the line, though madness and suicide always exist as a risk, since this would be to destroy all thinking and life. Rather, the task is to both 'cross the line' and make it endurable and workable; in short, this is the line of life cracked by death and conceived as *germinal*. The 'outside' is the line of life that links up random and arbitrary events in a creative mixture of chance and necessity. A new thought of the outside, and a new way of living on the outside, involves drawing new figures of thought and mapping new diagrams, in short, an intensive and vital topology that folds the outside into the inside. The passion of the outside is the passion of germinal life, releasing the forces of life from entropic containment and opening them up for a time to come.

The significance of this thought of germinal life can only be fully appreciated to the extent that the nature of Deleuze's movement beyond a philosophy of the subject has been understood and this requires, in turn, some understanding of his engagement with Kant's founding project of philosophic modernity. In *The Logic of Sense* Deleuze seeks to develop a new conception of the transcendental philosophy by approaching it as a topological field not inhabited by the 'I', the *cogito* or the synthetic unity of apperception, but populated by pre-individual singularities that constitute a 'Dionysian sense-producing machine' (Deleuze 1969: 130–1; 1990: 107). It is this 'surface topology' made up of populations (multiplicities that are not simply a combination of the one and the many) and pre-individual singularities which constitutes the 'real transcendental field' (see the next chapter for further insight into the notion of multiplicity in Deleuze). Singularities refer to 'ideal events' (ideal in the sense that they exist or endure beyond their specific individual manifestations and significations), such as bottlenecks, knots, points of fusion, processes of condensation, crystallization, and boiling, and which cannot be confused with the person or individual that is constituted by them and over whose genesis they preside. This surface topology cannot be restricted to the determinations of a centred subject (consciousness or natural perception); it is rather the surface of a 'skin' that acts as a membrane which allows for an interior and exterior to take shape and communicate, transporting potentials and regenerating polarities: 'Thus, even biologically, it is necessary to understand that "the deepest is the skin" ' (126; 103).

Although a transcendental field populated by impersonal and pre-individual singularities does not resemble a corresponding empirical field, this does not mean that it is devoid of differentiation. Deleuze does not pretend that determining the transcendental field is an easy task. In *The Logic of Sense* he accepts as decisive Sartre's objections to endowing this field with the 'I' or the synthetic unity of apperception[2] (a point repeated in *What is Philosophy?*, 1991: 49; 1994: 47, and in the 1993 piece on 'Immanence' with reference to Sartre's essay *The Transcendence of the Ego*, Deleuze 1997b: 4ff.). Before examining in more detail the character of Deleuze's encounter with Kant, it is important to note that the attempt to map the transcendental field as a field of pre-individual singularities resists the claim that unless the conditions of the real object of knowledge are conceived as the *same* as the conditions of knowledge, then transcendental philosophy becomes impossible since it has to establish autonomous conditions for objects and thereby is forced to resurrect the old metaphysics. Deleuze thus refuses to accept that the only option available to us is to choose between either an undifferentiated ground (an abyss without differences) or a supremely individuated Being and a personalized Form.

In Kant the transcendental turn is the only secure way of refuting the sceptical claims of empiricism and securing solid foundations for knowledge in terms of both its universality and necessity. The transcendental denotes the unity of self-consciousness but not *qua* a merely personal consciousness. The unity is in our own minds; it is not part of the objects apart from our knowledge of them. Experience can never give us anything that would be either universal or necessary. If knowledge remains empirical it can only ever be contingent and arbitrary. In applying a priori knowledge to experience we are already going 'beyond', notes Deleuze, what is given in experience. Now although this moment beyond the given is already present in Hume it is only with Kant that this movement beyond is provided with transcendental principles and arguments. In Hume, in other words, the principles which take us beyond the given are merely principles of *human nature*, that is, psycho-logical principles of association that concern our own representations of, and designs upon, the world. In Kant, however, 'the subjectivity of principles is not an empirical or psychological subjectivity, but a "transcendental" subjectivity' (Deleuze 1963: 21; 1984: 13; for Deleuze on Hume and Kant compare his 1991: 111–12).

Hegel frequently notes that expressions like 'transcendental unity of self-consciousness' have an ugly air about them, suggesting, he says, that there might be 'a monster' lurking 'in the background' (Hegel 1995b: 150). This is precisely what Deleuze will locate in his reworking of the transcendental in both *Difference and Repetition* and *The Logic of Sense*, namely, a thought of the monstrous and one that must become monstrous itself (as a thinking of difference 'and' repetition). This involves for Deleuze, whether one is speaking of these texts of the late 1960s or the later text *What is Philosophy?* of 1991 and the short essay of 1993 on 'Immanence', emancipating the transcendental field from the perceived stranglehold it undergoes at the hands of subjectivity and consciousness, whether this consciousness be psychological *or* phenomenological. Deleuze is committed to the seemingly extravagant claim that all phenomenology is *epi*phenomenology, since, his argument goes, it fails to penetrate the more profound individuations that are implicated in the creative evolution of difference and repetition.

What is an individual and how does it become what it is in a field of individuation structured by intensities and singularities? Deleuze's argument is incredibly complex and convoluted, and I can only here present a synoptic version of it, seeking to bring out those features that are pertinent to the concerns of this chapter. If the individual is inseparable from a world it is necessary to determine more precisely the character of this 'world'. A world can be viewed from two perspectives: that of an individual which actualizes and incarnates in its body the singularities which 'evolve' through forms of folding (envelopment, development, etc.), and that of the singularities which continue to persist and subsist over and above their particular incarnations and actualizations within an individuated body and self. There is, therefore, a 'continuum of singularities' that is distinct from the individuals that envelop it in variable degrees of determination. The world, says Deleuze, is not only actualized but also *expressed*, and while it is the case that such a world only comes into being in and through individuals, in which it exists only as a predicate, it is equally true to claim that it subsists in a highly different manner, namely, 'as an event or a verb, in the singularities which preside over the constitution of individuals' (1969: 135; 1990: 111). Put in Bergson's terms we can say that the singularities enjoy a virtual existence, and although they are actualized, and must be actualized, the actualization does not exhaust their powers of invention, and neither do the contours of an actualization ever resemble the potential that has been realized. These insights will become crucial when we approach the eternal return as a thought and willing of the event (how one becomes what one is as an event of becoming, which involves becoming the event of one's own actualization through a *counter*-actualization).

The error of attempts to define the transcendental with consciousness for Deleuze is that they get constructed in the image of that which they are supposed to ground, running the risk of simply reduplicating the empirical (Deleuze 1969: 128; 1990: 105). Metaphysics and transcendental philosophy are only able to think singularities by imprisoning them within the confines of a supreme self (*un Moi suprême*) or a superior 'I' (*un Je supérieur*) (129; 106). In *The Logic of Sense* the subject is transmuted into a free, anonymous, nomadic singularity which is said to 'traverse' humans, plants, and animals (131; 107). This 'nonorganic' life does this by being dependent neither on the matter of the particular individuations nor on the forms of their personality. This transversality, says Deleuze, constitutes the universe of Nietzsche's 'overman'. In later work this transversal field is identified as the plane of immanence, which, starting with Descartes and running from Kant to Husserl, gets treated as a field of consciousness: 'Immanence is supposed to be immanent to a pure consciousness, to a thinking subject' (Deleuze and Guattari 1991: 47–8; 1994: 46). So, while Kant denounces the employment of transcendent Ideas he discovers at the same time, Deleuze contends, the 'modern way of saving transcendence' through the auto-productive subject (the subject to whom immanence gets attributed and which is treated as an efficacious power).[3]

The character of Deleuze's peculiar transcendentalism can be further clarified by examining the discussion of Kant which figures in the chapter on 'The Image of Thought' in *Difference and Repetition*. In rethinking the 'image' of thought – the image by which thought becomes and goes astray – Deleuze argues that it is within our conception of thought that the empirical and the transcendental get distributed and related. He is concerned to show how within Kant and phenomenology the transcendental is never made truly transcendental. What does this mean? And why is it such a problem that philosophy fails to get right the installation of the transcendental? Deleuze's argument is that although it is Kant who discovers the prodigious powers of the transcendental he never escapes, in spite of his immense labours, to free transcendental structures of thought from the empirical acts of a psychological consciousness. We only have to witness in this regard, says Deleuze, the moves taking place between the first and second editions of the *Critique of Pure Reason*, and the fact that although psychologism is better hidden it still cannot be prevented from exposing itself. The reason why this is such a problem for Deleuze is that it means, in effect, that philosophy fails in the Kantian revolution to emancipate itself fully, if at all, from the domain of doxa, and which is always for him constituted by the two halves of common sense and good sense (we may also recall at this juncture that for Deleuze critique must always be violent and never at peace with established powers, a conception of philosophy which informs his reading of Nietzsche as the true philosopher of critique, 1962: 102–6; 1983: 89–93).[4] His argument here is quite specific:

No doubt philosophy refuses every particular *doxa*; no doubt it upholds no particular propositions of good sense or common sense. No doubt it recognises nothing in particular. Nevertheless, it retains the essential aspect of *doxa* – namely, the form; and the essential aspect of common sense, namely the model itself (harmony of the faculties grounded in the supposedly universal thinking subject and exercised upon the unspecified object)...so long as one only abstracts from the empirical content of *doxa*, while maintaining the operation of the faculties which corresponds to it and implicitly retains the essential aspect of the content, one remains imprisoned by it.

(Deleuze 1968: 175–6; 1994: 134)

Deleuze seeks to undermine the idea of knowledge that is implied in the transcendental model of modern metaphysics, which, he argues, is a model and form of *recognition* (between self and world, or subject and object, and self and other). Construed in terms of a model or form of recognition, philosophy is unable to open itself up to that which exceeds its faculties and the norms it imposes on their operation (the aberrant, the anomalous, the fuzzy, the indiscernible, and so on). It is for this reason that Deleuze refuses to distinguish the transcendental form of a faculty from its transcendent application: 'The transcendent exercise must not be traced from the empirical exercise precisely because it apprehends that which cannot be grasped from the point of view of common sense' (1968: 186; 1994: 143). Ultimately, this means that the transcendental must be answerable to a 'superior empiricism', which is obviously not the empiricism of common and good sense but that of another 'logic of sense' altogether: the empiricism of the unknown, the demonic, the anarchic, etc. There is something in the world that *forces us to think*. For Deleuze this something is not to become an object of recognition but rather assume the form of a 'fundamental *encounter*' (*rencontre*) (182; 139). It has to be an encounter with demons – as in Nietzsche's encounter with the demon who offers him the fateful and fatal task of undergoing the thought-experiment of eternal return, of taking up and passing on the gift and task of the overhuman (Nietzsche 1974: Section 335). The overhuman exceeds established philosophical modes of recognition and the reduction of becomings in the world to perceptual and affective clichés.

In the conjunction of Thought and Being that takes place on the plane of infinite movement (the movement peculiar to the plane of immanence, which is a movement not determined by the horizon of a subject or the position of an object), it is not, insists Deleuze, a question of 'fusion' (Merleau-Ponty's criticism of Bergson), but rather of a perpetual 'instantaneous exchange' (Nietzsche's image of the lightning-flash is deployed). If infinite movement is always double and reversible this is because 'there is only a fold from one to the other', and it is only in this sense that it can ever be said that thinking and being are one and the same (Deleuze and Guattari 1991: 41; 1994: 38). So when the thought of Thales (all is One) 'leaps out', it returns as water, and when the thought of Heraclitus emerges as *polemos* (all is strife), it is fire that retorts (ibid.). Deleuze acknowledges the dangers of thinking philosophically in terms of a plane of immanence. He speaks, for example, of its 'groping experimentation' and its recourse to 'measures' that many would deem unreasonable and unrespectable. These are measures that belong to the order of dreams, to pathological processes, to esoteric experiences, to drunken excesses, and so on (incontestably all part and parcel of the history of philosophy!). One is never quite sure, however, that Deleuze is fully cognizant of the nature or character of the selection he has made with the Thought-Being of immanence (although the plane of immanence is said to be 'pre-philosophical' its selection still involves a *philosophical* decision – see the conclusion to the next chapter for further discussion of this point). Deleuze himself is confident in the belief that it is 'we', human, all too human that we are, who suffer from the illusion that immanence is a solipsistic prison from which we need redemption by the Transcendent (one also thinks in this regard of Nietzsche's remark that the only redemption we need is from the need of redemption itself).

Individuation: Simondon and the difference of Darwin

In *Difference and Repetition* Deleuze links Simondon's reworking of individuation, which he sees as providing the basis for this new conception of the transcendental in that it presents the first properly worked out theory of an impersonal and pre-individual field, with Darwin's thinking on the origin of species, Weismann's thesis on the continuity of the germ-plasm, and the embryology of von Baer.

For Deleuze the importance of Simondon's reworking of ontogenesis lies in its demonstration that differentiation presupposes individuation as a prior intense field. Simondon argues that in spite of their differences both substantialism, a monism which conceives the unity of a living being as its essence, and hylomorphism, which regards the individual as a creation arising from the conjunction of form and matter, assume that a principle of individuation is operative independently of the activity of individuation itself (Simondon 1995 (originally 1964): 21; 1992: 297). What modern thought lacks, therefore, is a conception of ontogenesis as 'becoming'. In the hylomorphic model, for example, the actual process is deemed incapable of furnishing the principle itself and only puts it into effect. On Simondon's model, however, it is the process itself that is to be regarded as primary. This means that ontogenesis is no longer treated as dealing with the genesis of the individual but rather designates the becoming of being. Becoming is not to be thought of as a 'framework' in which the being exists, but as one of the dimensions of being which resolves an initial incompatibility that is rife with potentials. Becoming is not something that happens to being following a succession of events, since this is to presuppose it as already given and as originally substantial. Moreover, there is never a point of return in which the being would become fully identical with itself. A being never possesses a unity in its identity but only in its difference. This is because it exists as a certain kind of unity, namely a 'transductive' one that is capable of passing out of phase with itself, perpetually breaking its own bounds in relation to its centre. Individuation is not a synthesis requiring a return to unity, but rather the process in which being passes out of step with itself. This is to think invention as involving simply neither induction nor deduction but always only transduction.

Singularities and events belong to a system that evolves in metastable terms. In such a system energy exists as 'potential energy'. This, says, Deleuze is the energy specific to the 'pure event' (1969: 126; 1990: 103). Deleuze then draws attention to the fact that while an organism never ceases to assimilate and externalize through contraction and expansion, it is the membranes that play the crucial role of carrying potentials since they place internal and external spaces into contact with one another and without regard to distance. Singularities do not, therefore, simply occupy a surface but 'frequent' it and are bound up with its formation and reformation. Simondon himself writes:

> the individual is to be understood as having a relative reality, occupying only a certain phase of the whole being in question – a phase that therefore carries the implication of a preceding pre-individual state, and that, even after individuation, does not exist in isolation, since individuation does not exhaust in the single act of its appearance all the potentials embedded in the pre-individual state.
>
> (Simondon 1995: 22–3; 1992: 300)

Being is to be modelled neither in terms of substance nor matter and form, but rather 'as a tautly extended and supersaturated system, which exists at a higher level than the unit itself, which is not sufficient unto itself and cannot be adequately conceptualized according to the principle of the excluded middle' (23; 301). This means that the being does not exist, as traditionally conceived, as a stable but as a 'metastable equilibrium'. The former cannot capture the process of becoming since it requires that all the potentials in a virtual reality and becoming have been actualized. Such a being gets caught up in a system that has reached the lowest levels of energy, and so is unable to undergo any further transformations.

Simondon construes the becoming of individuation as a theatre in which there exists no pre-established sense of what is physically possible in terms of the rapport between the organism and the outside. Furthermore, and here Simondon trenchantly criticizes cybernetic modellings of systems, a living being does not exist functionally as a metastable entity in terms of being exclusively guided by the need to find compatibilities between its various requirements, in which it would merely follow a formula of complex equilibrium composed of simpler ones. Individuation is always 'amplified' for Simondon, meaning that it cannot be reduced to a mere functionalism. The process of individuation is not, therefore, akin to a *manufacturing* process, in which case it would not be a process involving *creative* evolution (it is on crucial points such as this that Simondon strikes me as being close to Bergson in spite of his own critique of Bergson; see Simondon 1995: 226–7). A living being is able to proceed in this way on account of the fact that it enjoys an 'internal resonance' in its interactions with its milieu, meaning that it never passively adapts. A relation between two terms, therefore, is never one of separate individuals, but always an aspect of the internal resonance which characterizes the system of individuation. This resonance requires permanent communication and the maintenance of a metastability as the precondition of a becoming.[5]

As noted above, the importance of Simondon's reworking of ontogenesis for Deleuze lies in its demonstration that differentiation presupposes individuation as a field of intensity. Deleuze deploys this insight in order to critically draw out what is at stake in Darwin's 'Copernican revolution' and to make it compatible with a philosophy of difference and repetition. His major claim is that individuation is not reducible to the determination of species. Individuation refers not simply to individuality but rather to intrinsic modalities of being that both constitute individuals and dissolve them (on the relationship between individuality and '*différenciation*' and between species and individuality, see Ruyer 1946: Chapter V, 133–77). Factors and processes of individuation precede the elements of a constituted individual such as matter and form, species and parts. This is to claim that they enjoy an independent 'evolution'. Deleuze begins his consideration of Darwin by noting that all differences are borne by individuals but not all these differences are individual, and then asks: 'Under what conditions does a difference become regarded as individual?' (Deleuze 1968: 319; 1994: 247). Biological classification has always been a problem of ordering differences by establishing a continuity of living beings, a problem of genus and a problem of species. According to Deleuze it was the novelty of Darwin's *Origin of Species* to inaugurate 'the thought of individual difference' by showing that we simply do not know what individual difference is capable of. The way Darwin posed the problem of individual difference is, Deleuze argues, akin to the way Freud would later pose it in relation to the unconscious: 'a question of knowing under what conditions small, unconnected or free-floating differences become appreciable, connected and fixed differences' (319; 248). An important utilization of Darwinism is taking place in *Difference and Repetition*, which inverts the emphasis placed on the origin of species. Deleuze rightly notes that in Darwin individual difference enjoys no clear status, to the extent, for example, that it is considered as the primary matter of selection and differenciation, for 'understood as free-floating or unconnected difference, it is not distinguished from an indeterminate variability' (ibid.).

In *Difference and Repetition* Deleuze reads Weismann in a manner entirely different from his late-nineteenth-century reception, where his doctrine on the continuity of the germ-plasm was placed in the service of a philosophy of pessimism. Weismann's contribution, on Deleuze's reading, is to have shown that sexed reproduction serves as the function of an incessant production of individual differences. To the extent, therefore, that sexual differenciation (variation) results from sexed reproduction, it is possible to claim that the three figures of Darwinism's Copernican Revolution consist in showing how the three major biological differenciations of species, organic parts, and of the sexes 'turn around individual difference, not vice versa' (1968: 320; 1994: 249). The species, therefore, a notion which caused Darwin so much anxiety since an accurate definition proves so elusive, is a transcendental illusion in relation to the virtual–actual movement of life, which is always evolving in the direction of the production of individuation.

> It is not the individual which is an illusion in relation to the genius of the species, but the species which is an illusion – inevitable and well-founded, it is true – in relation to the play of the individual and individuation.
>
> (321; 250)

It is on the basis of this privileging of a complicated 'evolution' through individuation that Deleuze is able to claim that: 'Dreams are our eggs, our larvae, and our properly psychic individuals' (ibid.), and, furthermore: 'Mute witnesses to degradation and death, the centres of envelopment are also the dark precursors [*les précurseurs sombres*] of the eternal return' (330; 256). This is to conceive evolution as the creation and production of lines and figures of differenciation, of difference through repetition, engineering excessively the futurity of germinal life over the continuity of the germ-plasm and the eternal return as the return of the same.

It is on the basis of this reading of the 'meaning' of Darwin's Copernican Revolution that Deleuze will situate von Baer's founding of a superior embryology which is able to show that every embryo is a phantasm of its parents and capable of undertaking forced movements, constituting internal resonances, and dramatizing the 'primordial relations of life', all in a way which would simply destroy an adult individual (von Baer 1973; Gould 1977: 52–63; Lenoir 1982: 72–95, 189–94). Development is individualization, proceeding 'from the general to the special', while the singular tendency towards differenciation precludes recapitulation (Gould 1977: 55, 63). The vital egg is a 'pure field of individuation' (Deleuze 1968: 322; 1994: 250). Von Baer's privileging of embryogenesis over phylogenesis – a new life does not simply recapitulate previous ancestral forms of other species but establishes new conditions of possibility – acknowledges that while an egg must reproduce all the parts of an organism to which it belongs (sexed reproduction taking place within the 'limits' of the species), it is also the case that an egg 'reconstitutes the parts only on condition that it develops within a field which does not depend on them. It develops within the limits of the species only on condition that it also presents phenomena of specific de-differenciation' (320; 249). In other words, von Baer's revolution was to show that there are conditions of life that are peculiar to embryonic life, conditions of possibility which *exceed* the limits of species, genus, order, or class. It is on the basis of von Baer's research, then, that Deleuze will claim in *Difference and Repetition* that the highest generalities of life 'point beyond species and genus...in the direction of the individual and pre-individual singularities rather than towards impersonal abstraction' (ibid.).

In declaring the world to be an 'egg' of germinal intensity (a position repeated in *A Thousand Plateaus*), Deleuze is arguing that it is the implicated egg which expresses the differential relations of the virtual matter that is to be organized. It is this intensive field of individuation that determines the relations incarnated in spatio-temporal dynamisms in both the differentiation of species and the differentiation of organic parts. It is intensity, therefore, which is held to be primary in relation to organic extensions and to species qualities. It is the notions of 'morphogenetic potential' and 'field-gradient threshold', elaborated by the embryologist Albert Dalcq, and discussed in strikingly similar terms to Deleuze by Merleau-Ponty, with reference to work in embryology since the time of Hans Driesch (Merleau-Ponty 1988: 192ff.), that are able to account for this complex ensemble and which significantly reconfigure our understanding of the relation between genic nucleus and the cytoplasm (the actualization of the differential relations is determined by the latter in terms of its gradients and fields of individuation).

The notion of individuation plays a crucial role in the unfolding of the psycho-biology of *Difference and Repetition* since it serves to mediate the virtual and the process of actualization. As in *Bergsonism*, Deleuze stipulates that evolution does not simply progress from one actual term to another, or from general to particular, and this is precisely because there is the intermediary of an individuation which creates a realm of difference between the virtual and its actualization. It is this emphasis on individuation as an autonomous realm of difference and repetition that leads Deleuze to claiming that any opposition between preformism and epigenesis becomes redundant once it is appreciated that there is no resemblance or simple correspondence between the intensive and enveloped pre-formations and the qualitative and extensive developed formations. What governs actualization, therefore, is primary individuation: 'the organic parts are induced only on the basis of the gradients of their intensive environment; the types determined in their species only by virtue of their individuating intensity' (1968: 323; 1994: 251).

The moves Deleuze is making here are crucial to his insight into germinal life and to his attempts to articulate a philosophy that conceives of the evolution of organisms as complex systems in terms of a field of individuation and intensities. What he is now calling the field of individuation populated by pre-individual singularities and intensities becomes what in the later work is called the plane of immanence (also the plane of consistency) that is populated by 'anonymous' matter and which develops through transversal communication across distinct phyletic lineages. The possibility of such communication is already encountered in the argument of *Difference and Repetition*. In demarcating so insistently individual differences Deleuze does not wish to deny that these differences partake of a certain confusion. On the contrary, he maintains that the clear is confused by itself and only in so far as it is clear. If Darwin's theory of 'descent with modification' put an end once and for all to the theological doctrine on the fixity of species, then Deleuze's thinking of radical difference also aims to expose the impossibility of the fixity of fields of individuation. Here his argument is complex and nuanced. On the one hand, he acknowledges that forms of animal life can only be considered as certain kinds of species in relation to the particular fields of individuation that express them. This means that we cannot readily say that individuals of a given species are distinguished by their participation in other species (as if, says Deleuze, there was an ass, lion, wolf, or sheep in every human being, although, Deleuze says, metempsychosis retains 'its symbolic truth', 1968: 327; 1994: 254 – something he will later articulate with his conception of 'animal becomings', but where such becomings are to be conceived as taking place on a molecular level having little to do with the animal on the level of species and genus). The realm of enveloped life, however, is a confused realm and individual differences, or composing souls, play the role of variables within fields of individuation. The 'universal individual', such as, says Deleuze, the thinker of the eternal return of all things (the repetition which takes things to their superior form and power) is one who makes full use of the power of the clear and the confused in order to attest to the 'multiple, mobile, and communicating character of individuality' and its 'implicated character' (ibid.). In other words, and however strange or paradoxical it may appear, the life of individuality is the life of the germinal and intensive, it is non-organic life:

> We are made of all these depths and distance, of these intensive souls which develop and are re-enveloped. We call individuating factors the ensemble of these enveloping and enveloped intensities, of these individuating and individual differences *which ceaselessly interpenetrate one another throughout the fields of individuation*. Individuality is not a characteristic of the Self but, on the contrary, forms and sustains the system of the dissolved Self.
>
> (ibid., my emphasis)

For Deleuze the 'problem of heredity' is never, therefore, simply one of reproduction since reproduction is tied to a complex and confused 'variation'. This is why, he says, all theories of heredity must ultimately open out onto a philosophy of nature: 'It is as if repetition were never the repetition of the "same" but always of the Different as such, and the object of difference in itself were repetition' (330; 256). In short, the differential and intensive factors of individuation never cease to be *implicated* in a complicated 'evolution'. For Deleuze, this 'evolution' becomes less and less that of exogenous mechanism (hence the need for scare quotes) to the extent that 'complex systems' achieve the tendency to interiorize the differences that constitute them. The more evolution involves the interiorization of difference the less 'repetition' comes to depend on external causes and the less guarantee there is that evolution can be simply understood as the reproduction of the same.

Ethics after individuation

It is individuation conceived as a field of intensive factors (haecceities) which informs Deleuze's rethinking of ethics 'beyond' the subject so as to move thought beyond the 'human condition' and to open it up to the inhuman and overhuman. The abyss of this Dionysian world, a world of free and unbound energy, is not, Deleuze insists, an impersonal or abstract Universal *beyond* individuation. For him this is precisely what defines Nietzsche's break with Schopenhauer (see Deleuze 1969: 131; 1990: 107). It is rather the 'I' and the self that are abstract universals that need to be conceived in relation to the individuating forces that consume them: 'What cannot be replaced is individuation itself' (Deleuze 1968: 332; 1994: 258). This is to think of life not as the expression of a *determined* world but as a *possible* one, replete with diffuse potentialities and virtual actualizations. The 'life' of the individual always exceeds that of the species, but the individuating factors that are implicated in a process of individuation assume neither the form of the 'I' nor the matter of the self (if the former represents the psychic determination of the species, the latter expresses its psychic organization). The formation of the 'I' is inseparable from a form of identity, while the self is always constituted by a continuity of family resemblances (the 'I' and self function as genealogical figures on a kind of Weismannian landscape). 'Individuation', Deleuze writes, 'is mobile, strangely supple, fortuitous and endowed with fringes and margins' (331; 257). It is correct to assert that individuality is always an indetermination. But this does not expose a lack or a deficiency at the centre of its being but rather denotes its full positive power.

The self, therefore, is always dissolved and the 'I' fractured: 'For it is not the other which is another I, but the I which is an other, a fractured I' (1968: 335; 1994: 261). Death can be situated from this life of the dissolved self within the folds of an involution. In the becoming of any system death has a double face: a face of implication and involvement, of involution, and a face of explication and evolution, of entropy. The most powerful face of death is that which is hidden, almost noumenal. It is the death which acts as the agent of dissolution and decomposition, being that 'internal power' which frees the process of individuation from the form of the 'I' and the matter of the self in which life has become imprisoned. Every death, therefore, can be deemed to be double, being both a cancellation of large differences through entropic extension, and the liberation of those little differences which swarm through an intensive involution.

Deleuze considered it a mistake to regard the late Foucault as making a straightforward return to a discredited subject of ethics. There are for Foucault, Deleuze argues, only processes of subjectification and relations of self to self that involve relations of forces and foldings of forces (intensities, densities, etc.). Ethics concerns the production of new creative lines of life that depend on how one folds the forces, producing not a new subject but a 'work of art' (Deleuze 1990: 127; 1995: 92), and through which one becomes master of one's particular speed and master of one's molecules (1986: 130; 1988: 123). What gets produced in this art of living is a field of forces (electrical, magnetic, kinetic), speaking of an *individuation* that evolves and involves through weak and strong intensities and through active and passive affects. The subject on this model is divested of both interiority and identity, transformed into an 'event' of individuation, so that a self always exists as a mode of intensity, never simply as a personal subject (134–5; 98–9). The cracked line is, more often that not, deadly, moving fast, and violent, transporting us into 'breathless regions'. The task, however, is never simply to surrender to the lethal character of the line, but always to seek to extract some surplus value from its productivity and fecundity, so as to pass something on that will continue to exist germinally: 'You write with a view to an unborn people that doesn't yet have a language' (Deleuze 1990: 196; 1995: 143).

Deleuze makes a rigorous and strict separation of the movement of 'desire', which is always transversal, from the experience of 'pleasure'.[6] Pleasure is 'on the side of strata and organization', serving only to interrupt the purely immanent process of desire (Deleuze 1997a: 189–90). He insists that desire does not have pleasure as its 'norm' because pleasure always wants, on Deleuze's austere reading, to attribute the production and distribution of affects to a subject or person, acting as a means for the person to find himself or herself again in a process that overwhelms them and which ensures that the act of repetition is always undergone again and again (Deleuze 1977: 119; 1987: 99–100). In making this point Deleuze is, in part, following Nietzsche in his incisive attack on the great errors of psychology. Nietzsche points out that our inability to allow for new experiences and new sensations lies in the tendency to stratify existence by always tracing the unknown back into what is already known and familiar, producing for ourselves effects that are alleviating, gratifying, soothing, and giving ourselves 'moreover a feeling of power'. This, he notes, is 'Proof by *pleasure* ("by potency") as criterion of truth' (Nietzsche 1979b: 51). The result is to exclude from experience the unexperienced, or the yet-to-be experienced, as well as the strange, the uncanny, and the new and the immeasurable. Deleuze's desire seeks the experience or time that is not a return of the repressed (a return of the same, as say, for example, in Freud's uncanny), but rather the repetition of the future in its difference.

This explains why in his reworking of Freud's positing of a 'beyond' of the pleasure principle Deleuze grants primacy to the phenomenon of repetition. The death-drive needs to be related not so much to destructive tendencies and aggressivity, but more to phenomena of repetition. If the pleasure principle is tied inextricably to a psychological principle, that of the 'beyond' is linked necessarily to a transcendental principle (Deleuze 1968: 27; 1994: 16). It is not so much a question of drawing something new from repetition as more a matter of making repetition a novelty, that is, liberating the will from all that binds it by making repetition the object of willing: 'The question in each and every thing: "Do you desire this once more and innumerable times more?" would lie upon your actions as the greatest weight' (Nietzsche 1974: Section 341). On this reading the 'subject' of this willing in eternal return is not the self taken as an abstract doer but the field of intensities and singularities, chance encounters and fortuitous circumstances, that it finds itself implicated in. As such, the demonic question invited by the eternal return seeks to transform mere action in the world into the event of a world, that is, it seeks to cultivate the event ('a life'). Deleuze cites the following important passage from a piece by Klossowski:

> the vehement oscillations which upset the individual as long as he seeks only his own center and does not see the circle of which he himself is a part; for if these oscillations upset him, it is because each corresponds to an individuality *other* than that which he takes as his own from the point of view of the undiscoverable center. Hence, an identity is essentially fortuitous [the innocence of becoming – KAP] and a series of individualities must be traversed by each, in order that the fortuity make them completely necessary.
>
> (Deleuze 1969: 209; 1990: 178)

The experience of repetition is never for Deleuze, as it is in Freud, simply implicated in a 'regression', such as found in the compulsion to repeat which determines the death-drive.[7] This would entail reducing death to the model of a purely physical or material repetition, to what Deleuze calls a merely 'brute repetition'. The reason why Freud has to construe the death-drive in terms of the step backwards, a desire to return to inanimate matter, is because of his commitment to a conception of the personal unity and integrity of the organism. On this model death must, therefore, always be conceived as a negative splitting and falling apart, a 'regressus' involving reactive, internalized violence of a self upon itself. In the repetition of return, however, we are exposed to 'demonic power' that is more complicated, living *between* life and death, at the border, on the edge of chaos: 'If repetition makes us ill, it also heals us; if it enchains and destroys us, it also frees us' (Deleuze 1968: 30; 1994: 19). Deleuze always insists on thinking beyond the death which is apodeictic knowledge (our sense of the mortality of the self is both empty and abstract, he maintains) and places the stress on the importance of coming to know the 'event' of death that always enjoys an open problematic structure (Deleuze 1969: 170–1; 1990: 145): 'Death is...the last form of the problematic, the source of problems and questions, the sign of their persistence over and above every response, the "Where?" and "When?" which designate this (non)-being where every affirmation is nourished' (Deleuze 1968: 148; 1994: 112).

Before we can fully explore this superior repetition of Nietzsche's, and the character of Deleuze's critical reading of Freud on the repetition of death, it is necessary to examine in general terms the manner in which Deleuze stages the question of repetition in the text of *Difference and Repetition*.

The phenomenon of repetition and the three syntheses of time

Deleuze's unfolding of the problematic of repetition in Chapter II of *Difference and Repetition* is convoluted but not imprecise or indeterminable. In fact, it is best approached in terms of the double movement it seeks to effectuate: first, to show that the need of the organism draws on repetition as a passive synthesis of time, and which links it to the pleasure principle; and second, to show that repetition, whether in habit or the passive synthesis of binding, is also involved in a 'beyond' of the pleasure principle. Indeed, Deleuze wants to claim that it is the habit of a passive synthesis that precedes the pleasure principle and actually renders it possible.

The opening sentence of Chapter II of *Difference and Repetition* ('Repetition for Itself'), which is concerned with the question of whether repetition changes anything in the object that is repeated or simply produces a difference into the mind that contemplates, returns us to the treatment of Hume and Bergson on repetition first broached in the 1956 essay. There Deleuze notes that for Bergson difference is introduced into the phenomenon of repetition as its spiritual component, producing a 'contraction' as well as a 'fusion' and an 'interpenetration' as the very movement of spirit (Deleuze 1956: 102–3).

In the 1968 text Deleuze locates in Hume's 'contractile' power of the imagination not simply the operations of memory or of human understanding, but the formation of a synthesis of time. The repetition of a sequence – such as the pattern of AB, AB, AB, etc. – produces something new in the mind, a kind of 'originary subjectivity', so that when A appears we expect the arrival of B and with a force that corresponds to the qualititative dimension of all the contracted ABs. It is this sense of anticipation that leads to the formation of a *synthesis* of time in the mind:

> A succession of instants does not constitute time any more than it causes it to disappear; it indicates only its constantly aborted moment of birth. Time is constituted only in the originary synthesis which operates on the repetition of instants.
>
> (Deleuze 1968: 97; 1994: 70)

The synthesis of time concerns a living present in which past and future do not designate separate instants but rather dimensions of a present that are involved in contraction. Deleuze stresses the 'passive' character of this synthesis because, in spite of its constitutive character, it is not carried out by the mind but rather takes place 'in' it prior to memory and reflection. If time is subjective then its subjectivity belongs to a passive subject, a subject that contracts matter, and moves like the arrow of time asymmetrically from past to future in the present that contracts its being-present.[8]

However, such a passive subject is already going beyond itself and its particular present since the past is no longer simply the past of retention but also that of reflexivity and the future is no longer simply that of anticipation but also that of prediction. These are the 'active syntheses' of memory and understanding which are both superimposed upon, and supported by, the passive synthesis of the imagination. Now examples of sensible and perceptual syntheses refer us back to a more primary sensibility that expresses itself in the form of 'organic syntheses' that constitute the elemental ground of living forms: 'We are made of contracted water, earth, light and air' prior not only to their recognition but also prior to their being senses. Every organism is a sum of contractions, retentions, and expectations so that at the level of a 'primary vital sensibility' it is the lived present of need that constitutes the synthesis of time. Need is the organic form of expectation, while cellular heredity accounts for the past that is always retained. When coupled with the perceptual syntheses that are built upon them these syntheses become part of the active syntheses of a 'psycho-organic' memory and an intelligence that take the form of instinct and learning. Deleuze will now refer to Hume in order to attack the illusions of psychology that would interpret the operations of habit from the perspective of action, rather than locating habit in the more profound contemplations of a contracting organic life: 'It is simultaneously through contraction that we are habits, but through contemplation that we contract' (1968: 101; 1994: 74; see Hume 1985: 78ff.). It is in this context of the contemplative organic life that Deleuze approaches a pleasure principle, noting that while pleasure is involved in contraction, tension, and the process of relaxation, it can only serve as the 'principle' of biopsychic life by not reducing it to a mere function within such life. This is the 'beatitude' that can be associated with the passive syntheses, speaking of a primary Narcissus in relation to the pleasure that is fulfilled in contemplation.

The organism, therefore, is a life constituted through the expression of need, habit, and contemplation. It is made of elements and cases of repetition (contemplated and contracted water, nitrogen, carbon, etc.). It is need that marks the limits of a variable present and which coincides with the duration of a contemplation. Repetition and need are inextricably linked since it is only through the repetition of an instance that need can express itself as the for-itself of a certain duration. It is only the need of the present which can impart signs to the past and the future as signs in need of interpretation and action. Deleuze does not, however, let us note, invoke the illusion of a perpetual present. The fact that the present is constituted through need and synthesized temporally through acts of repetition means that there can be no present which could be co-extensive with time. The organism exists outside itself precisely because it is constituted 'in' time. Action is constituted, therefore, only through the contraction of elements of repetition. Such contraction, moreover, takes place not in an action itself but only in a contemplative 'self' that serves to double any agency. Deleuze thus insists that underneath the self of action there is to be found those larval selves – molecular selves – that contemplate and that render possible the actions of an active subject. This results in *Difference and Repetition* in a significant reworking of Kant's model of the self found in the *Critique of Pure Reason*, granting primacy not to the receptive capacity that receives impressions and experiences sensations but to the contractile power of contemplation that constitutes the organism before it constitutes the sensations that affect it. The self is a multiplicity not on the order of representation, or even sensation, but on the level of the passive syntheses peculiar to contemplation. A self does not undergo modifications simply because it is nothing other than a modification.

This attention to the mute witnesses of change and becoming leads Deleuze to a highly novel conception of the 'activity' of need, which for him is not simply, in fact not at all, allied to a lack or a negative. Need already contains desire within itself as an experimentation of life through the exploration of possibilities, hence adaptation on this model can never simply be a merely mechanistic process: 'Need expresses the openness of a question before it expresses the non-being or absence of a response' (1968: 106; 1994: 78).

In his analysis of repetition so far Deleuze has introduced only the first synthesis of time associated with the time of the present as a time of 'need'. However, he now implicates this first synthesis in another time, in effect, a second synthesis that the first must presuppose. The time of the present is fundamentally paradoxical since in order to be 'present' it must not be self-present, that is, it must pass, and it is the passing of time that prevents the present from ever being coextensive with time. Habit is the domain of the first synthesis, that of the passing time. But it is memory that provides the 'ground' on which habit assumes a foundation for itself as a synthesis of time. Memory is the condition of possibility of habit. There is, argues, Deleuze a 'profound' passive synthesis that can be identified with memory constituting the 'being' of the past as that which precisely allows the present to pass. The past is never simply 'it was' but enjoys a virtual co-existence with the present. Deleuze's argument is confusing at points since it speaks of memory as both active and passive synthesis. The distinction is between memory as the synthesis of representation and as a synthesis that involves the 'pure past', that is, a productive unconscious of the past that exceeds the calculations of present needs and habits (it is the discovery of this pure past which makes Bergson's *Matter and Memory* such a great book, argues Deleuze). The paradox of this pure past is this: the present can only pass because the past is contemporaneous with the present that it was. The past is not simply past because a new present arrives. Moreover, the whole of the past coexists with the new present but it assumes only its most contracted state. It is for this reason that the past can be said to be something more than a dimension of time and declared to be its true passive synthesis. The 'ground' of time can never, as such, be represented. Instead Deleuze says that it can be granted a kind of 'noumenal character', so that 'what we live empirically as a succession of different presents from the point of view of active synthesis is also *the ever-increasing coexistence of levels of the past within passive synthesis*' (1968: 113; 1994: 83). This leads Deleuze to draw a distinction between two kinds of repetition. The first he names a 'material' repetition that is, in effect, the repetition of representation, the repetition of an active synthesis which is concerned almost exclusively with conservation and calculation. The second, however, is named a 'spiritual' repetition, and this is a repetition that unfolds in the being of the past independently of the identity of the present in reflection and, moreover, is unable to assume the form of a mere reproduction in the present. Again, one sees the critical force of Deleuze's working out of the importance of Bergson's notion of virtual duration, namely, that it involves actualization as invention, never simply reproduction.

From this we can perhaps best appreciate the character of Deleuze's reworking of Kant and the 'I think' that is fully implicated and complicated 'in' time. Kant's decisive break with Descartes's *cogito* consists in showing that the undetermined existence of the 'I am' can only be determined within time as the existence of a passive, receptive, and phenomenal subject that also appears within time. This means that the spontaneity of the self cannot be conceived as the attribute of any substantial being, but rather only as the 'affection' of a passive self that experiences its own thought – the intelligence which allows it to say 'I' – as being exercised in it and upon it but not *by* it. The self, therefore, Deleuze declares, always lives itself as fractured and as other on account of the 'pure and empty form of time'. Again this is, for Deleuze, the discovery of the 'transcendental'. Descartes can only save the 'I' by completely expelling time or by reducing the *cogito* to an 'instant and entrusting' time that belongs to the operation of a creation manufactured and controlled by God, to the extent that the identity of the 'I' has its only guarantee in the unity of such a God. Of course, Kant goes on in the *Critique of Pure Reason* to save the world of representation from the powers of this fractured self and its world of time, covering up the event of the 'speculative death of God' and the disappearance of both rational theology and rational psychology (1968: 117; 1994: 87). One of the principal ways this is done in Kant's *Critique* for Deleuze is by giving primacy to active synthetic identity, which also entails conceiving passivity as a simple receptivity lacking in synthesis.

It is the order of the pure, empty form of time (pure because it is free of any imperial designs of an active, foundational subject, and an empty form because it involves the movement of difference and repetition) which brings Deleuze to expounding the third and final synthesis of time, the time that is 'out of joint'. This is a time that ceases to be 'cardinal' (serial) and becomes 'ordinal' (intensive). It is the time of the caesura. The paradox peculiar to this order of time is that time is no longer subordinated to movement but is, itself, the most radical form of change as a form that does not itself change. Deleuze calls it, strangely perhaps, a 'totality of time'. But this is totality only in the sense of Nietzsche giving the name of 'eternity' to his 'moment'. It is total only to the extent that it draws together in the caesura the before and the after. It is the event 'of' time which creates the possibility for a temporal series to be instantiated. It is, says Deleuze, the time of the Overman (1968: 121; 1994: 90). The third order of time involves a repetition by excess, a repetition of the future as eternal return. It is a form of time which introduces into being a revelation of the formless, the order of chaos and metamorphosis, and in which the ground of time (memory) is super-seded by 'a universal ungrounding which turns upon itself and causes only the yet-to-come to return' ['*effondement*' *universel qui tourne en lui-même et ne fait revenir que l'àvenir*] (123; 91).

The critical and clinical question concerns how it is possible to live this life as the event and to live for it. We are reminded of Nietzsche's exhortation: 'this world is will to power and nothing besides! And you yourselves are this will to power and nothing besides!' The 'ethical' moment of Deleuze's thinking of difference and repetition emerges at this point, but it is perhaps more readily visible and intelligible in *The Logic of Sense*. The third order of time – the time 'of' the future – exceeds the measure of what we are, hence its overhuman character. In *Difference and Repetition* this excessive time of repetition, which comes from the future like a shock, is related to Nietzsche's doctrine of eternal return as a thought 'beyond good and evil' that entails a 'suspension of ethics'. But there is no contradiction here between the suspension of ethics in the one text and the affirmation of ethics in the other. This is simply because the 'ethics' Deleuze is seeking to articulate in *The Logic of Sense* are, however paradoxical, of an overhuman persuasion. They are an ethics of the event.

Before we can fully explore the character of this ethics we need to look at the way in which Deleuze's stress on the excess of repetition involves a reworking of the figuration of death in Freud's attempt to think the 'beyond' of the pleasure principle. Deleuze holds that life becomes 'poisoned' by moral categories, categories of good and evil, blame and merit, sin and redemption. In short, morality is totally lacking in a thinking of the event to the extent that it can only interpret the signs of life nonvitally as commands and punishments. In order to cultivate joyful passions and articulate the 'gay' science of life, it is necessary to denounce falsifications of life, including the values through which life gets disparaged. Do we live? Or are we merely existing within the semblance of a life? Why do we fear death so much? Might it be this fear which explains why civilization has only ever been conceived in terms of its 'discontents'? Has the history of humanity (or should I say 'man'?) been guided by the pursuit of a secret death *worship*? It is out of these Spinozist and Nietzschean concerns that we can, I believe, most fruitfully explore Deleuze's critical encounter with the Freudian death-drive.

The death-drive: Freud's reworking of Weismann

In *Difference and Repetition* Deleuze locates in Freud's idea of a working-through of memory in order not to repeat the past an important contribution to a thinking of difference and repetition. The compulsion to repeat contains a demonic power that needs to be linked not simply to regression, to a death that would be the realization of identity and self-presence, but to the moving-forward, that is, to the freedom of the future. Freud, at least on Deleuze's reading, maintains for the most part, and especially in his positing of a death-drive, the model of a brute repetition. This a purely material model in which the movement of desire is implicated solely in regression as in the desire to return to inanimate matter. In contrast to a material model Deleuze places stress on the difference produced by repetition in which its variations express differential mechanisms that belong to the 'origin' and 'essence' of that which is repeated. Repetition is always disguised, always covered by masks, since it involves the play of simulation and simulacra. It is in the disguised selection of repetition that we find conjoined the best and the worst, both good and evil, both terror and freedom (it is on the basis of this insight that Deleuze will produce his reading of modern literature, including the likes of Hardy, Kafka, Lawrence, and Zola).

However, more is at stake in Deleuze's encounter with the death-drive than simply showing the monstrous power of repetition. Because it is so crucial for grasping the details of his response to the Weismannian legacy I want to explore at length Freud's own reworking of Weismann in *Beyond the Pleasure Principle*. It is from understanding this encounter, and the alternative model of death Deleuze develops, that we will be able to gain the necessary insight into how repetition is to be linked to the movement of the 'living beyond'. The aim, in short, is to seek to understand the overhuman character of the Event of time. This requires, at the very least, understanding death as the pure, empty form of time itself. It is always important for Deleuze to engage with Freud's positing of the death-drive simply because in its biophilosophical aspects it can be seen to be diametrically opposed to his attempt to undertake the task of thinking 'difference and repetition'.

In *Beyond the Pleasure Principle* (first published in 1920) Freud approaches the compulsion to repeat from two angles: namely, those cases in which it ultimately conforms to the pleasure principle (as in the '*fort/da*' game of the child, the repetition of which transforms a passive situation into one in which s/he plays an active part); and those cases in which there is revealed a strange 'beyond' of the pleasure principle. Trying to determine the character of this 'beyond' takes us to the heart of Freud's biophilosophy. Freud notes that psychoanalysis was designed as 'first and foremost an art of interpreting' in which the chief emphasis is upon the resistances of the patient. The 'art' now shifts to uncovering these resistances and getting the patient to give them up. Under analysis a patient will repeat the repressed material, which is more often than not repeated as a contemporary experience, and demonstrate that resistance emanates not from the unconscious but rather from that higher strata of the mind which initially undertook the repression – the conscious or rather, to be more precise says Freud, the 'coherent ego'. What informs and motivates the repression, notes Freud, is the pleasure principle: seeking to avoid the unpleasure which would be produced if the repressed content was liberated. Under the compulsion to repeat we witness many instances that defy the pleasure principle, however, including recollections of past experiences, and their contemporary repetition, that include no possibility of pleasure. Freud then offers the hypothesis that this compulsion to repeat reveals a 'destiny' that is more elementary, more 'primitive' and instinctual than the principle of pleasure it 'overrides'. It is at this juncture in his argument that Freud enters the realm of what he calls 'far-fetched speculation'. It is important to appreciate that there is no fixed configuration of the death-drive in Freud's speculations, to the extent that it is not clear that he does, in fact, and as Deleuze alleges, subscribe to a merely material or 'brute' model of repetition. It is clear from the above, for example, in relation to the phenomenon of repetition, that repetition, including the death-drive, is involved in an emancipation of the subject from its own repressive character.

The point of departure for psychoanalytic speculation is said to be the discovery that, on the basis of an examination of the processes of the unconscious, consciousness is not the most universal attribute of the mental realm but only a particular function of it. Consciousness is implicated in perceptions of excitations that enter from the external world and is associated with feelings of pleasure and unpleasure that arise from within the mental apparatus itself. The 'space' of this perceptive consciousness is that of a 'borderline' between the outside and inside, turned towards the external world and enveloping the other psychical systems. Freud seeks to build this understanding of consciousness on the back of work in cerebral anatomy which locates the 'seat' of consciousness in the cerebral cortex, defined as the 'outermost, enveloping layer of the central organ'. What interests Freud, however, is why it should be that consciousness is lodged on the surface of the brain rather than being 'safely housed' in the inmost interior. He lays particular stress on the importance of the 'permanent traces' (memory-traces) that are left behind by the excitatory processes in the psychic systems. This is the 'foundation' of memory that is formed independently of any process of becoming-conscious. Before we can grasp the significance of this for Freud we need to understand the picture he offers of the living organism.

In its most simplified form, Freud argues, a living organism can be described as an 'undifferentiated vesicle of a substance that is susceptible to stimulation' (1991: 297). It is the 'surface' of this substance that is turned towards the external world in terms of receiving stimuli that constitutes the differentiated element. We may note that this is the first distinctive feature of Freud's biophilosophy of the organism – that differentiation relates not to the 'substance' but to the 'subject'. It is this feature that Deleuze's biophilosophy contests, along with others peculiar to Freud's positing of a death-drive, in *Difference and Repetition*. Freud revealingly utilizes research in embryology throughout the essay. On the two notable occasions he draws on its research he refers to it as the theory of 'recapitulation' of developmental history, showing clearly that he adheres to the theory critiqued by Deleuze (and now widely refuted by contemporary embryology) that ontogenesis simply recapitulates phylogeny. At this moment in the unfolding of his argument he refers to embryology to legitimize the view that the central nervous system has its origins in the ectoderm: 'the grey matter of the cortex remains a derivative of the primitive superficial layer of the organism and may have inherited some of its essential properties' (297). In other words, the advanced complexity of an organic life that enjoys a nervous system remains caught up in the primitive excitatory life of the more elemental life form from which it owes its origins. The destiny of the living organism, therefore, resides in its primitive past. The principal argument Freud seeks to sustain at this point is that it is protection against stimuli, rather than reception of it, which constitutes the most important function for any living organism. The receptive cortical layer provides such a protective shield, without which it would be killed by an overwhelming influx of stimulation that could be neither distributed nor dissipated and organized. The living substance evolves by acquiring this shield and developing it as an 'inorganic' material, which then comes to function as a special envelope or membrane that offers resistance to stimuli: 'By its death, the outer layer has saved all the deeper ones from a similar fate – unless that is to say, stimuli reach it which are so strong they break through the protective shield' (298–9). Freud acknowledges that in more complex organisms the receptive cortical layer has become enfolded within the depths of the interior of the organismic body, though portions of it remain behind on the surface in the form of sense organs whose task it is to deal with small quantities of external stimulation and regulate their impact through sampling.

The next stage in Freud's argument is to note that the sensitive cortex is not only exposed to external stimulation but also stimulatory excitations from within. It is here that the living organism lacks a protective shield and that feelings of pleasure and unpleasure predominate over all external stimuli. In addition, however, Freud notes that any internal excitations that produce too much unpleasure get treated as though they were acting not from inside but from outside, and defence mechanisms are erected against them, which is said to be the origin of the projection that plays such a key role in the causation of pathological processes. However, so far we remain within the element of the pleasure principle when the task is to think its 'beyond'. It is in terms of the 'order' of instincts that Freud now approaches the question, noting that it is the instincts of an organism that constitute its most abundant sources of internal excitation. Freud notes that in those instances when the manifestations of the compulsion to repeat exhibit a high degree of instinctual character, and when they appear to act contra the pleasure principle, then there is at work some 'daemonic force'. Freud is especially keen to comprehend those cases where the compulsion to repeat defies the insight that would claim that a condition of enjoyment is novelty, or perpetual creation of the new. Certain kinds of repetition as the re-experience of the identical, most noticeable in the play of children as well in their socialization, can be said to be sources of pleasure. The patient under analysis who is undergoing painful experiences from childhood, however, is clearly dwelling 'beyond' pleasure and shows signs of being in possession of the daemonic power.

It is at this point in his argument, and the attempt to comprehend the instinctual basis of the compulsion to repeat, that Freud introduces his speculative idea that organic life is formed by the instinct and urge to restore an earlier condition. This, says Freud, is a condition that the living organism has been 'obliged to abandon' only owing to the pressure of external circumstances, such as evolution. There is, he adds, a law of 'inertia' inherent in organic life. This is perhaps the most revealing moment in the unfolding of Freud's argument since it is clear that he is able to posit the death-drive – the fact that organic life is driven by a desire to return to the primitive conceived as a condition of stasis – only by constructing a reified dualism between the organism and its external circumstances, that is, speaking of the organism as if we could say that it exists independently of its evolution. It is not surprising, therefore, that his argument should get entangled in some untenable positions, most notably disregarding the mass of evidence from the life sciences that demonstrate that sexual reproduction (and with it, death) is a late acquisition in the evolution of life on the planet, and so positing a primordial death-drive that is said to be present, if only latently so, in the very first forms of life. Freud concedes that his views necessarily lead to the claim that the essential nature of the living substance is a 'conservative' one. The organism has no real desire to evolve, and if left to itself it would not. It is only on account of the contingencies of external circumstances, such as the environment treated by the theory of natural selection, that it is compelled to adapt and become more complex. This means for Freud that although instincts evolve and organisms can be shown to acquire new traits that enable them to survive through adaptation, what is never evolved out of is the primordial instinctual desire within all organic life to return to the first stages of matter (Freud provides an entirely unconvincing account in these terms of diverse phenomena such as the laborious migrations of fishes at spawning time to the migratory flight patterns of birds). Perhaps what is most important to note, however, is the extent to which his adherence to the theory of regression relies on the adoption of dubious and disputed – most would now say wholly discredited – theories of heredity and embryology, notably the recapitulation doctrine which contends that in the course of its ontogenesis the 'germ' of the organism has to recapitulate the embryological structures of all the forms of life from which it has sprung (1991: 309). It is not that Freud does not consider the empirical fact, and the theoretical claims that aim to support it, that life is characterized by the production of new forms; rather, he wishes to maintain that progress or development in evolution does not serve to cancel out the primordial and instinctual tendency to restore the earlier state of things. One crucial difference between the Freudian account and the Deleuzian one (shared by contemporary complexity theory) is that Freud wishes to explain organic development almost exclusively in terms of the accidents of what he calls 'external disturbances' and 'diverting influences' (310). It is worth quoting Freud on this point:

In the last resort, what has left its mark on the development of organisms must be the history of the earth we live in and its relation to the sun. Every modification which is thus imposed upon the course of the organism's life is accepted by the conservative organic instincts and stored up for further repetition. Those instincts are therefore bound to give a deceptive appearance of being forces tending towards change and progress, whilst in fact they are merely seeking to reach an ancient goal by paths alike old and new. Moreover it is possible to specify this final goal of all organic striving. It would be in contradiction to the conservative nature of the instincts if the goal of life were a state of things which had never yet been attained. On the contrary, it must be an *old* state of things, an initial state from which the living entity has at one time or other departed and to which it is striving to return by the circuitous paths along which its development leads.

(ibid.)

Freud's conception of an instinctual desire for death is only sustainable on account of the privileged status accorded to the organism. This conception of a curious 'creative evolution' (creating new avenues of life in order to find the road back to the earlier state of things) will be critiqued by Deleuze in his work in a number of ways: by a thinking of repetition 'and' difference, by a reconfiguration of the death-drive or death-instinct (the ambiguity over its precise formulation as an instinct or drive is in Freud), by a rhizomatic conception of creative evolution/involution that does not take the organism as 'given', and by an appreciation of the arbitrary character of limiting life to the desires or perspectives of the organism. As far as *Difference and Repetition* is concerned one of the principal moves Deleuze makes, and which leads him to a non-Freudian position, is that of exposing the transcendental illusion of entropy. The importance of this move is that it contests the view, articulated by Freud in the above passage, that the evolution of living systems is due essentially to the conservation and dissipation of energy. Deleuze does not neglect the importance of solar energy and radiation in activating life on the planet but he refuses to make the evolution of living systems reducible to external circumstances and accidents alone. Hence his stress on the vital role played by individuation and the endogenous powers of organisms, as they evolve into ever more complex interior systems, establishing their conditions of possibility relatively autonomously in relation to the environment, and to the extent that their repetitions are not simply the subject of, and subject to, the exogenous mechanism of natural selection.

Before moving on to exploring how Deleuze reconfigures a death-instinct in relation to Nietzsche, and also in terms of a reading of Zola, there is a further aspect of the essay *Beyond the Pleasure Principle* I want to examine. This concerns the way in which Freud responds to Weismann's theory of the germ-plasm. At one point Freud speaks of the 'striking similarity' between Weismann's distinction of soma-plasm and germ-plasm and his own separation of the death instincts from the life instincts, noting that Weismann's treatment of the 'duration of life and the death of organisms' is of the greatest interest to psychoanalysis (1991: 318, 322). Freud suggests that it might be possible for him to provide a 'dynamic corollary' to Weismann's morphological theory by distinguishing two kinds of instincts *within* the living substance: the drive towards death in the living and the drive towards the perpetual renewal of life. It is important to appreciate, however, that what causes Freud the greatest problem is not death but life. If, as he wishes to disclose to us, the aim of all living is death, then how is *life* possible? Freud responds to the challenge implicit in Weismann's account of organismic life, which stresses that death is a late acquisition in evolutionary life, peculiar to multicellular organisms and not characteristic of protista – so that death is not a drive or an urge but simply a matter of expediency, 'a manifestation of adaptation to the external conditions of life' – by asserting that such a death, death as a late acquisition, applies only to its *'manifest phenomena'* and does not, therefore, render impossible 'the assumption of processes *tending* towards it' (322). The creativity of evolution (variation, vital differences, divergence) is seen by Freud to amount to a circuitous detour from the path of death, where the organism fulfils its desire for self-preservation by attaining perfect, self-sufficient stasis (it is this which informs Deleuze's claim that Freud's model of death is individualist, solipsistic, and monadic). No matter how much life diverges from its original unity as substance through the complexity and creativity of evolution it remains the same subject – the life that desires death (311). This is how it is possible for Freud to state that the self-preservative instincts are not instincts of life but rather, ultimately, ones of death, to be regarded as 'component instincts whose function' is simply 'to assure that the organism shall follow its own path to death, and to ward off any possible way of returning to inorganic existence other than those which are immanent in the organism itself' (ibid.). This insight explains the paradoxical behaviour of all living organisms who struggle energetically against events that if embraced would help them to achieve their life's goal quite rapidly 'by a kind of short-circuit' (312). We are, then, granted the freedom and dignity to die in our own way and to take our time in doing it.

It is on the basis of this privileged life and death accorded to the organism – the freedom to die in its own fashion through the prolongation of life – that Freud judges germinal life (the life of Weismann's germ-cells). Clearly, Weismann's account of the germ-cells offers the greatest challenge to Freud's theory simply because they suggest that there is a 'life' *independent* of the instincts of the organism and its desire to return to the inanimate. Freud acknowledges that Weismann's theory suggests these cells of perpetual life are complemented in their original structure by 'inherited and freshly acquired instinctual dispositions' that lead a separate existence from the organism as a whole (1991: 312). However, in working against 'the death of the living substance' the germ-cells do not introduce any new life but merely gain for life a temporary reprieve from the final execution, simply attaining 'a lengthening of the road to death' (313). Any 'fresh start' in life, therefore, is to be treated as more than a prolongation of the journey, since all throws of the dice lead to the same fate. All moving-forward or progression, whether as deviation, degeneration, or variation, a cornucopia of good and bad, the monstrous itself, remains implicated in the ultimate moving-backward (regression). 'Fresh *vital differences*' exist to be 'lived off':

> The dominating tendency of mental life, and perhaps of nervous life in general, is the effort to reduce, to keep constant or to remove internal tension due to stimuli...a tendency which finds expression in the pleasure principle; and our recognition of that fact is one of our strongest reasons for believing in the existence of death instincts.
>
> (Freud 1991: 329)

Freud must retain death as a fundamental causal instinct or drive at all costs in order to sustain his argument that it operates regressively as the desire to 'return' to the early, inorganic state from which all life has evolved. However, while insisting that modern biology does not at all contradict the theory of the death instincts, and while openly stating that his argument could benefit from more information on the origin of sexual reproduction and on the sexual instincts in general, he concludes the essay by speaking of biology as a land of 'unlimited possibilities' which at some point in the near future may 'blow away the whole of our artificial structure of hypotheses' (1991: 334). But in the essay on the pleasure principle, and elsewhere, Freud insists upon all new life, evidenced, he notes, in the processes of embryology and germinal-cellular life, being merely a straightforward repetition of the 'beginnings of organic life' – including, and especially, a repetition of the original entropic desire of organic life (329; see also 1991b: 139–41; in this essay on 'Anxiety and Instinctual Life' Freud maintains that everything that is described as the manifestation of instinctual behaviour in animals, from the spawning migrations of fishes to the migratory flights of birds, takes place 'under the orders of the compulsion to repeat', meaning that all instincts express an essentially conservative nature). He remains keen throughout, therefore, even though it proves so difficult, to ascribe to the sexual instinct – to that which keeps alive the 'vital differences' and makes life live on – the 'repetition-compulsion' that so securely puts us 'on the track of the death instincts'.[9] The depiction of the death-drive as an instinctual tendency of organic life rests on a privileging of the 'ego-instincts' over the sexual instincts. This means, in short, that the organism is anti-life (in the sense of a creative evolution), being a fundamentally conservative substance that desires to remain, above all, a constant and coherent unity in its self-same, self-present identity, to the extent that the desire for death amounts to a complete externalization of multiplicity, heterogeneity, and difference. Freud has not provided a straightforward 'dynamic corollary' to Weismann's

111

morphology but rather introduced into its conception of the living substance a theory of forces that privileges the instincts of death (the ego) over those of life (sexual reproduction). (It is perhaps interesting to note that at one point Freud comes close to the theory of the 'selfish gene' when he accuses the germ-cells of behaving in a 'completely "narcissistic" fashion' (323)). As Freud himself recognizes, it is on the question of death – on its meaning and significance – that the appearance of any correspondence between his thinking and that of Weismann's disappears. Whereas Weismann's account of the origins of sexuality and death is strictly evolutionary, Freud's is not but resembles most, as he himself acknowledges, the metaphysics of Schopenhauer.

In a recent reading of the death-drive, as it figures in Freud and Lacan, Slavoj Zizek takes the drastic step of divorcing it from the unconscious altogether, marking it as the site of the symbolic order. On this reading the drive is not to be defined in terms of a simple opposition between life and death since its space is occupied by, on the one hand, the monstrous split within life itself between ordinary or normal life and 'horrifying "undead" life', and, on the other hand, that within death between the ordinary dead and the 'undead' machine. Like Deleuze and Guattari in *Anti-Oedipus*, Zizek wishes to separate the Freudian death-drive from any Heideggerean 'being-towards-death', but maintains that, although the drive itself is 'immortal, eternal, "undead" (the annihilation towards which the death-drive tends is not death as the unsurpassable limit of man *qua* finite being)', death belongs, along with mortality, to the domain of consciousness. For the unconscious there is no death-anxiety since such anxiety belongs only to consciousness (Zizek 1997: 89). Elsewhere he has stated that the death-drive is not a biological fact but 'defines *la condition humaine*' (1989: 4–5). Lacan himself held the theory to be neither true nor false, simply suspect (Lacan 1992: 213).[10]

Zizek's move succeeds in drawing attention to the curious role occupied by the death-drive in Freud's psychoanalysis. By associating the drive and its anxiety with consciousness we begin to appreciate the extent to which the drive is posited in Freud from the perspective of organismic life, in which even the unconscious is not free from the self-destructiveness that marks the consciousness of the organism. This is clear in Freud's remarks on his 'attitude towards death', which provides the context in which to think through the paradoxes and complications of Freud's position, a context I find lacking in Zizek's novel account (and also ignored or downplayed by Rose in her analysis of Freud, repetition, and death, 1992: 101–10, especially 107).[11]

Freud writes this essay, entitled 'Thoughts for the Times on War and Death' (1915) in order to 'overcome', or work through, the sense of disillusionment that has accompanied the onset of war among those who are not combatants and cogs in the huge machine of war. How could the unexpected, the worst, happen in the midst of European civilization? How could the 'world-dominating nations of the white race' regress to such tribal instincts and brutality? The aim is to learn something ancient and buried about death from this destructive experience. We are to discover that war gives way to dis-illusionment only because it is the result of a fundamental *illusion* regarding the real character of life, a truth hidden to us by the veneer of civilization. In this essay Freud does not simply declare the unconscious to be ignorant of death, but rather ascribes to it a lust for annihilation and war. On the one hand, Freud says, the unconscious does not believe in its own death but considers itself to be immortal. The deepest strata of our minds that is made up of only so-called 'instinctual impulses' is said to know 'nothing negative...no negation' (Freud 1987: 85). In this denial of death deep within our unconscious there resides, speculates Freud, the 'secret' and illusion of heroism. On the other hand, the unconscious houses the desires of the ultimate paranoid monad that is full of death-wishes towards strangers and outsiders. In our unconscious impulses we thus 'daily and hourly get rid of anyone who stands in our way....The expression "Devil take him!"...which really means "Death take him!", is in our unconscious a serious and powerful death wish' (86). Our unconscious, says Freud, murders over trifles and can be compared to the Athenian code of Draco which knows only death as the punishment for some perceived crime (in this case, the crime of being strange, a stranger, to the 'unconscious'). It is such an 'insight' which leads Freud to declare that 'if we are to be judged by our unconscious wishful impulses, we ourselves are, like primeval man, a gang of murderers' (ibid.). It is on the basis of these claims that Freud then arrives at his so-called 'positive' or affirmative conception of death. War, Freud tells us, strips away from us the 'later accretions of civilisation', so laying bare before each and every one of us our real primeval self. Moreover, it 'compels us to be heroes who cannot believe in our own death; it stamps strangers as enemies, whose death is to be brought about or desired' (88). The issue for Freud, therefore, is not one of abolishing war but, on the contrary, adapting ourselves to its inescapable and positive reality and affirming the primeval pleasures it allows the self that risks life to indulge in (the destruction of what is outside and strange to itself). This, Freud notes with due cruel irony, amounts to a *positive regression* (war brings about an 'involution' (74)), one which at least has 'truth' on its side. The illusion that we can live life without us all being, at some point in time, mass murderers is what makes life intolerable, Freud challenges.

We can now see why it is important for Deleuze to engage with Freud and to articulate a different model of death (something he does in almost every text). Freud's conception of organic life runs counter to the deepest tendencies of Deleuze's thinking of difference and to his affirmation of univocal Being *as difference*. It is to Nietzsche and his teaching of eternal return that Deleuze turns in order to generate a different model of death than is encountered with Freud where the *telos* of life is portrayed in fundamentally entropic terms and where repetition is conditioned by a drive that involutes simply in terms of regression. Deleuze is not positing a desire for the nonorganic but seeking to show how in life there is a creative evolution that involves a constant play between the organismic and the nonorganismic. On the human plane this involves a play between the self and the field of intensities and singularities it finds itself implicated in and which present it with new possibilities of existence. Death is not, therefore, merely the negation of life but a sign of the vital life that arrives from the future and which seeks to emancipate organic life from the fixed and frozen forms which entrap it. Organic life is invited to open itself up to the open system of the 'outside', not to resist the temptation of the demonic and the monstrous but to encounter those 'dark precursors', the difference in itself, which relate heterogeneous systems and the disparate.

The heredity of the crack and Nietzsche's superior return

In a meditation on Zola and the 'crack' that appears as one of the appendices to *The Logic of Sense* Deleuze draws a distinction between two types of heredity: the one small, historical, and somatic; the other epic and germinal. A heredity of the instincts and a heredity of the crack or fault-line. The somatic heredity might include a phenomenon such as alcoholism being passed down the generations from one body to another (the example given by Zola himself in *La Bête Humaine* when writing of the hereditary crack). This kind of transmission reproduces the return of the same. What is novel about Deleuze's reading of Zola is the attempt to implicate the somatic transmission in a different crack; where one kind of crack transmits something well-determined, reproducing whatever is transmitted, the other communicates in terms of a vital and virtual topology (dealing with thresholds, limits, transformations, connections and disconnections, implications and complications, closed and open systems). In the case of the heredity of the crack 'it is not tied to a certain instinct, to an internal, organic determination, or to an external event that could fix an object'. It is thus able to '*transcend life-styles*' (my emphasis). The movement of the heredity of the crack is for Deleuze 'imperceptible' and 'silent'. The crack cannot be 'replicated' since its mode of transmission is diffuse and inchoate; it proceeds via an 'oblique line, being ready to change directions and to alter its canvas' (Deleuze 1969: 377; 1990: 325). The crack, then, transmits only the undetermined crack. With the heredity of the crack characteristics are not simply acquired but have to be invented and are forced to undergo transmutation. The crack 'follows' only itself, like a runaway train destined for derailment. On the tracks of this germinal train of life there is neither beginning nor end, neither a given genealogy nor a given teleology, but only the broken middles that allow for novel intersections, cross-connections, and unpredictable growths, constituting a cornucopia of good and bad. The crack enjoys a capacity for self-overcoming, making possible creative 'evolutions', in which the creation involved offers not a simple redemption but allows for the germinality of the most destructive inclinations and tendencies. 'He knew that all his life he had been wrenching at the frame of life to break it apart. And now, with something of the terror of a destructive child, he saw himself on the point of inheriting his own destruction' (Lawrence 1995: 221). For Deleuze, to inherit one's own destruction is to inherit only the *crack* of that destruction.

In the crucial chapter of the novel *Doctor Pascal* (the final book in a cycle of twenty) entitled 'The Genealogical-Tree' Zola has the doctor lay out before himself and the young woman Clotilde the entire genealogical tree of the Rougon-Macquart family in order to present her with a 'terrible lesson in life' (Zola 1995: 109ff.). But this tree stretches way beyond and over this particular family, encompassing the strata of races and civilizations from the dawn of time to the present and branching out endlessly into unknown futures (126). This is 'the whole monstrous florescence of the human tree' (131). Moreover, this genealogy is presented in the terms of a highly complex, erratic, and unpredictable, monstrous descent, a descent subject, to employ Deleuze's later terms, to perpetual deterritorialization and reterritorialization. Zola is imaginative in depicting the unfolding of this tree of human life in terms of a complicated entanglement of sickness and health, of death and renewal, a vitality of life that is caught up in destruction, decay, and degeneration. For example, Pascal asks: once the floodgates have been opened by the overflowing river of life (creative evolution as monstrous), and as detected by the emergent science of heredity (as the science of difference *and* repetition), then is it possible to say that among the weeds and flowers on the banks there also mingles and floats by – gold? (132). This is, says Pascal, a world 'beyond good and evil'. The chapter concludes by asking whether it is necessary to burn the tree of genealogy or whether this can only be a matter of the future as its affair.

In the novel Zola presents Dr Pascal's work on the new science of heredity moving through Darwin's confused theory of pangenesis, the peri-genesis of Ernst Haeckel (whose works were avidly read by Nietzsche), the eugenics of Galton (also read by Nietzsche), finally arriving at an intuition of Weismann's major thesis on the continuity of the germ-plasm in which a portion of this 'delicate' and 'complex' substance is held in reserve and passed on without variation or mutation from generation to generation (Zola 1989: 36). Pascal, however, does not stop there, which is what makes his case especially interesting. He propounds his own theory, which he calls 'the failure of the cells theory', which consists in granting a high degree of freedom and novelty to plasmic evolution. It is not atavism that Pascal privileges but 'perpetual change', the change that denotes 'an increasing transformation, due to that transmitted strength and effort, that perturbation, which imbues matter with life, and which is, indeed, life itself in an abstract sense' (38). In short, there is no plasmic finality and it is at this point that Zola is able to resist the biological nihilism of Weismann's continuity thesis. It is in terms of such a nihilism that Weismann is appropriated in Hardy's tragic fiction where the anomalous and the aberrant (the 'unfit') are not allowed to survive but must face extermination (Tess, Jude etc.). This biological nihilism amounts, in short, to repetition without difference, to the eternal return of the same, since the germ-plasm is posited as evolving completely independently of perturbations (Deleuze's 'crack') and free of the endogenous powers of the organism itself which may exert an influence on the character of evolution.

In Deleuze's reading of Zola, Weismann's distinction between the two plasms, soma and germ, operates as a distinction between a love or a body that dies and a movement that creatively 'evolves' through germinal intensity. This is a movement from the organized body of the organism to the 'body without organs' which involves the release of singularities and intensities from entropic containment (Deleuze 1969: 384; 1990: 331). But the two exist in implication and complication; this is life and death lived and died from 'within the folds' or 'on the train' as in Zola's *La Bête Humaine* (published in 1890 as the seventeenth of the twenty novels that make up the Rougon-Macquart cycle), where the machine functions as the pure death instinct: 'The instincts or temperaments no longer occupy the essential position. They swarm about and within the train, but the train itself is the epic representation of the death-instinct' (Deleuze 1969: 384; 1990: 331). In the novel the train is undoubtedly depicted in terms of the demonic power of a death-drive:

> the train was passing, in all its stormy violence, as if it might sweep away everything that lay in its path....It was like some huge body, a giant creature laid out on the ground...past it went, mechanical, triumphant, hurtling towards the future with mathematical rigour, determinedly oblivious to the rest of human life on either side, life unseen and yet perennial, with its eternal passions and its eternal crimes.
>
> (Zola 1996: 44)

Unlike the murder of the husband in *Thérèse Raquin*, committed by Thérèse and Laurent simply because he stands in their way and is an inconvenience, Zola's aim in *La Bête Humaine* was to link murder with an ancient hereditary impulse buried by the sedimented layers of civilization, to show the 'caveman' dwelling deep within the civilized man of modernity, as he put it in a letter to a Dutch journalist. In the novel the 'hereditary crack' is not simply a matter of ill-health but is said to be involved in those 'sudden losses of control' that lie deep within our being 'like fractures, holes' through which the self seeks escape, losing itself 'in the midst of a kind of thick haze that bent everything out of shape' (Zola 1996: 53). At such moments as these, where the self is no longer the master of its own body but the obedient servant of its muscles and the 'rabid beast within', it is cast into paying back an ancient debt:

> paying for the others, for the fathers and grandfathers who had drunk, for the generations of drunkards, of whose blood he was the corrupt issue...paying the price of a gradual poisoning, of a relapse into primitive savagery that was dragging him back into the forest.
>
> (Zola 1996: 53)

The extent to which Zola's novel, with its stress on a hereditary regression and atavistic instincts, anticipates both Freud's conception of death (1915) and his positing of the death-drive (1920) by several decades is indeed remarkable. On Deleuze's reading, however, the complicated investment of the erotic instincts in destructive ones – Zola's novel was read in the precise terms of this complication on its publication – expresses not simply the noise of primal instincts caught up in an involution but rather the silent echoes of a repetition that drives us ever forward and upward. This is why for him the key actor or agent in the novel is the train itself (a field of action, a body without organs distributing intensities and producing transformations). The train is a creation of modern civilization but it is also the crack which derails it, making sure that it is the 'great health' which lives on in humanity (the dissolutions of the novel, it should be noted, take place against the backdrop of the dying days of the Second Empire). Michel Serres is incisive in detecting in Zola's epic series of novels with its cycle of destruction, waste, dispersion, irreversible ebbing towards death, disorder, and degeneration, revolving around in *La Bête Humaine* a veritable thermodynamics of the train, an 'epic of entropy' (see Serres 1975: 78; 1983: 39ff.). Such an insight, however, discloses on Deleuze's reading only half of the story. Deleuze's conception of germinal life aims to demonstrate that entropy is never the final word. Thermodynamics always needs to be linked up with good sense since they share the same characteristics: the single direction from the most to the least differentiated, from the singular to the regular, and from the remarkable to the ordinary, so orientating 'the arrow of time from past to future, according to this determination' (Deleuze 1969: 94; 1990: 76).

In his reading of literature, therefore, Deleuze is proposing a vital rework-ing of heredity, in which it is shown that it is not heredity that passes through the crack, which would fix desire to a morbid ancestry, but that *heredity is the crack itself*. Hence the claim: 'In its true sense, the crack (*la félure*) is not a crossing for a morbid heredity; it alone is the hereditary and the morbid in its entirety' (Deleuze 1969: 373; 1990: 321). For Deleuze everything depends on grasping the significance of this paradox, confusing this heredity with its vehicle, that is, the 'confusion of what is transmitted with its transmission' (the transmission which transmits only itself). This is what he means when he declares the 'germen' to be the crack *and* nothing but the crack.

The way Deleuze links the heredity of the crack with the death-instinct is crucial since it wants to implicate death in the dimension of 'difference and repetition'. This explains the momentous role Nietzsche's eternal return comes to play in Deleuze's reworking of the death-drive, subjecting all questions of descent and heredity (origin and genealogy) to deterritorialization (a change of form and function). In Freud the desire of the organism to return to inanimate matter assumes that there is an original model of death to be returned to (the One death, the single truth of death). As such, the death-drive is a desire for identity (Nirvana). So far as the possibility of difference enjoying its own concept, as opposed to being laid out under the dominion of a concept in general that already presupposes a founding identity, Nietzsche's thought-experiment of the eternal return amounts to something of a 'Copernican revolution'. Identity now operates as a secondary power and speaks only of differences *qua* differences (it speaks of repetition). 'The true subject of the eternal return', Deleuze contends, are the intensities and singularities which refer to the mobile individuating factors that cannot be contained within a self, and in which the One does not return (that which aspires to recentre the acentred or decentred circle of desire). Rather, what returns is the divergent series and inclusive disjunctions: 'In the circle of Dionysus, Christ will not return; the order of the Antichrist chases the other order away' (1969: 348–9; 1990: 300–1). The system of the Antichrist convulses a world where the identity of the self is lost – 'not to the benefit of the identity of the One or the unity of the Whole, but to the advantage of an intense multiplicity and a power of metamorphosis' (345; 297). Thus if there is an 'essential relation' to be encountered between eternal return and death it is because the repetition of eternal return implies the death of that which is one 'once and for all'. Likewise, if it enjoys an essential relation with the future it is because the future is the explication and complication of the multiple, the different, and the fortuitous, 'for themselves and "for all time" ' (Deleuze 1968: 152; 1994: 115). The eternal return thus selects against the One, against that which will not select.

The crack is not involved in a movement that would ensure the repetition of the same since it overcomes itself by always exceeding the directions it invents. Death effects the transmutation of instincts, turning death against itself, so creating 'instincts' that do not involve simply a repetition of the same but are allowed to live and grow germinally. On Deleuze's reading Zola's socialist optimism derives from the recognition that the proletariat makes its way forward and onward through the crack. It is in the finding a way through the crack that it becomes endowed with the future. This is to read Darwin and *social* Darwinism against the grain so that life now serves not the strongest conceived solely by means by heredity, but is a swarm that gets created epically through revolutionary change by a new generation. As Zola the author dramatizes:

> Was Darwin right, then, was this world nothing but a struggle in which the strong devoured the weak so that the species might advance in strength and beauty?...if one class had to be devoured, surely the people, vigorous and young, must devour the effete and luxury-loving bourgeoisie?...in this expectation of a new invasion of barbarians regenerating the decayed nations of the old world, he rediscovered his absolute faith in a coming revolution...whose fires would cast their red glare over the end of this epoch even as the rising sun was now drenching the sky in blood.
>
> (Zola 1954: 496)

The crack announces a series of jolts that will assault bourgeois society and lead to its dissolution and transformation:

> This time their old, tottering society had received a jolt and they had heard the ground crack beneath their feet, but they felt other jolts on the way, and yet others, and so it would go on until the old edifice was shaken to pieces and collapsed and disappeared into the earth...a black avenging host was slowly germinating in the furrows, thrusting upwards for the harvests of future ages. And very soon their germination would crack the earth asunder.
>
> (497, 499)

For Deleuze, the epic novels of Zola provide the creative evolution of life with the open, pagan space that is beyond tragedy (Deleuze 1969: 384–6; 1990: 331–3). Deleuze utilizes Zola's appropriation of work on heredity, which identifies the 'difference' between a homologous and determined heredity and a dissimilar or transformational heredity, in order to render impossible any stable duality of the hereditary and the acquired. Deleuze concedes that the small type of heredity may well indeed transmit acquired characteristics, simply because the formation of instincts is inseparable from a social field and historical conditions. But in the case of the grand heredity the acquired characteristics are brought into a quite different relationship since here we are dealing with a 'diffuse potentiality' (377; 325). Moreover:

> if it is true that the instincts are formed and find their object only at the edge of the crack, the crack conversely pursues its course, spreads out its web, changes direction, and is actualized in each body in relation to the instincts which open a way for it, sometimes

mending it a little, sometimes widening it, up to the final shattering....The two orders are tightly joined together, like a ring within a larger ring, but they are never confused.

(ibid.)

The crack is simply the emptiness, the great void; it is death – 'the death instinct' [*l'Instinct de Mort*] (378; 326). For all the 'noise' they make, the swarming instincts cannot hide the fact that they belong to this more profound silence from which they burst forth and to which they return. When death happens the stillness in the room is such that one can hear a 'Fly buzz'; it is 'like the Stillness in the Air – Between the Heaves of Storm' (Dickinson 1969: 31). The death instinct is *'not merely one instinct among others*, but the crack itself around which all of the instincts congregate' (Deleuze 1969: 378; 1990: 326). On this model of life entropy is never to be allowed the final word or the last say; rather, its 'transcendental illusion' must always be exposed. Deleuze wants to argue that the death-instinct constitutes the complicated 'grand heredity' of the crack (it's not tragic but epical, since it concerns revolutionary movement). But the death-instinct is retrieved from any global entropic determination or telos:

Is it possible, since it [the death-instinct or drive] absorbs every instinct, that it could also enact the transmutation of the instincts, turning death against itself? Would it not thereby create instincts which would be evolutive rather than alcoholic, erotic, or financial...?

(1969: 385; 1990: 332)

For Deleuze it is enough that we dissipate ourselves 'a little', 'that we are able to be at the surface, that we stretch our skin like a drum', liberating singularities that are neither general nor individual, neither personal nor universal, from their imprisonment in the self and the 'I': 'All of this is traversed by circulations, echoes, and events which produce more sense, more freedom, and more strength than man has ever dreamed of, or God ever conceived' (1969: 91; 1990: 73). In Freud the 'daemonic power' of the 'beyond' of the pleasure principle (the compulsion to repeat that characterizes masochism and the death-drive) is too readily domesticated by being reduced to an Oedipal determination and articulation, namely the influence of early infantile experiences which are invariably used to explain the apparent malignant fate we feel possesses us, driving us on and tormenting our minds. Deleuze's reading of Zola endeavours to remove the death-instinct from Freud's tragic model in order to open it up to the pagan space of the *epos*. The crack comes from the future as a sign *of* the future. In the 'exaggeration' of life as composed by the great writer or artist, inspiration comes not from the *logos* but the *epos*, which discloses to us that we can never go far enough in the direction of 'decomposition', 'since it is necessary to go as far as the crack leads' (385; 332). Art is, therefore, implicated in degeneration. Civilization is haunted by its discontents and malcontents. Degeneration and decay not only disclose signs of the future but show that the future is only possible to the extent that it grows out of the great health (the health that has incorporated the crack as that which allows for repetition *and* difference). Zola's 'putrid literature' is a literature of the future: 'Zola's optimistic literature is not anything other than his putrid literature. It is in one and the same movement' (385; 332). The death-instinct on this model is not a manifestation of buried repressed desires, a primeval lust for murder, but is to be read as signs of the future, of new life, new affects, and new bodies of desire.

In learning to will the event one undergoes a 'volitional intuition and transmutation' (1969: 176; 1990: 149). One does not will *exactly* what occurs, but rather wills *germinally*, willing something '*in* that which occurs' and as 'something yet to come'. This is said to be a willing of life that is one 'with the struggle of free human beings'. A thinking of the eternal return as the time of the event seeks to make comprehensible, therefore, much of the paradoxical aspect of the doctrine, such as how, for example, it involves an affirmation of everything as it is but yet entails no Stoic resignation. Zarathustra's 'Thus I willed it!' is above all, on Deleuze's reading, a battlecry, not an acceptance of the past but an affirmation of the arrival of the future. Perhaps it is in the thought of the event that we can discover and invent the 'meaning' of Nietzsche's cryptic and enigmatic statement that in order to understand anything at all of his 'Zarathustra' one must have one foot *beyond life* – not one foot in the grave but, I would suggest, one foot in the event.

For Deleuze the crucial question pertaining to the actualization of the event concerns the possibility of sustaining its *non-actualization*: Is it possible to will the event without also not willing its full actualization in a corporeal mixture? If the surface is cracked how can it not out of necessity break itself up in wanton destruction, losing life itself in the process? Is it possible to maintain the 'inherence' of the incorporeal crack while not incarnating it in the depth of the noisy body? Is it possible, Deleuze asks, and risks asking (taking the risk of making oneself appear as a philosophical fool), to live, love, and die 'a little alcoholic', 'a little crazy', 'a little suicidal'?

Staging the event

The eternal return, Deleuze argues, is not a theory of qualities but of pure events. Before we can comprehend the meaning and significance of this major claim we need to understand better both the conception of ethics and the notion of the event he is working with.

In *The Logic of Sense* Deleuze deploys the thought of the event through a reading of Stoic doctrine, as well as the philosophy of Leibniz, in which it serves to denote neither an attribute nor a quality of subject but rather the incorporeal predicate of a subject of the proposition (so instead of saying 'the tree is green' one would say 'the tree greens'). In *The Logic of Sense* events are presented in terms of a Stoic theory of bodies. If all bodies are causes in relation to each other, and causes for each other, then it is possible on Deleuze's reading of Stoic doctrine to say that the effects of these causes are not bodies but incorporeal entities. These entities are not physical qualities and properties, things or facts, but events that do not exist but rather subsist and inhere. Thus, the thickness of bodies can be contrasted with the incorporeal events that play on the surface of bodies and animate them, such as 'growing', 'becoming smaller', and 'being cut'. Such events have become free of the states of affairs to which they are reduced when they are only mixtures that exist deep inside bodies. Deleuze goes on to argue that it is the world of incorporeal effects which makes language possible, drawing sounds from their simple state of corporeal actions and passions. Pure events are said to 'ground' language in the sense that they enjoy a singular, impersonal, and pre-individual existence inside the language which expresses them. In other words, events enjoy an independence of expression in relation to their actual incarnation in bodies and states of affairs. Sounds do not belong to bodies as physical qualities but acquire their 'sense' (signification, denotation, etc.) only as events. As Paul Patton notes, for Deleuze this means that the relation between language and the world is no longer one of representation but of effectivity, with language directly intervening in the world and creating it in novel ways (Patton 1996: 13; see also Foucault 1977: 172–6).

In his book on the 'Fold' Deleuze attempts to unfold the 'meaning' of the event once again in terms of Leibniz's chief principles, namely, the principle of sufficient reason and the principle of the identity of indiscernibles. He seeks to show that the predicate is, above all, a relation and an event. Predication is not attribution but rather a movement and a change: 'That the predicate is a verb...irreducible to the copula and the attribute, mark[s] the very basis of the Leibnizian conception of the event [*l'evénement*]' (Deleuze 1988: 71; 1993: 53). The predicate, Deleuze argues, is to be thought as an 'execution of travel' (a movement and a change), and not a state of travel. Such a conception of thought and event necessarily leads to a major reworking of the notion of substance, since it no longer denotes the subject of an attribute but rather the inner unity of an event and the active unity of change. Substance is here being conceived as a 'double spontaneity' involving movement as event and change as predicate. Badiou notes that for Deleuze the event denotes 'an immanent activity over a background of totality, a creation, a novelty certainly, but thinkable within the interiority of the continuous. *Un élan vital*' (Badiou 1994: 60). Deleuze goes so far as to describe the world itself as an event and as an incorporeal predicate that can be included in every subject 'as a *basis* from which one extracts the manners that correspond to its point of view' (Deleuze 1988: 71; 1993: 53) (for other readings of Leibniz on substance and predication see Russell 1992: 42ff., and Woolhouse 1993: 54ff.; for Deleuze on Russell's reading see 1988: 71–2; 1993: 53–4).

Deleuze's reading of the event in these terms is far from being entirely idiosyncratic. We may note that at the same time as Deleuze published *The Logic of Sense* (which, as Paul Patton has noted, could equally have been called *The Logic of the Event*), Donald Davidson published in 1969 an essay on 'The Individuation of Events', which sought to challenge the claim that the category of events is conceptually dependent on the category of objects and that there is no symmetrical dependence the other way round (a position he attributes to P. F. Strawson). In challenging the view that objects are more fundamental than events Davidson argued that a crucial insight resides in showing that the category of substance and the category of change are inconceivable without each other, even though he is unclear, as he readily admits, 'what substance it is that undergoes the change' (see Davidson 1980: 174). Clearly, Deleuze is wanting to go much further than this both in granting primacy to dynamic predication as constituting the subject and in conceiving the event in terms of the incorporeal and virtual. In other words, Deleuze holds that one cannot even think of a substance independently of the inner unity of an event and the active unity of change. Deleuze's thinking on the event is distinguished from most of the literature that has been generated on the subject in the last few decades in that it does not make the 'event' reducible or equivalent to the 'state of affairs' (for a helpful overview of this body of work see Lombard 1998). Let us now move on to see how this conception of the event, which gets worked over in three major texts of Deleuze (*The Logic of Sense*, 1969, *The Fold*, 1988, and (with Guattari) *What is Philosophy?*, 1991), and each time gets recombined with new elements and placed under new emphases, is worked through in relation to the question of ethics. We shall then attend to the difficulties of Deleuze's conception of events.

'Either ethics (*la morale*) makes no sense at all', Deleuze writes in *The Logic of Sense*, 'or this is what it means and has nothing else to say: not to be unworthy of what happens to us' (1969: 174–5; 1990: 149). This, says Deleuze, is ethical life beyond the *ressentiment* of the event, in which we use moral notions like just and unjust, merit and fault immorally by rendering them personal. The event is greeted by resentment whenever we treat whatever happens to us as unjust and unwarranted. 'There is no other ill will', Deleuze says, than this revenge against time itself, time's 'it was'. The event is the event of time (the moment) that both happens to us and which lives beyond us (eternity): 'The event is not what occurs (an accident), it is rather inside what occurs, the purely expressed. It signals and awaits us' (ibid.). Learning to live the crack (and crack-up) as the event requires perfecting one's complex inheritance, both its misfortunes and its promises, becoming what one is as a piece of necessity and fate. Every event can be said to have a double structure. On the one hand, there is necessarily the present moment of its actualization: the event 'happens' and gets embodied in a state of affairs and in an individual ('the moment has come'). Here the time of the event, its past and future, are evaluated from the perspective of this definitive present and actual embodiment. On the other hand, the event continues to 'live on', enjoying its own past and future, haunting each present, making the present return as a question of the present, and free of the limits placed upon it by any given state of affairs. My life appears as too weak for me, slipping away at the present instant, or I appear as too weak for life, the life that overwhelms me, the life that scatters singularities all around me without any relation to what I take to be me myself I (this is the life, says Deleuze, akin to Blanchot's death that has two faces: the death that has an extreme and definite relation to me, grounded in my mortal coil, and the death that has no relation to me at all, being incorporeal and impersonal, grounded only in itself).

The concept of the event is not to be confused with any actualized state of affairs, but rather refers to the 'shadowy or secret part' that can always be subtracted from or added to an actualization as an 'infinite movement' which gives life *consistency*. This is the 'virtual': not a chaotic virtual but one that has acquired consistency. And while the event may *appear* transcendent to the state of affairs to which it relates, it must be conceived to be an entirely immanent movement ['*L'événement est immatériel, incorporel, invivable: la pure "réserve"* ']. Ironically, it is the state of affairs in which the event becomes actualized that is transcendent, as well as 'transdescendent'. The time that is implicated in the becoming of the event is not that of the determination of isolable points or the measure of discrete instants, but that of the 'meanwhile' (*un entre-temps*), the *durée* of the interval or 'dead time'. The event names not the passing of time but the vectorial passage of time: 'When time passes and takes the instant away, there is always a meanwhile to restore the event' (Deleuze and Guattari 1991: 150; 1994: 158). The germinal life enjoyed by the event refers to a becoming that takes the battlefield and the wound as components and variations that live in excess of the subject's own body, including its rapport with other bodies and the physico-mathematical field that determines them.[12] The 'dignity' of the event is lived and over-lived by philosophy and its ethic of *amor fati* on account of the fact that philosophy becomes equal to the event by being able to 'disembody' it as a state of affairs; that is, to free it for a germinal existence. Paradoxically, it is only through this disembodiment that the event can become truly *embodied* (150–1; 159). This is how, for example, it is possible to 'treat' death as an event: 'Death may be assimilated to a scientifically determinable state of affairs, as a function of independent variables or even as one of the lived state, but it also appears as a pure event whose variations are coextensive with life' (153; 161). Deleuze makes an important distinction between action and event in order to liberate action from the embodied state of affairs: 'Nothing more can be said, and no more has ever been said: to become worthy of what happens to us, and thus to will and release the event, to become an offspring of one's own events, and thereby to be reborn, to have one more birth, and to break with one's carnal birth...' (Deleuze 1969: 175–6; 1990: 149–50). In 'becoming what one is' the task is to become the offspring of one's events and not one's actions, since the action is produced only by the offspring of the event.

In his encounters with Fitzgerald and Artaud, Deleuze wants to know, as already indicated, whether it is possible to maintain the silent trace of the incorporeal crack at the surface without letting it descend completely, where it is lost forever in a black hole, into the coagulating thickness of the noisy body. Surely if we are to go to the outer limits we must go all the way and never come back? Does the philosopher not find himself dancing close to the edge of ridiculousness, speaking from the safety of the shore? Why not just take gallons of alcohol and have done with all this futile philosophizing on matters of life and death? Does not the truth of Fitzgerald speak from the depths of alcoholic excess and does not the truth of Artaud speak from the depths of self-mutilation and the abysmal pit and pendulum of schizophrenia? However, to take just the one case, it is not a question of saving Nietzsche from madness for the reason that Nietzsche saved himself from complete descent into its black hole, giving himself over to a future germinal life by making sure that he would be born *posthumously*.

> The eternal truth of the event is grasped only if the event is also inscribed in the flesh. But each time we must double the painful actualization by a counter-actualization [*contre-effectuation*] which limits, moves, and transfigures it. We must accompany ourselves – first, in order to survive, but then even when we die. Counter-actualization is nothing, it belongs to a buffoon when it operates alone and pretends to have the value of *what could have happened*. But, to be the mime of *what effectively occurs*...like the true actor and dancer, is to give to the truth of the event the only chance of not being confused with its inevitable actualization. It is to give the crack the chance of flying over its own incorporeal surface area, without stopping at the bursting within each body; it is, finally, to give us the chance to go farther than we would have believed possible.
>
> (1969: 175–6; 1990: 149–50)

It is impossible simply to opt for the good healthy life with its denigration of the consuming passions of bodily life. Attention must be drawn to the confusion of lines, of flight and death, of paranoia and schizophrenia, 'the central point of obscurity which raises endlessly the problem of the relations of thought to schizophrenia and depression' (Deleuze 1969: 243–4;

1990: 209). All life involves a process of demolition, even and especially the speculative life. Which is greater, the more monstrous, the more extreme, the more experimental – corporeal, individual life (somatic life) or the incorporeal life of the event (germinal life)? It's a question of being superficial *out of* profundity, which involves having the courage to stop at the surface, to play at the fold. The depths are always to be navigated in terms of a geography of thought and existence, for in the 'entire biopsychic life' it 'is a question of dimensions, projections, axes, rotations, and foldings. Which way should one take? On which side is everything going to tumble down, to fold or unfold?' (272; 222–3). Germinal life does not live and die in the name of victims, of the tortured, oppressed, and depressed; rather it gives them the living line *as* a broken line. In dying we come to know that 'death is neither the goal nor the end, but that, on the contrary, it is a case of...passing life to someone else' even in destruction, murder, and the worst (Deleuze 1977: 76–7; 1987: 62).

To return to the question of eternal return as a theory of pure events. The problem of a 'logic of sense' is, says, Deleuze, one of knowing how an individual is able to 'transcend' its form and its syntactical link with a world so as 'to attain to the universal communication of events' (which involves affirming disjunctive syntheses that are beyond mere logical contradictions).[13] This involves the individual learning to recognize itself not in- and for-itself but as the fortuitous case, as just like everyone else (perhaps as all the names in history). This would amount to the individual recognizing itself solely as the event, the event that is 'actualized within her as another individual grafted onto her' (1969: 209; 1990: 178). The task of the eternal return is to effect this transformation of the individual – its qualities and actions – into the event:

> we raise each event to the power of the eternal return in order that the individual, born of that which comes to pass, affirms her distance with respect to every other event. As the individual affirms the distance, she follows and joins it, passing through all the other individuals implied by the other events, and extracts from it a unique Event which is once again herself, or rather the universal freedom.
>
> (ibid.)

Is there any basis in Nietzsche's own texts for interpreting the eternal return in terms of a notion of events? One could claim that the notion of the event implicitly informs every crucial articulation Nietzsche gave of the eternal return, such as in *The Gay Science* and *Zarathustra*. But there is one crucial passage, dated 10th June 1887, from the 'Will To Power' *Nachlass* – to my knowledge never treated in the literature in respect of the question of the event – where Nietzsche explicitly makes a link between eternal return (posited as a 'terrible' form of thought) and the event. It takes place in the context of a long section in which the eternal return is considered as both an expression of nihilism and as an intensified solution to it (as the self-overcoming of nihilism). Many of the remarks made are enigmatic, so the entire passage requires the most attentive and exact of readings. Only certain key aspects of it can be deciphered and presented here. It is, I would contend, one of the most significant passages in Nietzsche's later *oeuvre* (in the original German the passage bears the title of 'European Nihilism' and is composed of sixteen numbered sections, Nietzsche 1987: Volume 12, 211–17; in the English translation by Kaufmann, *The Will To Power*, the passage is split into Sections 4, 5, 55, and 114).

The passage begins with Nietzsche reflecting on the peculiar point at which nihilism emerges and the particular style it assumes, namely, as a psychologically necessary affect of the decline of belief in God. The displeasure with existence has not become any greater than in previous times, it is simply that we moderns have come to mistrust any 'meaning' in suffering and in existence itself. One extreme position is now succeeded by another equally extreme position, one that construes everything as if it were *in vain*. It is this 'in vain', and in particular its tie with *duration*, which constitutes the character of 'present-day nihilism' – or this, says Nietzsche, is what requires a demonstration. The problem is twofold, therefore: it involves a mistrust of all previous valuations that borders on the pathological (all values are now seen as lures that 'draw out the comedy of existence but without bringing it closer to a solution'), and, second, 'duration "in vain", without goal [*Ziel*] or purpose [*Zweck*], is the most paralyzing idea'. In short, we understand we are being fooled and yet lack the power *not* to be fooled. I take this to mean that we have become the victims of an excess of knowledge that does not enable us to orient and comport ourselves in newfound conditions and we currently lack the resources to create new modes of knowing and acting. Hence we find ourselves trapped in paralyzing forms of nihilism.

It is this quite specific situation that the eternal return is a reponse to. Hence Nietzsche's formulation of its experimental character in the following manner: 'Let us think this thought in its most terrible form: existence as it is, without meaning or purpose, yet recurring inevitably without any finale of nothingness: "*the eternal recurrence*" [*die ewige Wiederkehr*]'. The thought, therefore, is designed as a response to, and a responsibility in the face of, the problem of the 'in vain' by which we now interpret duration. This means that taken away from the 'finale', posited without the aid of notions of goal or aim and purpose, we are left confronted with the 'nothingness' and the prospect of the *Sinnlose* returning eternally. Nietzsche considers this to be 'the most extreme form of nihilism'. It is a peculiarly *European* form of Buddhism, in which all the *energy* of knowledge compels us to arrive at this position. It is, therefore, on this level, 'the most *scientific* of all possible hypotheses. We deny final goals [*Schluss-Ziele*].' But the aim is not to cultivate a will into, or for, nothingness (one cannot now make nothingness the end goal of all activity). Neither is it to remain stuck within the horizon of nihilism as some absolute horizon. Rather, one might propose, the task is one of traversing the negativity of our current condition, the pathos of nihilism and the stasis of the nothingness, in order to generate a new positivity of being. This is the link that needs to be made with the event.

Nietzsche's turn to the event emerges in this long passage out of a consideration of the problem of pantheism and the affirmative position of Spinoza. He asks whether it would make any 'sense' to conceive a god 'beyond of good and evil', which suggests that the task is not simply one of deifying a godless world, becoming without end-goals and final purposes, but of living each 'moment' as if it were constituted by the 'event'. What does this mean for Nietzsche? He notes that Spinoza's ethics involve removing the representation of purpose [*Zweckvorstellung*] from the process and then affirming the process in which something is attained at every moment and as always the same (the beatitude of the perfection of each moment). It is this idea that each moment possesses a logical necessity, Nietzsche argues, which provided Spinoza with a sense of triumph over the world. Nietzsche then says that Spinoza's case is 'only a single case' and now unfolds his thinking of the event:

> Every basic character trait that is encountered at the bottom of every event, that finds expression in every event, would have to lead every individual who experienced it as his own basic character trait to welcome every moment of universal existence with a sense of triumph.
> (1987: 214; 1968: Section 55, 36)

What is the link between this conception of the event and the 'single case' of Spinoza? It lies, I suggest, in the way in which the thought-experiment is being offered by Nietzsche as a cultivating thought, one that enjoys its own 'scholarly presuppositions' (active nihilism is said to presuppose a certain degree of 'spiritual culture', while our feeling of spiritual weariness that has culminated in 'the most hopeless skepticism regarding all philosophy' is also a sign of our by no means low position as nihilists: to be a nihilist is to be in a 'well off' position, spiritually and culturally speaking). For Nietzsche it is not a question of the 'logical necessity' of the moment, or even of a pantheism (a god 'beyond good and evil'), but rather a question of the style in which the event is thought and constituted as an event of being: it is a matter of 'impressing' [*aufzuprägen*] upon becoming 'the character of being' as the 'highest will to power', which is to recognize that being has no other being than that of 'becoming' (Nietzsche 1968: Section 617; 1987: Volume 12, 312). In the passage on the event just cited we need to note the precise link being forged between the event and the character trait, namely, that it is the 'event' – 'in time' – which gives style to the trait. It is this kind of play, I would suggest, that is informing Deleuze's staging of the relation between the action and the event.

Eternal return enjoys a link with nihilism simply because it arises out of the specific context of the problem of the nothing and the 'in vain' (nihilism, Nietzsche insists, is a consequence of our *idealism*). This explains why Nietzsche holds that eternal return will assume different forms and articulations, implicated in differing styles of nihilism. If nihilism is a symptom of the physiologically underprivileged who live without comfort, and whose horizon of being is determined by a 'lust for destruction', then these types will encounter the eternal return as a 'curse' (they will be confronted by the ridiculous character of this lust). This nihilism exposes the will to destruction as the instinct of self-destruction, 'the will into nothingness' [*Willens ins Nichts*]. This is the European style of Buddhism, actually *doing* 'No after all existence has lost "its" meaning [*Sinn*]'. Active nihilism is related to the 'moderate' types who do not require 'extreme articles of faith'. They both concede and love a fair amount of nonsense and accident and are able to think of a reduction in the value of the human without becoming small and weak themselves on this account, or turning the world into something ugly and reprehensible, and so unworthy of living in terms of the event.

Conclusion: Deleuze beyond Weismann and the problem of the event

Deleuze has reconfigured the Weismannian legacy that informs biophilosophical modernity in a particular and highly novel way. The germinal life is not fixed to the continuity of the germ-plasm, but allows for cracks and fissures through which new life is possible and the monstrous power of difference and repetition affirms itself. Germinal life is fundamentally antientropic in relation to both the entropy of the river of DNA (the contemporary rendition of the germ-plasm) that is immune to external perturbations and the entropy of the Freudian death-drive (both, in effect, examples of potentially closed systems). Deleuze places the stress on the field of pre-individual singularities that allows for the powers and forces of life to express a creative evolution beyond entropic containment.

Of course, one can still perhaps deploy in connection to it the neo-Darwinian language of vehicles and replicators (organisms and genes on the well-known model of Richard Dawkins), although the drama of life and death has undergone a fundamental reconfiguration. In Deleuze's conception of germinal life, for example, organisms become vehicles for the transmission and communication of intensities and singularities or haecceities. 'The event does not relate the lived to a transcendent subject = Self', he writes with Guattari, 'but, on the contrary, is related to the immanent survey of a field without subject' (Deleuze and Guattari 1991: 49; 1994: 48) (the notion of self-survey is taken from the work of the French philosophical biologist Raymond Ruyer). Subjects in this logic of life, then, are the bearers of singularities and intensities, created by them and destroyed by them.[14] However, singularities and intensities do not subordinate the invention of life to either a 'reproduction' of the same or guarantee the continuity of an unbroken line of descent. One inherits only the crack. This is Bergson, for example, working out the relation between his conception of creative evolution and the legacy of Weismann:

> There is no doubt that life as a whole is an evolution, that is, an unceasing transformation. But life can progress only by means of the living, which are its depositaries. Innumerable living beings, almost alike, have to repeat each other in space and in time for the novelty they are working out to grow and mature....Heredity does not only transmit characters; it transmits also the impetus in virtue of which the characters are modified, and this impetus [*élan*] is vitality itself.
> (Bergson 1962: 232; 1983: 231)

The significance of the thinking of the event in Deleuze is that it is with this notion that he aims to steer a conception of germinal life beyond the problem of heredity. This move is consonant with his later attempt, to be examined at length in the next chapter, to subject questions of genealogy and filiation, as well as the matter of natality, to processes of deterritorialization and decoding. Deleuze utilizes developments in embryology and ethology to provide insight into the nonorganic life and as a way of showing how individuals participate in this life through germinal intensity. The character of this process and the nature of the participation will become even clearer in the examination of Deleuze's later work to be pursued in the next chapter.

We now need to encounter a set of critical problems regarding Deleuze's thinking of the event. How coherent is the notion of an event in terms of naming a realm of incorporeal transformations? Does it make any sense to declare that 'the tree greens' without reference to a tree that *is* green? Does this imply that everything is an 'event' in Deleuze's sense? But if everything is an event, is this not to deprive the event of its genuine event-like status and character? Moreover, why do 'singularities' not belong to the most general order of things there is? Regarding this latter point, Deleuze himself argues that when a predicate is attached to an individual subject (such as a tree that greens or an Adam who sins) it does not enjoy any degree of generality since the generality of predicates 'blend' with their analytic character:

> To have a colour is no more general than to be green, since it is only this colour that is green, and this green that has this shade, that are

related to the individual subject. The rose is not red without having the red colour of this rose. This red is not a colour without having the colour of this red.

(Deleuze 1969: 136; 1990: 112)

We can make this less difficult by recognizing that Deleuze is not so much thinking in terms of concepts and mediations, but rather in terms of an order of mixtures that coexist and become through succession. He writes, for example, 'Analytic predicates do not yet imply logical considerations of genus and species or of properties and classes; they imply only the actual physical structure and diversity which make them possible inside the mixture of the body' (ibid.).

In essence, Deleuze is attempting to rethink through the notion of the event Leibniz's doctrine of the world as a *possible* world. The aim is to show that there is no determined world with fixed singularities, but only the aleatory point of singular points and the *ambiguous* sign of singularities. This means that the predicates used to define a subject are ultimately synthetic ones that open up different possible worlds and individualities (an Adam who sins, an Adam who does not sin). This reading is sustained and continued further in relation to Leibniz in Deleuze's book of 1988 on *The Fold*. One of the most telling moments in the text is when Deleuze brings into encounter Leibniz's question about the world with the questions raised by the likes of Bergson and Whitehead regarding an open-ended evolution. Deleuze suggests that in posing the question of the world the aim is not to arrive at eternity but rather determine the conditions under which the objective world allows for a *subjective production of novelty*. This is a production which, in short, allows for a *creative* 'evolution'.

Justin Clemens has questioned Deleuze's separation of fact and event, or what we might designate as the objective world and subjective world (Clemens 1997: 188). This is also a major concern of Alain Badiou, for whom the event does not come into being from the world, whether ideally or materially, but from not being attached to it. The event is an 'interruption' that is always separate from the world. Instead of a world defined by 'creative continuity' there is the 'founding break' (Badiou 1994: 65, 68). In his recent book on Deleuze, Badiou once again distances his own thinking on the event and this time in the context of a treatment of the doctrine of eternal return (Badiou 1997: 101–16). The implicit criticism being made by Badiou is that Deleuze is not, in fact, thinking the event since the 'affirmation' of return is always taking place as an articulation of the contingency of the 'One' (Being as univocal). For Badiou, however, the contingency of the event belongs to *each event*. Where for Deleuze chance is the play of all things [*Tout*], '*toujours re-joué tel quel*', for Badiou there is only the 'multiplicity and rarity' of chance, and it is only *by chance* that we come to the chance of an event, and not according to the expression of a univocal One (115).

Badiou, however, fails to understand the work being done with Deleuze's conception of the event, namely, that it seeks to provide an account of how rupture and discontinuity are explicable and possible. Badiou's point that we need to consider a different question to Deleuze – 'what are the conditions of an event for *almost nothing* to be an event? – is as misguided as the charge of Platonism he brings against Deleuze is misplaced (*'Au fond, le deleuzisme est un platonisme ré-accentué'*, 1997: 42). The event is both actualized *and* possesses a virtual reality (it enjoys a germinal life). It is the dimension of the virtual that is missing from Badiou's account, such as moments of tension, inactivity and inertia that have incorporeal potential. To simply declare the event to be the rupture of a continuous state is not only to grant an unwarranted normative status to that state, but also to posit the break with it in terms that are both blind and transcendent, simply because one does not have, in fact, a theory of the event.

Clemens argues that in collapsing the distinction between facts and events Deleuze has deprived the 'meaning' of the event of any functional effectiveness. Not only is this criticism too harsh it, too, fails to comprehend the nature of Deleuze's thinking of the event. For Deleuze any opposition between the state of affairs and the event is meaningless simply because it is the meta-physical (the *extra*-Being) that his thinking on the event aims to show as fundamental to the becoming of Being. The world only becomes a 'world' on account of the incorporeal event that subsists over and above the depths of individual bodies. In turn this means that the event becomes the analytic predicate of a subject but only as a synthetic event and a possible world. 'To green', Deleuze says, indicates a 'singularity-event' that subsists in the vicinity of which the tree is constituted. Individuals, therefore, are constituted within the vicinity of singularities that they envelop.

In thinking the event *qua* a field of individuation and pre-individual singularities Deleuze is seeking to move thought beyond the terrain established by phenomenology. As Foucault noted, *'Logique du sens* can be read as the most alien book imaginable from *The Phenomenology of Perception'* (1977: 170). In turn, in his own book on Foucault, Deleuze develops a reading of Foucault 'beyond' phenomenology. He argues that the notion of 'intentionality' was devised by phenomenology as a way of surpassing naturalism and psychologism, but it only resulted, he contends, in restoring a psychologism that synthesized consciousness and significations and in a naturalism that appealed to a 'savage experience' of the thing. In short, he argues, the phenomenological *epoche* failed, in its bracketing out of the world, to attain the *event* of thought and Being, the event that exists beyond subject and object, beyond things and states of things. A 'statement', for example, is neither reducible to the expressive capacities of a subject nor explicable with reference to a single object. Rather, statements refer only to a 'language-being' that enjoys its own independent existence, and which produces subjects and objects for themselves as so many immanent variables.

Any and all intentionality, Deleuze argues, collapses in the gap that gets opened up in the relation between 'seeing' and 'speaking': we do not see what we speak about and neither do we speak about what we see (e.g.: in the statement 'this is not a pipe' intentionality can be seen and shown to deny itself).

In the production of a statement it is not a question of origins in which one would trace it back to a *cogito* or transcendental subject. Rather, different individuals can intervene in each case to the extent that a statement 'accumulates' into a specific object, which is then preserved, transmitted, and repeated. This complex production of an object through statements, and removed from the domain of origins, is compared by Deleuze to the 'Bergsonian memory' in that a statement can be said to constitute itself in its own space as a space that both 'endures' and gets reconstituted anew. Any group or family of statements is formed by rules of change and variation that guarantee that the 'family' exists only as a medium for dispersion and heterogeneity, and never for homogeneity (a thought of repetition and difference, again). If everything is 'knowledge', and if everything has the status of an 'event', then there can be no savage or brute experience. The claims of knowledge and of the event become, Deleuze goes on to argue, the 'essence' of the task of thinking for the later Heidegger and Merleau-Ponty, in which the standpoint of the intentionality of being gets surpassed in favour of an openness to the 'fold of Being': 'It was Merleau-Ponty who showed us how a radical, "vertical" visibility was folded into a Self-seeing, and from that point on made possible the horizontal relation between a seeing and a seen' (Deleuze 1986: 117; 1988: 110). Deleuze will insist, however, that Foucault remains the more radical since in his work there is no desire to refound intentionality and that, in this regard, he was right to reproach Heidegger and Merleau-Ponty.

It might be argued that problems still remain with Deleuze's thinking. As a theory of 'pure' becoming, one might argue that the thinking of the event does not acknowledge the peculiar violence it inflicts upon individuals and wish to criticize it for depriving them of their singular embodiment. Real alcoholics and addicts, as well as genuine schizophrenics (for whom Deleuze had little time, considering their contamination to be of an insidious kind), become portrayed as mere vehicles through which the philosopher is able to devise a conception of germinal life that allows for the experience of 'living on' at the edge of chaos (not descending into the murky depths or falling straight into a black hole). For Deleuze, however, it is the peculiar vocation and responsibility of the philosopher to show that it is possible to go further in life – and death – than we thought possible (compare Nietzsche 1966: Section 212). The issue of living a 'little' alcoholic, a 'little' suicidal, and a 'little' crazy, is no doubt both a brave and foolish one to raise. The difficulty, however, lies in the attempt to articulate an ethics of germinal life, of the event, around the notion of measure, and it is here, as Deleuze fully recognizes, that the philosopher risks appearing foolish. Depending on the conception one has of it, one can accuse Deleuze either of conceding too much to 'life' or of not conceding enough.

The philosopher finds himself on awkward, if not perilous, ground in trying to follow the line of the crack. Why does s/he not simply admit, respectfully, that there is a fundamental difference in nature between the two processes, the impersonal and personal, and abandon the need for any attempt to work through the relation between them? Let us consider briefly an example that inspires Deleuze, that of the crack-up depicted by Scott Fitzgerald. 'All rather inhuman and undernourished isn't it? Well, that, children, is the true sign of cracking up' (Fitzgerald 1965: 44). But where exactly is the crack of the crack-up to be found, Fitzgerald asks: inside or outside? The blows which strike one down seem to come from the outside; but then there's a tremendous blow which comes from within, the one which comes too late for you to do anything about, and then you realize 'with finality that in some regard you will never be as good a man again' (39). The twenties go by in a blissful ignorance, but by the mid to late thirties the dark secrets begin to expose themselves, the curse of family madness, the problem of the small heredity, catches up with one, and one finds oneself seriously afflicted (40). One has simply eaten up all one's Spinoza (44). It is a life based on punctuated equilibria, famine or feast, the boom and bust of the Jazz Age (56). As Lawrence asks, is not this cycle a more felicitous rendition of the movements of life than the weary, twisting rope of 'Time and Evolution' that requires the revolving hook of a *causa sui*?: 'I like to think of the whole show going bust, *bang!* – and nothing but bits of chaos flying about. Then out of the dark, new little twinklings reviving from nowhere, nohow' (Lawrence 1986: 10–11). Fitzgerald himself insists upon being able to see things as *both* hopeless and yet remaining determined to make things otherwise. Dying as the art of living, bearing testimony to an execution, that of the disintegration of one's personality, but in which the task has become one of a germinal life: to find the crack within the crack-up (Fitzgerald 1965: 46).

The descent into alcoholism reveals the ambiguous power of all instincts. For while the instinct is never to be confused with the crack, it is also the case that an alcoholic instinct passes on a complex, 'foul *germen*', so bringing about an important linkage with the crack itself. 'Instincts', such as the alcoholic instinct, become what they are in historically determined environments. They may express, for example, ways of coping and surviving in an environment that is unfavourable and alienating, even if their practice entails destruction. Destructive instincts, such as alcoholism, perversions, styles of masochism, and so on, serve to *conserve* existence, enabling an individual to will nothingness rather than not will at all. These socially mediated and historically informed instincts establish variable relations with the crack, at times covering over and mending it through the resilience of the body, at other times widening the crack and giving it a new orientation. In each case it is never simply a question of causality but always a matter of geography and topology. For Deleuze the problem begins when there develops a complete identification with the objects of horror or compassion. As Fitzgerald commented on his own bouts of alcoholic depression, such identification signals the death of all accomplishment, which is nothing short of the death of time. Identification is not so much a reaction to the loss of an object – whether that object be love, fame, fortune, or native land – as the very determination of that loss.

For Deleuze there can be no ultimate philosophy of suicide, of drugs, or even of life and death, but only philosophical learning about these things, a learning that for philosophy takes place in the encounter with the event 'of' time. Thought and life are neither innate nor acquired but 'genital' – 'desexualised and drawn from that reflux which opens us onto empty time' (Deleuze 1968: 150; 1994: 114). It always remained the case that for Deleuze only philosophy could assume the responsibility for this vision and riddle of germinal life. Germinal life, however, does not issue commands or imperatives; events are signs of a certain vitalism in which one can only have the *freedom* to become what one is. One must insist upon the point that Deleuze's ethics of the event in no way reproduces the moral law. This is what Deleuze means when he writes 'My wound existed before me' (1997b: 6). The wound does not refer simply to an empirical determination but to the event of a life. Nietzsche's 'thus I willed it!' signals not a Stoic resignation but a battlecry. It does not merely 'accept' the past but gives time to the future. If resignation is only another articulation of *ressentiment*, then *amor fati* is the struggle of free human beings. Indeed, Deleuze will describe the battle as the very essence of the event, and he selects Stephen Crane's novel *The Red Badge of Courage* (1895), with its exploration of demonic imagery and its navigation of a battlefield of intensities, as the most important book on the event. It is this 'will' affirmed by Nietzsche, and created through the event, which gives expression to the 'eternal truth' of the event.

In one of his last pieces of writing Deleuze names the life of immanence as a life of 'beatitude' that is beyond ephemerality. Once again, he returns to the issue of the transcendent and transcendental. A transcendental field is to be distinguished from the realm of experience in that it involves no reference to an 'object' and does not belong to a 'subject'. It does, however, *appear* as an 'a-subjective current of consciousness' but this is a consciousness 'without self', entirely impersonal and pre-reflexive, it is a 'qualitative duration' without beginning or end (1997b: 3). Deleuze insists that this is still a transcendental *empiricism* even though it has no reference to an object or a subject. Indeed, it is the very lack of reference that makes this empiricism 'wild and powerful'. He further insists that the relation of this field to a consciousness is only *de jure*, arguing that what makes it *de facto* is when a subject is produced that then relates itself to an object (the point at which the transcendental is said to be rendered 'transcendent'). The 'true' consciousness that is co-extensive with this field is said to withdraw itself from 'all revelation'. What is this 'consciousness'?

It is nothing other than the movement of a 'plane of immanence' (a notion we are about to encounter and explore at length in the next chapter), in which immanence enjoys its own existence as movement without being immanent either 'in' something or 'to' something. Immanence is not 'immanence to life' but speaks only of 'a life'. This is to refer to 'an absolute immediate consciousness whose very activity no longer refers back to a being but ceaselessly posits itself in a life' (Deleuze 1997b: 4). This, says, Deleuze, is the immanence of the late Fichte, a Fichte, it is claimed, who 'reintroduces Spinozism into the heart of the philosophical operation' (the text of Fichte's Deleuze referred to is *Die Anweisung zum seligen Leben, oder auch die Religionslehre* of 1806, appearing in French as *Initiation à la vie bienheureuse*). It is the impersonal but singular life of the individuating *haecceity* (the 'beatitude' in the title of Fichte's work). This is a *germinal* life since it is not positing life in a 'simple moment' confronting a 'universal death', but rather a life that is 'everywhere', contained 'in all the moments' that a 'living subject passes through', a life of virtualities, events, and singularities (5).

In the 'event' the life of the individual gives way to the life that is impersonal and yet nevertheless singular. It involves releasing the 'pure event' from the accidents of corporeal life, 'from the subjectivity and objectivity of what happens' (Deleuze 1997b: 4). This is the life of a 'haecceity', which speaks of a life of pure immanence that is 'beyond good and evil'. It is only the subject which incarnates this haecceity in the world that makes it either good or bad, in the sense that these categories aim to fix and determine things once and for all. It is crucial, therefore, that the two realms or planes be kept separate; that is, although there is a coexistence of the singularities of a 'life' and accidents of the 'corresponding life', the two do not come together and should not be confused. Deleuze provides the example of small children to illustrate his point, noting how although they all resemble one another and lack individuality, they nevertheless are traversed by singularities which animate but do not inhere in them – a smile, a gesture, a grimace, and so on.

The reference to Fichte in this late essay will strike many readers of Deleuze as problematic. It is certainly odd that Deleuze should locate in the later Fichte a renaissance of 'Spinozism' given Fichte's own declared hostility towards Spinoza and his influence (see Fichte 1994: 98–9, where he argues that Spinoza could not have believed in his own philosophy, but could only have 'thought' it). Nevertheless, in this series of popular lectures Fichte does present Being in terms akin to Spinoza's immanent substance as that 'which *is* absolutely through itself, by itself, and from itself...a self-comprehensive, self-sufficient, and absolutely unchangeable Unity (*Einerleiheit*)' (1962: 53; 1848: 48–9). The description of this unity and being as 'unchangeable' should not deceive or lead us astray. This is duration in Deleuze's sense that has neither beginning nor end, and it only ever becomes what it 'is' – pure becoming.

The spiritual life of blessedness for Fichte is the life filled with consciousness, love, and self-enjoyment (the latter becomes an important feature of the argument on thinking Being beyond phenomenology in *What is Philosophy?*). It is through the laws of reflection which govern its operations that consciousness creates a system of separate and independent individuals and so confronts itself with numerous paradoxes concerning the reality of time and change. Fichte's text is bound up in the moves it makes with classical metaphysics, evident in the distinction it draws between a 'true world' (life/blessedness) and an 'apparent world' (death/unblessedness), but this is a distinction that gets carried over into Deleuze's affirmation of life as pure immanence in contradistinction to the human, all too human satisfaction with the ephemeral and quotidian – a life that is, ultimately, beyond death. This is a distinction that Fichte presents as one between the merely sensuous world and the higher suprasensuous world that is available to, and attainable only by, *Thought*. (It is Lecture VIII which is the most philosophically serious and which would be the decisive one for staging an encounter between Fichte and Deleuze, 1962: 122ff.; 1848: 144 ff.) In the *Vocation of Man* beatitude is related to the achievement of supersensible death '*in life* and *through life*', which speaks of the praxis of a rational self as its peculiar 'sublime vocation'. The self-expressive life of the spiritual is one in which the universe can no longer be thought in terms of the circle 'returning to itself, that endlessly repeating game, that monster which devours itself so as to give birth to itself again as it already was'; rather, there is 'constant progress to greater perfection in a straight line which goes on to infinity'. This is because 'All death in nature is birth...in dying does the augmentation of life visibly appear....It is not death which kills, but rather a more living life which, hidden behind the old life, begins and develops' (Fichte 1987: 122). *My* death can only be a festive passing. It is not, therefore, for Fichte a question of living 'according to nature' simply because this is not even what nature does. The only 'law of life' is, in Nietzsche's almost Fichtean language, 'self-overcoming' (on the world as a 'monster of energy', 'without beginning or end' but as ceaseless transformation, a world of repetition and difference – the 'joy of the circle' as the only goal – compare Nietzsche 1968: 1067; 1987, Volume 11, p. 610).[15]

In this piece of 1993 (1997b) Deleuze has perhaps drawn the logical conclusions from the position he advanced in the texts of the late 1960s: affirming a quasi-Weismannian world of haecceities and individuations beyond the embodied subject, a world beyond good and evil, the expression of the repeated singularity of the One [*L'Un*]. Deleuze, the philosopher of univocal being, of the world as perpetual difference and repetition, the infinite play of simulacra, is keen to avoid two things: one, denaturing the transcendental and thereby reduplicating the empirical; and two, reducing the transcendental to the activity of a subject that synthesizes things and constitutes the world. One might contest that this 'destruction' raises the stakes of germinal life to a level that is too high and that it performs it at too great a cost, so that this thinking of the event does not change us as we are but simply *crushes* us. For Deleuze, however, it is only ever a question of *becoming* those that we are, a piece of germinal life. In *The Logic of Sense*, which sets out to think the event beyond both the banalities of everyday life and the sufferings of madness, the attempt is made, however philosophically 'foolish', to think the relation between the impersonal and the personal. The focus of the next chapter is on *A Thousand Plateaus*, a work that is perhaps Deleuze's most innovative in showing how a novel Weismannianism can be made to work in terms of bringing into relation the germinal life of the field of intensities and the embodied life of the subject, or what now gets played out as the 'body without organs' and the organism.

3

THE MEMORIES OF A BERGSONIAN

From creative evolution to creative ethology

The initial elementary events which open the way to evolution in the
intensely conservative systems called living beings are microscopic,
fortuitous, and utterly without relation to whatever may be their effects upon teleonomic functioning.

(Monod 1971: 118)

It is a poor recipe for producing monsters to accumulate heteroclite
determinations or to over-determine the animal. It is better to raise
up the ground and dissolve the form....There is no other sin than
raising the ground and dissolving the form.

(Deleuze 1968: 44; 1994: 28–9)

Introduction

To move from Deleuze's 'Bergsonism' and thinking of difference and
repetition to the rhizomatics of *A Thousand Plateaus* (abbreviated hence-
forth to *ATP*) is to travel some distance. This is not to say, however, that
there are not important continuities in Deleuze's thinking. The seeds of
Deleuze's attempt to map out the field of nonorganic life with Guattari in
terms of rhizomatic becomings can be seen to already exist in a dormant
state in the work of the late 1960s, as the above citation from *Difference and
Repetition* shows. Deleuze remains preoccupied with the 'transformation of
substances' and the 'dissolution of forms' in which there is 'a passage to the
limit' in favour of forces that are fluid and at the site of which there is
witnessed 'the incorporeal power of that intense matter' (Deleuze and
Guattari 1980: 138; 1988: 109). This is a matter that finds itself enveloped in
continuous variation and caught up in an absolute deterritorialization,
where the 'absolute' indicates neither perfection nor undifferentiation.
Differences 'infinitely small' now come to matter.

Perhaps most significant for the task of understanding the developments
in Deleuze's thought is the attempted novel reconstruction of 'creative
evolution' that is put forward in the 'Becoming-Animal' plateau of *ATP*

139

under the heading 'Memories [*souvenirs*] of a Bergsonian'. We need to try and understand what these 'memories' are and whether they are, in fact, *just* memories and ones that Deleuze wishes to cast off. It cannot be without significance that Deleuze here no longer speaks of 'creative evolution' but instead prefers to speak only of 'creative *involution*'. This indicates, I think, that not only is Deleuze keen to distance himself from a possible Darwinian evolutionism but equally from a Bergsonian one with its residual humanism and perfectionism. I think a decisive factor at play in this complicated and problematic move is that Deleuze's 'Spinozism' has changed its character, notably in its ethological dimension. This is not without significance, I believe, for understanding the kind of ethological 'ethics' (nonhuman becomings of the human) that Deleuze now wishes to put into effect in *ATP*. Whereas in the 1968 text on Spinoza and ethics the aim was to demonstrate a 'superior human nature', and which made it highly compatible with Deleuze's Bergsonian endeavour to think beyond the human condition, by the time of *ATP* Spinoza's ethology is being deployed to attack the gigantic memory of 'man' and to destratify the human organism as a particular kind of stratified organization.[1] But something more is at stake and has been put into play, and this concerns the introduction, beginning with *Anti-Oedipus* and continuing in *A Thousand Plateaus*, albeit with some significant modifications that cannot be traced here, of a *machinic* thinking. Why is this significant for an understanding of the later Deleuze and for understanding his break from the Bergsonism that characterizes the work of the 1960s?

Its significance largely resides in the fact that, in his collaborative work with Guattari, Deleuze is no longer addressing the 'becoming' of the human as a question of its 'evolution' as an individuated *biological organism* (any attempt to apply the charge of biologism to his work is based on a colossal misunderstanding and the most cavalier of readings). The 'human' is now understood solely and strictly in terms of it being a component in a machinic assemblage. Let me try and clarify this, and then say something on why this might be said to amount in some key aspects to a departure from the earlier Bergsonism.

Deleuze and Guattari critique two possible reductions of the question of the machine. First, they attack psychoanalysis for treating machines – which they often conceptualize in terms of the compound 'desiring-machines' – as phantasies, in which fragments of 'real machines' are made adaptable to, and conformable with, symbolical processes. Second, they attack the Marxian assumption of an abstract evolutive line in which man figures as the tool-making animal *par excellence* and where the machine derives from the tool, the tool, in turn, is made dependent on the biological organism and its needs, and, finally, the machine is posited as evolving more and more independently of its maker or designer (man). It is the latter that I will focus attention on for now, since I examine the character of their quarrel with psychoanalysis later in the chapter and at some length when I look at their conception of 'becomings-animal'. Deleuze and Guattari insist on treating the tool/utensil and the machine as a difference of nature (1973: 465; 1995: 122). This is a crucial move to make for them simply because they want to show that the machine enjoys a primacy over the tool and that to commit the error of privileging the latter is to make impossible the move to a machinic, and nonanthropocentric or nonhumanist, model of 'evolution'. This is evident in the Marxian approach, which moves from tools to machines and within which the human means of production imply social relations of production that are held to be merely 'external' to the means and are said to constitute their 'index'. As a result, social relations must necessarily appear as external to the tool and the machine, both of which have had imposed upon them an abstract biological schema that is related solipsistically to the 'needs' of the isolated human organism.

This approach has a number of limitations and problems attached to it. First, it is trapped within a representationalist model of technics and machines in the sense that they are being reduced to the level of extensions and projections of the living being. Second, the evolution of the human gets configured on this schema as if it could be simply understood in terms of a self-directed evolution, the realization of an estranged essence that rests on a naive narrative of freedom conceived as the self-realization of the species-being of the human. Third, and finally, such an approach blocks off access to the 'machinic phylum', that which radically undermines the abstract humanism of the Marxian approach and which is by no means peculiar to it. The idea of the 'machinic phylum' plays various roles and serves a number of functions in Deleuze and Guattari. A definition can be provided here, which is the one they give, but its 'meaning' and significance will only become clearer as the concerns and analyses of this chapter unfold. It names a single phylogenetic lineage that can be said to be 'ideally continuous' since it concerns 'materiality', whether natural or artificial, and frequently both. This materiality in turn speaks of a matter that is in 'movement' and flux, and the subject of 'continuous variation'. It is matter as a 'conveyor of singularities' (1980: 509; 1988: 409). Such an 'operative and expressive flow...is like the unity of human beings and Nature' (506; 406). There is, in addition, a relation of positive feedback between the (machinic) assemblages and the machinic phylum, since while the assemblages cut up the phylum into distinct, differentiated lineages, the machinic phylum cuts across these lineages and makes them coexist. It should be evident that this conception of the machinic phylum is really a reworking of Bergson's conception of creative evolution.

A 'machinic' approach, then, will not treat machines as projections of the human but rather in terms of 'monstrous couplings' involving heterogeneous components that 'evolve' in terms of recurrence and communications ('ergonomics', they note, comes close to this position in not treating the relation between human and machine as simply a question of adaptation).

Humans are both component parts of a machine and combine with other forms of organic and nonorganic life to constitute a machine (or, better, machinic assemblage since there exists no isolated and monadic machine). The difference between the tool and the machine is that one functions as an 'agent of contact' and the other as a 'factor of communication'. The machine enjoys a primacy in the sense that we can only understand *how* human and tool become, and already are, distinct components of a greater machine by referring to an actual *machinic* agency. They thus argue that the question of the machine needs to be posed not in terms of the event of the human as a biological organism but rather as directly conceivable in relation to a full social body, and it is this body that acts as an 'engineering agency' (for example, the full body of the steppe, the full body of the Greek city-state, the full body of the industrial factory, and so on). The attraction to the 'machine' is that it allows for the idea of an 'evolution' proceeding in terms that do not pre-judge what constitutes the character of open living systems, which is customarily done by privileging in a priori fashion organismic conceptions of unity and totality.

How this does new 'machinic' approach bear on the matter of Deleuze's earlier Bergsonism? In spite of his superior treatment – the stress on the virtual and inventive character of a creative evolution of these technics – the criticism that Deleuze and Guattari make of Marx can also be levelled against Bergson, namely, that his staging of questions concerning biotechnics in terms of the standpoint of consciousness, in order to 'privilege' the human, albeit in a nuanced and complex manner, remains too evolutionist, too biological, and too humanist. Bergson's schematics of creative evolution are insufficiently machinic in their appreciation of the deterritorialized and transversal movements of molecular matter. What remains Bergsonian in their approach is the claim that 'machinic' becomings participate in a mode of 'creative evolution' in the sense that they enjoy a capacity [*pouvoir*] for a potentially 'unlimited [*l'infini*] number of connections, in every sense and in all directions' (1973: 469; 1995: 126). In other words, Bergson's virtual, dynamical and durational conception of 'creative evolution' is now being modelled along decidedly 'machinic' lines.

We should also note further that Deleuze's attempt to distance himself from his former Bergsonism, which is never explicitly stated or acknowledged in his later writings, cannot be taken to amount to a complete break simply because in the texts he publishes under his own name in the 1980s, such as the two *Cinéma* books, one finds a return to, and the return of, this earlier Bergsonism. In some instances the language used in the *Cinéma* books is identical to formulations and expressions found in the 1966 text on Bergson. Moreover, it is in *Cinéma 1* that Deleuze appeals to Bergson to clarify the character of the plane of immanence, perhaps the pivotal thought-experiment of his work with Guattari. Bergson's critique of mechanism is utilized as a way of mapping closed and open systems. The plane of immanence is depicted as the 'facet of movement' that gets established between one system and another and which crosses all systems, so preventing them from becoming absolutely closed (Deleuze 1983: 87; 1986: 59). In an important footnote in this work Deleuze justifies this reading of Bergson as a thinker of open systems by arguing that the plane is itself mobile in the specific sense that it has duration as a variable *of* movement (87; 226). Furthermore, in his book of 1988 on Leibniz we also witness a return to the question of the organism that would appear to be quite different to the 'destruction' of the organism executed in *ATP*. This involves Deleuze returning to the question of the irreducible character of types of foldings in which he declares that 'we cannot be sure if preformism does not have a future'.[2] In decidedly Bergsonian fashion, however, he also speaks of the 'unity of movement' (mentioning Bergson in fact in this regard) and notes that 'whether organic or inorganic, all matter is one'. A detailed analysis of this text cannot be undertaken here, regrettably, since it would take us too far afield from the focus of the present study. Some salient features of Deleuze's return to the fold will, however, be dealt with in the conclusion of the study. I shall seek to show in what follows that the ethological move of Deleuze's later rhizomatics is consonant with his earlier thinking if it is unfolded carefully and critically, and taken as *expanding* the horizons of our conception of a possible 'inferior' and 'superior' human becoming – a becoming that involves both the 'less' than human and the 'more' than human.

We can note other lines of continuity in Deleuze's project. Bergson's pluralism, involving the difference between differences (of kind and degree) is, in fact, Deleuze argues, a monism. All the degrees of difference exist in a single nature that expresses itself differentially in terms of kinds and degrees. These degrees coexist in a single time that can be seen to be nature itself. Now to a certain extent this is what *ATP* also seeks to show but this time by conceiving nature in terms of a 'machinic phylum' that eschews any 'ridiculous cosmic evolutionism' and seeks to reveal the deterritorialized dance of the most disparate things. It would appear that it is no longer possible or desirable for Deleuze to appeal to a single principle of evolution, such as the *élan vital*, in which the virtual actualizes itself along lines of divergence that continue to diverge and effect a high degree of closure for the becoming of the virtual. It is as if Deleuze bestows so much significance on becomings-animal that proceed via the transversal communication of affects, and in terms of the establishment of rhizomatic relations, as a way of combatting the tendency within nature to proceed via reified forms of difference and divergence on the level of organisms and species.[3]

In *Difference and Repetition* Deleuze's stress is always on a plurality of modes of evolution to the extent that an organism can be defined in differing ways, including genetically in terms of the dynamisms that determine an internal milieu, and ecologically in terms of the 'external movements' that inform its constitution in an extensive field. In addition, Deleuze notes the importance of a 'kinetics of population' that adjoins, but does not resemble, the kinetics of the egg. In other words, a process of speciation through geographic isolation can be as formative of a species as internal genetic variations (Deleuze 1968: 280; 1994: 216–17). And in one crucial respect *ATP* is continuous with *Difference and Repetition*, in that the dissolution of forms and functions is viewed as a necessary and integral part of any conception of creative 'evolution': 'When representation discovers the infinite within itself, it no longer appears as *organic* representation but as *orgiastic* representation: it discovers within itself the limits of the organised; tumult, restlessness and passion beneath apparent calm. It rediscovers monstrosity [*le monstre*]' (61; 42). This concern with the monstrous continues in *ATP*. Now, however, the monstrous involves 'anti-natural' couplings, such as symbiotic complexes and anomalous animal becomings, that come from nature but through which nature can be seen to be working against itself. I take this claim to mean that nature has a tendency to produce reified forms of life. The emphasis placed on 'unnatural participation' in *ATP* is designed to expose styles of becoming that serve to reintroduce novel aspects into evolution (this preoccupation with novelty in 'evolution' remains fundamentally Bergsonian in inspiration). Such monsters already figure in Darwin's account of evolution but they are always used as examples to illustrate the character of natural selection. For example, in his work on 'The Various Contrivances by which British and Foreign Orchids are Fertilized by Insects and on the Good Effects of Intercrossing' of 1862, Darwin argues that it is a *universal law of nature* that 'organic beings require an occasional cross with another individual' (Darwin 1993: 224ff.). 'Monstrosity', therefore, is an essential component of the complex machine called 'natural selection'. How radical, therefore, is Deleuze and Guattari's stress on monstrous couplings and symbiotic complexes? We may note that they utilize to great effect the wasp–orchid symbiosis in order to illustrate their points, but this is the example examined at length by Darwin himself in his 1862 elaboration of certain ill-understood aspects of the theory of natural selection. The difference is that Deleuze and Guattari use the example in order to show how living matter can evade or transform 'selective pressures', whereas for Darwin such evolutions always serve to confirm the power of natural selection. The important point to note, however, is that the novel alliances created through symbiotic complexes do not take place in terms of simple, predictable additions but involve noncumulative new formations in which the 'extraordinary composite' does not resemble each of the parts of forms of life from which it has been generated (see Margulis and Sagan 1995: 146, who discuss the example of lichens, which are the result of cross-kingdom couplings between fungus and green alga or blue-green bacterium). Whether there is anything further at stake in their turn to symbioses is something we shall examine below.

Before we can adopt a critical perspective on the ethics of *ATP* it is first necessary to unfold the innovations Deleuze and Guattari are seeking to make in the domains of evolutionary thinking and ethology. In the first case this involves moving away from genealogical and filiative models of evolution to rhizomatic or machinic ones. In the second case it involves a move away from an ethology of behaviour to one of assemblages. What these moves mean in more precise terms, and how they relate to current developments in the life sciences themselves, will be examined in detail in this chapter. Now it might be thought that I am making a spurious and non-Deleuzean separation in holding apart ethology and ethics since the two are configured in Deleuze in terms of an intrinsically immanent relation as a question of affect. However, I want to insist on the need both to appreciate the immanence *and* to place it in suspension. It is precisely the manner in which Deleuze stages his claims about the possibility of nonhuman becomings of the human that reveal the necessity to maintain the distinction. This will enable us to question the nature of the leap that is involved in moving from a construction of nature as a plane of immanence and consistency, which proceeds in accordance with the transversal communication of molecular affects, to the articulation of an ethological ethics that concerns nonhuman and animal becomings of the human.

Complexity and the organism

Deleuze's relation to Weismann and to the Darwinian tradition is highly complex. If Deleuze is in some fundamental sense a Bergsonian thinker, then he is close to Weismann simply because, as was shown in Chapter 1, the 'germinal life' Deleuze develops from his reading of Bergson has its 'origins' in the reception of Weismann we find in Bergson's *Creative Evolution*. Bergson is typically treated as part of the tradition of vitalism which, at least in one of its principal aspects, maintains that 'life' is irreducible to the mechanisms of physics and chemistry. But, as we have seen, this conception of Bergson as a straightforward vitalist neglects the extent to which his text on creative evolution is inconceivable without the influence and inspiration of the first neo-Darwinian revolution performed by Weismann. Like the Weismannian tradition, both Bergson and Deleuze give primacy in their philosophies of life not to the organism or the species but to the flows of vital intensities and the becoming of durational forces. However, Deleuze never subscribes to a genetic determinism, whether in the shape of the continuity of germ-plasm or the incessant and imperturbable river of DNA. Moreover, for him 'evolution' involves a great deal more than hereditary transmission and reproduction. However, in affirming the vitality of nonorganic life and, thereby seemingly demoting the organism in the manner of the Weismannian tradition, Deleuze shows himself, on one level at least, to be in sympathy with the modern revolution of neo-Darwinism, genetics, and molecular biology.

And, as we shall see, in *ATP* Deleuze and Guattari do draw heavily on this post-Bergsonian tradition in order to promote a rhizomatics. But they also have an important relation to complexity theory and its attention to issues of self-organization. We need to understand better how Deleuze and Guattari freely draw on these different trajectories in biology.

Stated in very general terms complexity theory denotes a movement within contemporary biology that has sought to go beyond the alleged genetic reductionism and determinism of neo-Darwinism, in which the emphasis is on the 'co-evolution' of organism and environment. The attempt is made to dissolve the opposition of the 'organism' and the 'environment' and to dereify it as a subject–object split. In other words, the claim is that it is erroneous to view the organism as an entity entirely separate from, and evolving independent of, its environment, or to reify the environment by treating it as something given and fixed, and which, it is alleged, produces only a 'passive' model of adaptation. Organisms cannot be treated as closed systems simply subjected to external forces and determinations; rather, they have to be understood in more dynamical terms as open systems that undergo continual flux. The process of 'adapting' involves, Bergson argued anticipating this approach, not a mere 'repeating' but an active 'replying' [*répliquer*] (Bergson 1962: 58; 1983: 58). On the model of natural selection it is the environment which simply selects the organism, exterminating the ill-fitted and picking off the fitter ones. On the model of co-evolution, however, it is equally the case that an organism 'selects' its environment. Brian Goodwin, a leading figure in these so-called 'post' Darwinian developments, argues that organisms have the potential for effecting appropriate responses to their environment, with the result that the variation available for evolutionary change and adaptation cannot be seen as simply arising from random genetic mutation, but rather 'from the intrinsically regulative and plastic responses of the organism to the environment during its life-cycle' (Goodwin 1994: 104–5). Organisms for Goodwin are agents of 'immanent, self-generating, or creative power' that participate actively in 'a flowing unity, a creative river of life' (108). George Kampis, another complexity theorist, argues that if one adopts a co-evolutionary view, in which the relationship between an organism and an environment is seen to rest on a series of feedback loops, then it has to be recognized that the problems to be solved by evolutionary adaptation are themselves products of an evolutionary process (this is an essentially Bergsonian point about *creative* evolution). In other words, the process and its products are meaningful only with reference to one another (Kampis 1991: 16). Kampis contends that neo-Darwinism remains wedded to the transformationalism of Darwin's theory of natural selection, in which evolution is conceived as the unfolding of subsequent stages that already exist, separately from the process, from the beginning on and determined by an initial set of problems posed by an 'environment' once and forever.[4]

It is also possible to see Bergson's statement in *Matter and Memory* that 'the brain is part of the material world; the material world is not part of the brain' as being in communication with this line of thought. This is not the metaphysical opposition as it might first appear on an initial reading. For Bergson the body is an aggregate of the material world, an 'image' in his terms, which both receives and gives back movement and which is capable of *selection*: 'Our perception of matter is, then, no longer either relative or subjective...it is merely dissevered by the multiplicity of our needs' (Bergson 1959: 49; 1991: 50). The molecular movements of the 'cerebral mass' that is the brain are bound up inseparably with the rest of the material world. This is why he argues that to ask whether the universe exists only in our heads or outside of them is to pose an insoluble problem for ourselves. The living body can be construed as a kind of 'centre' but not in terms of mathematical point; such a body does not merely reflect action from without but also struggles with and absorbs it (this is the basis of the distinction he makes between perception and affection) (see 55–7; 56–8).

The 'new' biology, therefore, emphasizes the fact that an environment cannot be separated from what organisms are and what they do (see also Whitehead 1978, originally 1927–8, Chapter IV 'Organisms and Environment', 110ff., for an early conception of this view). Francisco Varela, one of the leading thinkers of the school of autopoiesis, has argued that living beings and environments stand in relation to one another through the activity of 'mutual specification' and 'codetermination' (Varela *et al.* 1995: 198). One promising idea in the search for a paradigm of 'complexity', one commentator writes, is the notion that complexity of dynamic structure increases the more it is 'endogenously' driven, with the result that change cannot be simply located in external compulsions (Wesson 1991: 36). It is precisely such an understanding of complexity that Deleuze was seeking to articulate in *Difference and Repetition* in 1968 (note again that the word 'evolution' is placed in scare quotes):

> What is the formula for this "evolution"? The more complex a system, the more the *values peculiar to implication* appear within it. The presence of these values is what allows a judgement of the complexity or complication of a system, and determines the preceding characteristics of biological systems...we claim that complex systems increasingly tend to interiorise their constitutive differences.
>
> (Deleuze 1968: 329; 1994: 255–6)

This view is taken up in the later work where the word 'complexity' has been replaced by the notion, which we shall examine more closely later on, of 'deterritorialization': 'The more interior milieus an organism has on its own stratum, assuring its autonomy and bringing it into a set of aleatory relations with the exterior, the more deterritorialized it is' (Deleuze and Guattari 1980: 70–1; 1988: 53–4).

In fairness to Darwinism, however, it should be noted, the idea that it is only with the arrival of complexity theory that the co-implication of the organism and environment has been recognized within biology is erroneous and misleading. Such a view constitutes an essential element of Monod's *Chance and Necessity*, a book that exerted a substantial influence on Deleuze's understanding of biophilosophy. As a committed Darwinian Monod rebuts the idea that the theory of selection places the entire responsibility for selection upon changes in the external environment. Monod insists that 'selective pressures' cannot be said to exist independently of specific interactions of different organisms within similar ecological niches. The organism is said to play an 'elective' role in these interactions, determining, in part, the nature and orientation of the selective pressures that inform its behaviour. In addition, the orientation of selection is said to become greater the higher the level of autonomy an organism attains with respect to an environment (Monod 1971: 126). In his insights into the subterranean and 'molecular' events and intensities that exist alongside and independently of the teleonomic activity of conservative living systems, as well as 'beneath' the macroscopic level of natural selection, Monod also makes possible and inspires the trajectory that Deleuze and Guattari will pursue in both *Anti-Oedipus* and *A Thousand Plateaus*.

A 'complexity' view of the rapport between an organism and its environment is articulated to great effect in Deleuze's oft-cited and celebrated essay on 'Spinoza and Us', which utilizes the ideas of Uexküll in order to unfold and enfold the plane of nature as a plane of immanence that distributes affects and which cannot be conceived as operating in terms of an arbitrary distinction between nature and artifice. On this ethological model an animal can never be made separate from its relations with the world since 'The interior is only a selected exterior, and the exterior, a projected interior' (Deleuze 1981: 168; 1988: 125) (it is this insight that echoes those of Merleau-Ponty's, drawn from Uexküll, mentioned in Chapter 1). However, in *ATP* the complexity of the organism is mapped out in terms of a deterritorialized movement of assemblages which significantly reconfigures our appreciation of the organism as an autopoietic agent. In other words, one of the crucial issues at stake is precisely how the organism is to be conceived in terms of its boundaries and its boundedness as well as its components. The view that rhizomatics takes to task is the conception of the organism, identifiable in Kant and within the recent theory of autopoiesis, which posits in a priori terms the unity, stability, and identity of the organism. On this model the whole is superior to the parts or components – the parts exist for the sake of the whole – and the organism is posited holistically as a self-regulating and self-perpetuating unity and integrity. The problem with this conception is that it pre-judges questions of evolution and involution by individuating organisms without due regard for the 'complexity' of their becoming, such as, for example, ignoring the extent to which living systems enjoy dynamical boundaries subject to constant perturbations and mutations, and incorporating different structures, and life-cycles, of other forms of life. Both the boundaries and actual components of organisms are subject to machinic modes of communication which call into question the kind of unity and integrity attributed to living systems by the tradition of autonomy from Kant to autopoiesis.

It is only in recent years, largely as a result of the shifts brought about by complexity theory that the organism has, in fact, begun to be taken seriously within theoretical and investigative biology: that is, treated as a self-organizing system in which 'life' cannot be made reducible to a phenomenal level and so treated purely epiphenomenally (say as a direct expression of DNA sequences, a manifestation of primary structures which presuppose a simple one-to-one relationship between relational, functional organizations and the structures which actualize them; see Rosen 1991: 280). In other words, life and biology are not reducible to physics or chemistry, at least as typically conceived. Rather, the life peculiar to organisms is seen to be an emergent property of complex autopoietic systems involving nonlinear feedback processes. Goodwin contends that for neo-Darwinism organisms have no agency since they are not treated as 'real entities' in their own right but reduced to an expression of genes and their products (1994: 159). Complexity theory, by contrast, seeks to show that organisms display innovative capacities for self-organization and self-regulation that need explaining. The creative life of organisms cannot be understood in terms of 'modern' mechanistic conceptions of causality since these have been shown to be clearly deficient for any adequate mapping of living systems. For complexity theorists this entails a number of things: one, treating organisms as natural kinds and not as historical individuals; two, placing the emphasis in biology on morphogenesis in which complex order in space and time is to be understood as an emergent property that arises from organizational principles of an essentially simple kind (173). Life is not DNA but 'a rich network of facilitating relationships' (174). The nature and function of genes are not disregarded in complexity; rather, the emphasis is placed on understanding their activity within the dynamic context of the morphogenetic field, a field that is to be granted a high degree of relative autonomy in relation to genetic activity and molecular life. Robert Rosen has followed the complexity turn and sought to completely invert the Weismannian model by claiming that the question can only pertain to the soma and the organism: life *is* nothing other than an emergent phenomenon of complex systems, emerging through a process of morphogenesis (these systems are relational and involve 'entailment') (Rosen 1991: 275). This means for him that while it is relatively easy to conceive life and biology without evolution, the same is not true if we try to think of evolution without life. In other words, living things do not simply *instantiate* evolutionary processes; rather they and their specialized somatic activities provide the conditions of possibility for evolution (255).

The problem with the Weismannian account is that it reifies the role of the germ-plasm, or of DNA, in the various cycles that inform life. DNA, for example, does not function as an independent 'replicator' but is dependent on the environment provided by the dividing cell. Similarly, the replicating molecules, that is the nucleic acid templates, need an energy source, the building blocks provided by the nucleotide bases, as well as an enzyme, to facilitate the process of polymerization which is involved in the self-copying of the templates (see Goodwin 1995: 34). Goodwin argues that in all unicellular organisms and plants, as well as in many species of animal, there is no separation of germ plasm from somatoplasm,[5] meaning that there is no 'hereditary essence', such as the germ or DNA, which is responsible for the replication of life.[6] The new biology of complexity theory argues that the molecular mechanisms which make up DNA enjoy a versatility, making it a fluid rather than a stable polymer (35). All the molecular components of cellular life, such as the DNA, the RNA, and proteins, undergo a 'molecular turnover' in which their constituent parts get replaced (ibid.). The hereditary material of life, therefore, plays an important role in stabilizing certain aspects of the spatial and temporal order of the dynamic field in which life evolves and involves. For complexity theorists such as Goodwin, however, what is decisive is the 'generative field' which provides the dynamical context in which the organization of the egg cell or the organism takes place. Note the emphasis is not on a physical 'essence' but on a dynamic *field* of forces. It is not the actual nature of components involved in a physical field, such as molecules and cells, that is decisive in the production of spatial patterns, but rather the manner of interaction with one another in time (the kinematics of the egg) and in space (their relational order: that is, how the state of one region depends upon the state of neighbouring regions) (Goodwin 1995: 49). 'Nude DNA', therefore, does not replicate itself but requires a complex assemblage of protein enzymes: 'the cellular dance is mediated by a host of protein enzymes' (Kauffman 1995: 39).

Deleuze and Guattari attack DNA dogma and mythology in both *Anti-Oedipus* and *A Thousand Plateaus*. In the latter, for example, they suggest that genetic linearity is above all spatial even though the segments are constructed successively. It is best not to compare the genetic code to language, they argue, since it has neither emitter nor receiver, and concerns redundancies and surplus values rather than comprehension and translation[7](Deleuze and Guattari 1980: 81; 1988: 62) (on the genetic code and language see Jacob 1974: 306). As a protein molecule that reorders molecular components DNA cannot be said to 'translate' in the way language translates. Coded within the genetic material is a scheme by which particular sequences of nucleotides (the four bases of adenine, cytosine, guanine, and thymine) correspond to specific amino acids. But, as Brian Massumi notes, the transformations effected as a result of DNA mechanisms are 'not a function of the code itself, but of the successive syntheses its reorderings are taken up by', such as metabolism, natural selection, reproduction, and viral transfer (Massumi 1992: 187).

Orthodox biology would contest these claims by pointing to the invariant character of DNA replication and reproduction. Deleuze and Guattari accept, of course, that invariance is not simply a feature of molar organizations but constitutes the essential nature of nucleotides and nucleic acids whose 'expression' is independent of molecules of 'content', such as proteins, and of any directed action in an external milieu.[8] At the same time proteins can be seen to be equally independent of nucleotides in the form and substance of their 'content'. In the linear determination of a nucleic sequence 'expression' takes on a form that is relative to 'content', and involves a 'folding back upon itself of the protein sequence of amino acids', thus yielding the characteristic three-dimensional structures. However, the critical point Deleuze and Guattari insist upon is that while this configuration of expression and content determines an organism's capacity for reproduction, it also enhances the capacity for *deterritorialization*. To quote their key point again: 'The more interior milieus an organism has...assuring its autonomy and bringing it into a set of aleatory relations with the exterior, the more deterritorialized it is' (Deleuze and Guattari 1980: 70–1; 1988: 53–4). This is a radical inversion of standard Darwinian dogma since recognition of the invariant structure of DNA does not necessarily lead to a reification of reproduction of the species as the *telos* of life, but rather is able to point out that reproduction is dependent upon a primary deterritorialization (see 77–8; 59–60).[9] If reproduction is always inseparable from deterritorialization then 'evolution' must involve something more than the simple reproduction of species *qua* species. If codings are inseparable from intrinsic and extrinsic processes of decoding (through supplementation and 'side-communication'), and if territorialities are equally involved in these processes, then the key insight is that it is *populations* that are the 'subject' of these mutually implicated processes of coding, decoding, and deterritorialization. On this model 'evolution' does not simply entail the passage from one pre-established form of life to another, involving only the translation of one code into another. Rather, there are the phenomena of mutations,[10] genetic drift, and the transferral of cells of one species to another that takes place not in terms of 'translation' but in rather in terms of a surplus value of code ('side-communication') (viruses, for example, do not translate code but side-communicate via fragments of code).[11] Ultimately, this signals a break from the molar norms of orthodox Darwinian theory with its emphasis on large numbers and statistical aggregates, allowing for anomalous animal-becomings, and affording a significant insight into how Deleuze and

Guattari are able to break free of the grip of Weismann's germ-plasm. On a rhizomatic model it is not so much a question of 'evolution', with its perfectionist and progressivist values, but more a question of passages, bridges, and tunnels; a question neither of regression nor of progression but of 'becomings': 'There is no plasmic finality....Life knows no finality, no finished crystallisation' (Lawrence 1993: 182).

The body without organs and the organism

Given the proximity of a great deal of Deleuze's 'Bergsonism' (whether of 1966 or 1968) to the arguments of complexity theory, and given that in *ATP* Deleuze and Guattari can be seen to be developing a rich conception of self-organization, why do they in *ATP* lay so much stress on attacking the organism? In short, why in *ATP* do Deleuze and Guattari seem to have it in for the organism? 'The enemy', they declare, 'is the organism' (1980: 196; 1988: 158). This is not to ask after in an abstract fashion the nature of Deleuze's strange attraction to the 'powerful' and 'intense' life that for him is anorganic, but rather to ask critically whether the philosophy of germinal life adumbrated in *ATP* results in an unmediated opposition between organismic life and nonorganic life, as well as a merely abstract negation of the organism. In *ATP* the organism appears to be treated as something that is intrinsically molar, trapping life in formal organization, and effecting stratification and rigidification on the anorganic life in which the vital and virtual power of the disparate and the demonic, as that which haunts the organization of living systems, is nullified: 'The *judgement of God*...the theological system, is precisely the operation of He who makes an organism, an organization of organs called the organism' (196–7; 158). Indeed, it is often noted, as well as being clearly stated by Deleuze and Guattari themselves, that their curious notion of 'the body without organs' is poorly phrased in that the attack is not on the organs as such but on their organization by the organism.

Deleuze and Guattari's conception of the organism as a highly stratified molar organization is not, in fact, entirely idiosyncratic or peculiar to them. On the contrary, a similar conception of the organism has been advanced by the celebrated founders of the contemporary theory of autopoiesis, Humberto Maturana and Francisco Varela: the former one of the world's leading biologists, the latter one of the world's leading neuroscientists. They argue that considered as a 'metasystem' the organism is made up of components that enjoy only a minimum autonomy. Because the organism is driven by the desire to preserve itself as a unity in the environment it tends to evolve as a highly conservative system. They come very close to Deleuze and Guattari when they argue that 'The organism restricts the individual creativity of its component unities, *as these unities exist for the organism*' (Maturana and Varela 1992: 199, my emphasis).

I want to put forward the argument, however, that in *ATP* Deleuze and Guattari neither abstractly negate the organism nor disregard the vital contribution it makes to 'life'. The way in which the relationship between the body without organs and stratification gets configured at the outset of the plateau significantly titled 'The Geology of Morals' is important in this regard. The 'body without organs' refers to the 'body' of the energies and becomings of the Earth that gets permeated by matters which are highly unformed and unstable, characterized by free-moving flows, 'free intensities' and 'nomadic singularities'. Deleuze and Guattari then argue that there also forms on the planet, at the same time, the phenomenon of 'stratification' that can be regarded, in the effect it has of providing life with organization and direction, as both 'beneficial' and 'unfortunate'. It is the system of strata that provides matters with forms, imprisons intensities into systems of redundancy, and organizes small and large molecules into 'molar aggregates' (see note 3 for insight into the distinction between molecular and molar). These strata, ranging from the energetic, the physico-chemical, and the geological, to the social and cultural, function like acts and apparatuses of 'capture' and operate in two principal ways: coding and territorialization. The problem of *organic* stratification is that of how to make the 'body' – the body as constituted by free intensities and nomadic singularities – into an organism. It is this problem that is laid out and addressed by Monod, in which natural selection draws the intensities of life out of the realm of chance and operates as a force of necessity on the macroscopic level of organisms (Monod 1971: 118–37). We do not need to follow in any further detail the rich contours of Deleuze and Guattari's cartography. Many of the crucial expressions and formulations they come up with, such as the 'plane of consistency', 'deterritorialization', etc., will be explored in more detail as the chapter progresses. Let us simply note a crucial aspect to their presentation of the problem of the organism in relation to the body without organs: namely, that it is a question of understanding how the organism is formed on the level of an organic stratum, how it works, what functions it performs, what deterritorializations and decodings it is implicated in, and what we may make with it as the affair and praxis of a germinal life. It is crucial, therefore, how we construe the 'ethics' involved in the complex relation between the body without organs and the organism.

Deleuze and Guattari stress that the 'body without organs', the immanent field that is 'desire' producing and distributing intensities, does not simply come 'before' the organism, as if it was some kind of preorganic, amorphous soup; rather, it has to be thought as adjacent to the organism and as 'continually in the process of constructing itself' (1980: 202–3; 1988: 164). The problem is compounded by the fact that for Deleuze and Guattari there is not just the one type of body without organs but several. The principal contrast, however, which is designed to have critical import, is between the body without organs that constitutes the 'glacial reality' where the foldings, sedimentations, and coagulations involved in the composition of an organism take place, and the organism that constitutes the stratum existing on this body, imposing upon it forms, functions, hierarchized organizations, and transcendences (197; 159). Deleuze and Guattari, however, insist that no abstract opposition is to be set up between the strata and the body without organs. This is for two reasons: first, the body without organs comes into play both in the strata and on the destratified plane of consistency. This means that in the case of the organism conceived as a stratum one finds a body without organs which opposes the rigid organization of organs, but also that there is a body without organs *of the organism that belongs to its stratum.* The aim is not, therefore, to negate the organism but to arrive at a more comprehensive understanding of it by situating it within the wider field of forces, intensities, and durations that give rise to it and which do not cease to involve a play between nonorganic and stratified life. Creative processes inform *both* the body without organs and processes of stratification. Stratification, for example, is said to involve creating order out of chaos, and artists, in whose work this process becomes visible, are compared to 'God', making a world by organizing forms and substances, codes, milieus, and rhythms, and producing a harnessing of cosmicized forces (1980: 627; 1988: 502). By contrast, the 'becoming' of bodies without organs involves a play of individuation through haecceity/singularity, a 'production of intensities beginning at a degree zero' (633; 507). This is the 'powerful nonorganic life' that escapes the strata and is implicated in transversal modes of communication, which are modes that cut across the evolution of distinct phyletic lineages.[12] The organism that Deleuze and Guattari are attacking, I would contend, is not a neutral entity but rather the organism construed as a given hierarchized and transcendent organization. It can only be represented in such terms by being abstracted from its molecular and rhizomatic conditions of possibility.

Is it possible to understand this complex working out of the organism as implicated in a body without organs that is always adjacent to it in other than ethical terms? The question 'how do you make yourself a body without organs?' is addressed to the organism as an ethical question in Spinoza's sense that knowing what a body can do can never be determinable once and for all since it involves an experimental becoming. On my reading the ethical question is the only way to make sense of Deleuze and Guattari's statement that 'dismantling the organism has never meant killing yourself'. This is because the point is not a moral one, involving some entirely abstract negation of the organism. The ethical question concerns the theory and praxis of opening up the body to connections and relations 'that presuppose an entire assemblage' made up of 'circuits, conjunctions, levels and thresholds, passages and distributions of intensity, and territories and deterritorializations' (1980: 198; 1988: 160). Deleuze and Guattari's emphasis on caution, in which they warn against not destratifying too wildly, is entirely consonant with this ethics that has to do with a knowledge of the distribution and operation of intensities. When they argue that 'staying stratified – organized, signified, subjected – is not the worst that can happen', since the worst involves collapsing the strata into some demented or suicidal state, we are reminded of the ethics of the event (the event of a germinal life) that Deleuze was seeking to articulate in *The Logic of Sense*. Killing oneself as an organism does not amount to 'dismantling' the organism, therefore, but, ironically, preserves and mummifies it.[13] It is a question of becoming an intense body and of finding potential movements of deterritorialization *as an organism* (this is the great Spinozist paradox of the book).

But we also need to note the problems attendant on any attempt to simply or straightforwardly apply insights gained from the evolution of biological populations, and from the becomings-animal implicated in ethological assemblages, to the field of 'nonhuman becomings of the human'. The problem stems from the fact that we are dealing with inordinately lengthy geological timescales and highly complex evolutionary processes. This is, in fact, a fundamental problem with Deleuze and Guattari's attempt to rewrite ethics with a '*geology* of morals' (in the plateau entitled '10,000 years B.C.: A Geology of Morals' the date names as an 'event' the end of the last Ice Age and the accelerated process of deforestation that marks it).

What is a multiplicity? Bergsonism and neo-Darwinism

There is one issue on which Deleuze's utilization of Bergson remains entirely the same between the works of 1966–8 and *ATP*, and this is the issue of multiplicity. In *Bergsonism* Deleuze argues that the usage of multiplicity in Bergson is not part of the traditional vocabulary of philosophy when it is taken to denote a continuum. For Bergson, he insists, it is never a question of opposing the one and the multiple but of drawing a distinction and a difference between two kinds of multiplicity, namely, between a dicrete multiplicity that can be represented spatially and a qualitative multiplicity that can only be conceived durationally (Deleuze 1966: 30–1; 1988: 38; and see Bergson 1960: 121–3 for a clear presentation of the two types). In *Bergsonism* the difference between the two is presented as one between a difference in degree and a difference in kind. The crucial difference is that in the first case multiplicity is being represented purely extensively in terms of a homogeneous ordering (it is strictly numerical); in the second case, however, there is an internal fusion and heterogeneity that cannot be made subject to a simple numeration. Conceived purely as a unit of arithmetic, Deleuze argues, the number is always that which divides without changing in kind. By contrast, a continuous or durational multiplicity always changes in kind when it divides and so can be considered to be nonnumerical (one can speak of 'indivisibles' but only at each stage of the division).[14] Here number does exist but only as a 'potentiality', meaning that its duration is of the order of the *virtual* that must divide in kind through differential actualization.

The notion of multiplicity constitutes a crucial component of Deleuze's Bergsonism and allied critique of dialectics. To map a 'becoming' in terms of an opposition between the one and the many is not to reach multiplicity, since, it is argued, one does not attain the concrete by combining the inadequacy of one concept with the inadequacy of its opposite (a procedure based on the illusion that abstract mediation results in real determination). It is not for Deleuze a question of saying that a subject or self is one and also several, resulting in a unity of the multiple; neither is it a question of declaring the one to be already multiple. Multiplicity does not, therefore, designate a combination of the one and the many, but only an organization of the heterogeneous that does not require an overarching unity in order to operate as a system. If we take the organism as an example, whether considered as one or as many, we can say in the terms of Deleuze's Bergsonism that it is constituted by, and implicated in, an assemblage made up of heterogeneous components. This assemblage functions as an acentred multiplicity that is subject to continuous movement and variation. Deleuze will go on to insist in connection to systems thinking that there are types of multiplicities and that these different types enjoy an immanence to each other – the point being that there *can* take place in material terms an arborification of multiplicities, just as a rhizome can become subjected to segmentation and stratification. A rhizome never exists in some 'pure' rhizomatic state but rather oscillates between genealogical lines that segment and stratify and 'lines of flight' and the dissemination that rupture the tree lines (on multiplicity contra dialectics compare Derrida 1987: 45, where dissemination is discussed in terms of 'an irreducible and *generative* multiplicity').

There is for Deleuze, therefore, a 'topology of multiplicities' which marks 'the end of dialectics' since the multiple is now uprooted from its state as a mere predicate and has become a noun ('multiplicity'). The important difference is now conceived as a difference between 'distances' and 'magnitudes', with the former being inseparable from a process of continuous variation and the latter always distributing constants and variables (Deleuze and Guattari 1980: 603–6; 1988: 483–5) . An intensity, for example, is not made up of addable and subtractable magnitudes, in which it would be possible to describe a temperature as the sum of two smaller temperatures, just as a speed cannot be described as the total of two smaller speeds. Rather, an intensity is a 'difference', and while differences can form an order of smaller and larger differences they are enveloped in one another in such a way that it is not possible to judge their exact magnitudes.[15]

A multiplicity is not to be understood extensively, say in terms of the elements that compose it in a certain space, but rather intensively by 'the lines and dimensions it encompasses in "intension" ' (Deleuze and Guattari

1980: 299; 1988: 245). A multiplicity is always potentially subject to mutation or transformation simply because of a change in the dimensions that deterritorialize it (we have noted that it exists as acentred). If the dimensions change, whether through the loss of a dimension or the gain of one, then the multiplicity will change. Multiplicity is another word for 'becoming' with no centre of unification and is simply definable by the number of dimensions it enjoys:

> Since its variations and dimensions are immanent to it, *it amounts to the same thing to say that each multiplicity is already composed of heterogeneous terms in symbiosis, and that a multiplicity is continually transforming itself into a string of multiplicities, according to its thresholds and doors.*
>
> (305; 249)

The multiple can only be thought as a becoming-multiplicity, that is, as a substantive and not a predicate of a subject (such as the 'One'). It is never a question of a subject bringing a multiplicity back home into itself simply because the subject, conceived as an act of consciousness which gathers and unifies matter, is never the ground or source of its conditions of possibility.

It is the notion of multiplicity which informs Deleuze and Guattari's conception of the rhizome. The emergence of novel modes of 'evolution' presuppose communication is taking place between bodies and material forces that are already made up of heterogeneous components. An emphasis on the composite character of living entities informs Lynn Margulis's work in symbiogenesis, which has much in common with the stress placed on rhizomatic styles of evolution in Deleuze and Guattari, especially in its challenge to standard genealogies of evolutionary life and its focus on transversal assemblages in which genes are seen to cross taxonomic boundaries 'because DNA travels easily in the form of small replicons: plasmids, viruses, transposons, and so forth' (Margulis in Margulis and Fester 1991: 9). Similarly, in Deleuze and Guattari's mapping of germinal life a rhizome functions through transversal communication and serves to bring about creative and novel links across phyletic and informational lineages. A rhizome is an underground sprout, such as a bulb (which is best thought of as a stem, not a root), in which parts constantly die off in the same measure as the rhizome conceived as a multiplicity rejuvenates itself. It is a subterranean 'network of multiple branching roots and shoots, with no central axis, no unified point of origin, and no given direction of growth' (Grosz 1995: 199). In effect, the rhizome constitutes the surplus value of evolution, always coming into being without origin, and only conceivable when evolution is understood as functioning transversally, that is, as cutting across distinct lineages.[16]

A rhizome refers to an assemblage of heterogeneous components, a multiplicity which functions beyond the opposition of the one and the many. It is neither a One which becomes Two (which would be to presuppose evolution as involving little more than a linear accumulation) nor a multiple that is either derived from a One or to which One might be added: 'It is composed not of units but of dimensions, or rather directions in motion' (Deleuze and Guattari 1980: 31; 1988: 21). The rhizome is 'anti-genealogy' since it operates in the 'middle' without *arche* or *telos*, operating not through filiation or descent but via 'variation, expansion, conquest, capture, offshoots' (ibid.). It is the rhizome that can be said to provide the most inventive domain of 'creative evolution' simply because it functions in terms of a potentially infinite open system.

It is on the issue of multiplicities that a rapprochement between Bergsonism and neo-Darwinism (population thinking) can be established. Deleuze and Guattari credit the latter with a 'double deepening' of biology consisting of, on the one hand, an attention to populations as multiplicities in the Bergsonian sense, and, on the other hand, an equally important attention to degrees of development in terms of speeds, rates, coefficients, and differential relations (Deleuze and Guattari 1980: 63–4; 1988: 48). Taken together, these contributions move help to move biology in the direction of a nomadology or 'science of multiplicities'. In the former case, it means that forms of life cannot be taken to pre-exist a population: 'The more a population assumes divergent forms, the more its multiplicity divides into multiplicities of different nature' (ibid.). In the case of substituting rates for degrees the effect is to show that the degrees of development do not pre-exist and await perfection through a process of realization (as in any conception of preformationism), but rather function as global and relative equilibriums within the context of the advantages they bestow on particular multiplicities operating with a particular variation in particular milieus. This means

> Degrees are no longer measured in terms of increasing perfection or a differenciation and increase in the complexity of the parts, but in terms of differential relations and coefficients, such as selective pressure, catalytic action, speed of propagation, rate of growth, evolution, mutation, etc.
>
> (ibid.)

The utilization of neo-Darwinism in *A Thousand Plateaus* extends Deleuze's reading of Darwin in *Difference and Repetition*, where Deleuze sought to enlist natural selection in support of a philosophy of internal difference by transforming its molar articulation as a theory of species. Now, however, the reading of Darwin is conducted largely in terms of the modern synthesis of neo-Darwinism and the introduction of population thinking. As we saw in the previous chapter, in *Difference and Repetition* the utilization of von Baer's work in embryology follows the discussion of Darwin, but in *A Thousand Plateaus* the order is now reversed (Deleuze and Guattari 1980: 62ff.; 1988: 47ff.). This shift in emphasis is not without significance for understanding the complex movements Deleuze and Guattari are forging in relation to modern biology. The attraction is still to a *molecular* reading of Darwinism (a 'molar' Darwinism is one which gets fixated on the question of *species*), but now it is made clear that the neo-Darwinism of population thinking is able to think a way out of the problem of what they call the 'irreducibility of the forms of folding'. If forms of life are irreducible, as the work of von Baer and his disciples suggested, then how can we explain novel becomings such as creative involutions and communications that take place across phyletic lineages? The answer for Deleuze and Guattari lies in the insight that it is through populations that one is formed and assumes forms. The suggestion is that one can only understand a molar population, such as a species, in terms of a different kind of population, a molecular one, which is the subject of the effects of, and changes in, coding. This molecularization of a population is obviously contingent since it is dependent upon the ability of a code to propagate in a given milieu or create for itself a new milieu in which any modification is caught up in a process of population movement (1980: 69; 1988: 52). 'Change', therefore, cannot be conceived as the passage from one pre-established form to another but rather in terms of a process of decoding. To support this view they appeal to the 'modern theory of mutations' and its claim that a code enjoys a margin of decoding that provides supplements which are capable of free variation. Moreover, and as already noted, innovation in evolution takes place not simply through 'translation' between codes but equally in terms of the phenomenon of what they call, drawing heavily on the work of François Jacob, the 'surplus value of code' or 'side-communication'.

Nature as a plane of consistency

This plane is neither one of organization nor of development, but refers to a quite different mode of 'evolution' which has the character of a 'becoming'. Such a 'becoming' involves neither the development of forms nor the constitutions of substances and subjects, but rather modes of individuation that precede the subject or the organism. This plane is said to consist only of 'abstract' and 'nonformal' elements, such as relations of speed and slowness between them and the composition of intensive affects. Becomings that take place within it are rhizomatic, taking place nongenealogically in the 'middle', and, as such, do not require for their comprehension principles of finality.

The emphasis on an 'anti-nuptial nature' through anomalous 'animal becomings' explains why in *A Thousand Plateaus* Deleuze continues, as he had done in *Difference and Repetition*, to side with Geoffroy Saint-Hilaire over his great rival Cuvier: even on the level of the foldings it is a question of *'Monsters'* (Deleuze and Guattari 1980: 62; 1988: 47). The contrast between the two is presented as one between a topological conception of the fold in Geoffroy and a Euclidean conception of space in Cuvier, but the crucial point of difference is that one allows us to gain access to nature as a plane of consistency and the other does not (see also Deleuze 1968: 238–9; 1994: 184–5).[17] Cuvier restricts biological definitions to a logic of organs, of their relations and functions. On his schema the unity of the plane of nature can only assume the character of a unity of analogy, a transcendent unity which 'realizes' itself by fragmenting into distinct branches and lineages, and which then evolve in terms of irreducible and uncrossable heterogeneous compositions. By contrast, Geoffroy goes beyond organs and functions to the 'abstract' anatomical elements, to nature as an immense abstract machine, in which the key operations and evolutions are 'pure materials' entering into combinations and recombinations. In this schema, therefore, it is speeds and slownesses, relations of movement and rest, that assume priority over the forms of structure and over types of development. As Deleuze and Guattari note in line with new research developments in fertility, as in other examples such as sex-hormonal determination, it is less a question of form and functions and more a question of speed: 'do the paternal chromosomes arrive early enough to be incorporated into the nuclei?' (Deleuze and Guattari: 1980: 312; 1988: 255). It is this mapping of the 'plane of nature' as a plane of consistency that provides valuable insight into the nature of the 'body without organs'. This is a plane of immanence in which the molecular order of nature is one of composition (of particles and their speeds) as much as it is a molar one of organization (of forms and functions). Such a plane is populated by all kinds of successes and failures, by jumps and rifts between assemblages, not because of some essential formal or functional irreducibility, but precisely because there are always elements that are slow or late to arrive. The plane of consistency, therefore, is at one and the same time a plane, or flat multidimensional surface, of absolute movement and absolute immobility, which gets 'traversed by nonformal elements (*éléments informels*) of relative speed that enter this or that individuated assemblage' (312–13; 255).

The plane of immanence marked by transversal communications provides an alternative model of 'creative evolution' to the genealogical or filiative model that is dominant in modern biology. The particles and speeds it draws our attention to are not to be confused with atoms simply because atoms remain finite elements still endowed with a form that is too determinate. As Deleuze himself asks: can one even speak of atoms as distinct from worlds and interatomic influences? (Deleuze 1983: 86; 1986: 58; compare Nietzsche 1968: Section 636). The particles of relative speed do not 'exist' *independent of the plane of consistency*. One must learn to think the immanent nature of this plane: 'each individual is an infinite multiplicity, and the whole of

Nature is a multiplicity of perfectly individuated multiplicities' (Deleuze and Guattari 1980: 311; 1988: 254). The plane of nature enjoys *as a plane of consistency* a real unity, but this unity encompasses both the animate and inanimate, the natural and the artificial. Moreover, the 'evolution' of this unity has nothing to do either with a deeply buried ground or with a transcendent creator, such as the agency of God and principles of design. The plane of nature–artifice is conceived as a plane of univocity in order to combat forms of modern evolutionism that would restrict nature to a model of *analogy*, within which anti-nuptial nature, modes of involution, and transversal communications would become unthinkable and unmappable.

Deleuze and Guattari insist that within this plane of immanence and consistency the univocal *is* the multiple: 'The One expresses in a single meaning all of the multiple' (1980: 311; 1988: 254). This is not the unity of a subject or a substance – the plane of nature is said to be peopled only by *anonymous* matter – but rather that of a potential infinity of modifications and combinations of particles and forces of matter that are implicated on this unique plane.

Creative involution

In *A Thousand Plateaus* Deleuze and Guattari do not place the emphasis on complexification through more or less differenciation, but on forms of creative *involution* and modes of transversal communication that produce an 'anti-nuptial nature' ('monstrous' couplings) that involve subterranean becomings. The 'memories of a Bergsonian' in the book herald, in fact, the 'becomings' associated with this move to involution. The book resists any attempt to conceive of evolution in terms of a logic of stages. 'It is difficult', they write,

> to elucidate the system of the strata without seeming to introduce a kind of cosmic or even spiritual evolution from one to the other, as if they were arranged in stages and ascended degrees of perfection. Nothing of the sort....There is no biosphere or noosphere, but everywhere the same Mechanosphere.
>
> (Deleuze and Guattari: 1980: 89; 1988: 69)

The phylum of the plane of consistency is described as 'machinic' partly because it involves becomings that are taking place without fidelity to relations of species and genus. The 'plane of consistency' upon which they wish to map these becomings claims to eschew distinctions of natural and artificial, of lower and higher, and of differences in level or orders of magnitude.

It is in the plateau entitled 'Memories of a Bergsonian' that Deleuze and Guattari introduce the notion of 'blocks of becoming' as a way of breaking out of the grip of evolutionism which configures natural history in terms of a grand mimology. Within germinal life there is a 'subterranean' level of animal becomings. As in the actualized movement of the virtual, a process of becoming involves neither resemblance nor imitation. It is not, therefore, a question of playing at being or imitating an animal since becoming 'produces nothing other than itself' (1980: 291; 1988: 238). In speaking of this reality that is specific to becoming Deleuze and Guattari refer to Bergson's notion of the coexistence of different durations which subsist in communication (ibid.). This is to ingeniously bring Bergsonism – the emphasis on virtual times, multiplicities, etc – into play with the non-evolutionist field of transversal communication. Becoming is of a different order than filiation simply because it concerns alliances which cut across phyletic lineages: 'If evolution includes any veritable becomings, it is in the domain of *symbioses* that bring into play beings of totally different scales and kingdoms, with no possible filiation' (ibid.). 'Originality in neoevolutionism' is to be mapped out as creative involution. Involution has to be removed from its association with regression – regression always involves a movement in the direction of something less differentiated (as in Freud's positing of the involution of the death-drive seeking to return to simple inorganic matter) – and conceived in terms of the formation of a block that creates abstract lines which allow for passage beneath any assignable relations. The domain of involution can be considered to be 'creative' because its field of production is not differenciation (whether more or less) but the formation of blocks which create their own lines of invention and which allow for non-filiative becomings. Involution produces a dissolution of form and freeing of times and speeds and so makes possible certain transversal modes of becoming (326; 267). Of course, an orthodox or strict Darwinian would claim that this is only to modify or enhance the model of natural selection, in which the survival of the fittest becomes configured in terms of the survival of the most symbiotic in particular instances. In other words, there exist strong selective reasons as to why the evolution of organisms and species should take place, in certain cases and contexts, in term of symbiotic complexes. However, the 'becomings' taking place on the destratified plane of consistency are doing quite different work for Deleuze, contributing to the creation of a nonorganic life that is not captured by natural selection and its values of adaptation and survival (of the fittest).

Communication between living systems, therefore, occurs not simply through filiative relationships but also through transversal modes that always involve heterogeneous populations and assemblages. This is a point also made by Richard Dawkins, the well-known advocate of the theory of the selfish gene; namely, that evolution cannot be restricted to individual organisms since at the molecular level one is dealing with populations of genetic material the movements of which are not governed or fixed by organismic boundaries. This is the only way to understand how evolution can become contagious and epidemic (hence the emphasis on rhizomatic modes of 'evolution' and the importance of viruses; for Darwin's view on the inexplicability of epidemics see Darwin 1987: 397). These phenomena involve terms which are heterogeneous, such as a human being, an animal, a bacterium, a virus, a molecule, micro-organisms, and so on (compare Jacob 1974: 291, 311). Such combinations are 'neither genetic nor structural' but rather 'interkingdoms' that involve 'unnatural participation', which they claim is the way nature operates, namely, 'against itself' (Deleuze and Guattari 1980: 296; 1988: 242). Populations are not to be treated as inferior social forms; they are affects and powers, in short, involutions and multiplicities (295; 241).

In his important collection of essays entitled *The Extended Phenotype* Dawkins contends that the existence and character of organisms needs to be treated as a phenomenon meriting an explanation (Dawkins 1983: 263). Too often it gets treated unproblematically as an obvious unit of selection with Darwin himself conceiving natural selection as working for the benefit of each individual. Dawkins's innovation is to come up with the notion of extended phenotype. He introduces the idea in the context of the classic Darwinian debate over the object of evolution by natural selection, and contends that if it is legitimate to speak of adaptations as being about benefits, then these accrue to germ-line replicators and not to individual organisms.[18] The most important replicator is the gene, or, to be more precise, the gene fragment or network. We need to bear in mind that for this model genes are not selected directly but by proxy, that is, by their phenotypic effects. It is only for reasons of convenience that we think of these effects as being packaged or contained in discrete bodies or entities, such as individual organisms. Dawkins, however, holds that the replicator should be thought of as having *extended* phenotypic effects which is made up of all its effects on the world, and not just on the individual body in which it happens to be sitting (4). He is referring here to the phenomenon of 'genetic action at a distance'. All bodies are made up of parasites, symbionts which infiltrate the systems of the host. A very similar line of thought to the one pursued by Bergson and Dawkins can be found in Gregory Bateson, who argued that the network of a system of life is not bounded by 'skin' but includes 'all external pathways along which information can travel' (1973: 290). Dawkins himself writes:

> These phenotypic consequences are conventionally thought of as being restricted to a small field around the replicator itself, its boundaries being defined by the body wall of the individual organism in whose cells the replicator sits. But the nature of the causal influence of gene on phenotype is such that it makes no sense to think of the field of influence as being limited to intracellular biochemistry. We must think of each replicator as the centre of a field

of influence on the world at large. Causal influence radiates out from the replicator, but its power does not decay with distance according to any simple mathematical law. It travels wherever it can, far or near, along available avenues, avenues of intracellular biochemistry, of intercellular chemical and physical interaction, of gross bodily form and physiology. Through a variety of physical and chemical media it radiates out beyond the individual body to touch objects in the world outside, inanimate artefacts and even other living organisms.

(1983: 237–8)

Dawkins situates the whole biosphere on the plane of this intricate network of fields of influence, speaking of a 'web of phenotypic power' (238). The key insight is that interactions are taking place among not only different gene-pools, but also different phyla and different kingdoms (245). As to why replicative life assumes organismic form, Dawkins answers: 'An organism is the physical unit associated with one single life cycle. Replicators that "gang up" in multicellular organisms achieve a regularly recycling life history, and complex adaptations to aid their preservation, as they progress through evolutionary time' (259). The important insight is that organisms need to be treated as 'complex assemblages' (263). But it is this conception of life as made up of 'assemblages' which is under-theorized in Dawkins and which also requires extending further in order to articulate a general ethology.

It should be noted that for Dawkins the extended phenotype is still subject to laws of natural selection: 'Evolution is the external and visible manifestation of the differential survival of alternative *replicators*....Genes are replicators; organisms and groups of organisms are best not regarded as replicators; they are *vehicles* in which replicators travel about' (82). Let us note, however, before moving on, that a Bergsonian-inspired critique can be made of Dawkins on this point. One such critique has, in fact, been outlined by Kampis, who begins by noting the paradox of evolution: evolution can only act on the phenotype but the real subject of evolution is the genotype (the distinction is simply between the genetic endowment or constitution of an organism – the genotype – and the observable characteristics of that individual organism – the phenotype – which result from the interaction between organism in its environment and its inherited genotype). Kampis is close to Dawkins in his insistence that phenotypes only evolve in specific contexts, such as the context provided by an ecosystem. Here he utilizes the work of von Uexküll and his claim that different species perceive the world differently, to suggest that if in a predator/prey system the prey develops new traits that are not perceived by the predator, then nothing happens. However, he goes beyond Dawkins in recognizing that a more structured appreciation of the evolution of the phenotype, over the reductive view provided by the classical one-dimensional concept of fitness, does not provide one with the

Bergsonian insight required: namely, that the properties of any given component-system, or assemblage, cannot be defined 'independently from and prior to the forces that realize them' (Kampis 1991: 264–5). In other words, the notion of fitness tells us little about the creative dimension of evolution, while the stress on an ecological appreciation of the phenotype is insufficient if it explains the becoming of that phenotype in terms of the tautologous principle of the survival of the fittest (as Depew and Weber note, 1996: 327, if the fit are by definition those individuals that prove to be the most reproductively successful, then the idea is devoid of any empirical content since it will not be verified or falsified by referring to any facts that might have been otherwise).

The importance of population thinking for Deleuze and Guattari lies in the way that it moves biology away from a typological essentialism in which individuals in a species are not 'real' but merely variations or deviations from an ideal type. As Ernst Mayr has put it: 'For the typologist, the type (*eidos*) is real and variation an illusion, while for the populationist the type (average) is an abstraction and only the variation is real' (Mayr 1994: 158; for a comprehensive introduction to population thinking see Sober 1994: 161–89). There is no typical member of any species, since 'phenotypical variation within populations is the same in kind as phenotypical variation between species' (Depew and Weber 1996: 310). Species refer to ensembles of organisms that are held together by both genetic and ecological bonds, and which populate adaptive landscapes. These landscapes are biogeographic distributions of adapted populations of organisms (populations that inhabit the peripheral points of a species' biogeographic range tend to differ most radically from the other populations of the group). Moreover, it is not so much the case that within these landscapes species are not static units but stages in a process, but rather that species are the *result* of geographic and ecological processes and *not* the *stages* of them. To treat species in terms of stages is too teleological, and runs the risk of mixing the reality of species *as populations* with the categories of taxonomy (312).

In its attempt to steer beyond the antinomical poles of organismic holism and neo-Darwinian reductionism, through the emphasis on a rhizomatics of evolution, the work of Deleuze and Guattari comes close to the novel ideas espoused by Lynn Margulis, first in her pioneering study *The Origin of Eukaryotic Cells* (1970) (now published as *Symbiosis in Cell Evolution*) and then over the course of the last few decades in a series of original texts. Margulis has never questioned the fact that natural selection plays an important role in evolution in acting upon variation, but she has insisted that the source of variation – a source of evolutionary innovation – cannot be accounted for by standard neo-Darwinian thinking. Her radical thesis of symbiogenesis amounts to the claim that the most significant inherited variation comes from mergers and associations which span across different phyla. Such a complex, acentred 'genesis' involves the 'incorporation of microbial genetic systems into progenitors of animal or plant cells. The new genetic system that evolves out of this merger is different from the ancestral cell that lacked the microbe. She contends that symbiogenesis is the major evolutionary innovator in all lineages of larger nonbacterial organisms.

The thesis is a radical one because it is providing biology with something it has lacked since the time of Darwin, a serious theory of speciation, one which contends that new species do not simply arise from random mutation and other random DNA arrangements such as recombination and gene duplication. Perhaps most worthy of note about Margulis's work for the purposes of this exploration is that it results in a reconceptualization of population thinking. Deleuze and Guattari follow Margulis in not making 'individuals' the units of a population mapping, since these individuals are themselves highly heterogeneous bodies. It is worth quoting Margulis at length to show how radical and far-reaching her thesis is:

> these lessons from symbiosis research and from molecular biology directly contradict the assumptions of mathematical biology (and its stepchild sociobiology)....The 'individuals' handled as unities in the population equations are themselves symbiotic complexes involving uncounted numbers of live entities integrated in diverse ways in an unstudied fashion. In representations of standard evolutionary theory, branches on 'family trees' (phylogenies) are allowed only to bifurcate. Yet symbiosis analyses reveals that branches on evolutionary trees are bushy and must anastomose; indeed, every eukaryote, like every lichen, has more than a single type of ancestor. Such analyses also reveal rampant polyphyly (e.g. more than eight independent origins of parasitism in dicotyledonous plants...). The fact that 'individuals' – as the countable unities of population genetics – do not exist wreaks havoc with 'cladistics', a science in which common ancestors of composite beings are supposedly rigorously determined. Failure to acknowledge the composite nature of the organisms studied invalidates entire 'fields' of study.
> (Margulis in Margulis and Fester 1991: 10)

Deleuze's rhizomatic conception of evolution is highly relevant to contemporary developments within the field of biology where a number of practitioners are adopting an open systems approach. A recent study by Alan Rayner stresses the importance of an appreciation of the fluid character of life in which evolution is seen to take place within 'dynamic boundaries'. Rayner refers to an entrenched theological tradition, emanating from Aquinas, in which the ability of living things to proliferate is viewed as a necessary evil in an imperfect world where absolute boundaries cannot be maintained. In other words, there is a bias within the Western tradition of ontotheology in favour of self-sufficiency and closed boundaries (a conception of life that is not without its political articulations and implications). As Rayner points out, however, absolute boundaries are radically anti-evolutionary since they entail stasis (Rayner 1997: 67). An appreciation of dynamical living systems, therefore, requires a reconceptualization of classical notions such as adaptation, survival, growth, reproduction, the organism, and so on. Our conceptions of 'growth' and 'reproduction' are dependent on the definition we produce of the contextual boundaries. The focus needs to be on environmental heterogeneity and evolutionary indeterminacy, while an emphasis on the dynamical movement of boundaries must result in an appreciation of the limited character of standard systems of classification and the mapping of evolution in terms of phylogenetic trees. The evidence of symbiosis and horizontal gene transfer, to give two examples, shows the extent to which the representation of evolution of species and larger groupings as a tree is highly misleading. Rayner comes close to suggesting that tree models are far too molar in their mapping of evolution, showing only 'gross relationships between major groups of organisms' (88). What has been left out of the picture is the occurrence within living systems of 'fully nomadic or free-floating assemblages' (85).

This novel conception of evolution amounts to a significant critique of the Weismannian tradition within modern biophilosophy simply because, in its emphasis on the dynamical contexts of evolution, it challenges the depiction of evolution in terms of the functioning of fully determinate and discrete units (whether genes or organisms) in which production is always bound up with an alleged reproduction of self-contained genetic units along pre-formed evolutionary lineages. On the Weismannian model, evolution is reduced to a predetermined genetic programme proceeding via the elimination and extermination of the unfit. But such a notion of evolution is bound up with a particular conception of individuals as determinate and discrete, functioning as closed systems. As Rayner notes, such individuals are 'ineducable' with a built-in obsolescence since they are unable to alter their (genetic) information content in response to novel circumstances. These individuals are then depicted as acting as 'self-centred objects' or agents enmeshed in the Darwinian jungle of the survival of the fittest through a ruthless competition which requires the maintenance of closed boundaries and entropic systems. This conception results, not surprisingly, in a reified appreciation of notions like 'adaptive fitness' and operates simultaneously on the theoretical level and the political/ideological level:

> Societies are thought to be organized most efficiently around central administrations. Evolution is interpreted as a calculational process that uses individuals, like beads on an abacus or bytes in a computer programme, to come up automatically with the best solutions to survival problems. Genes are cast in the role of potentially immortal

'central controllers' and bodies as temporary contraptions assembled bit by bit (or byte by byte) according to a strict code of conduct.

(Rayner 1997: 70)

Deleuze and Guattari lay important stress on the insight that it is never a question in rhizomatics of establishing a dualist opposition between the molecular and the molar, but rather of providing an account of their co-evolution within any given assemblage (a social machine, an organized mass). Moreover, a 'return' to the 'body without organs' as the site of creative becomings is not a matter of regression, but 'a resurfacing of the virtual field of intensities and singularities which offer possibilities for breaking out of rigidified structures and organizations. Not regression: invention' (Massumi 1992: 85). The full body without organs which sustains the germinal life of intensity and individuation, as opposed to the imploding body without organs which becomes vitreous or cancerous, involves a different composition, caught up in Brownian-styled motion, it can be conceived in terms of a desert populated by molecular multiplicities (that is multiplicities which do not receive their organization from the order of a transcendent One). In short, the aim of rhizomatics is not to abstractly negate the organism but to open up creative dimensions of 'evolution' through an insight into its machinic character.

Autopoiesis and machinic heterogenesis

The innovation of this reworked Bergsonism can perhaps best be approached by understanding what is at play in the hybrid phrase 'machinic heterogenesis'. Although this phrase has been developed separately by Guattari it is consonant with the machinism that characterizes *ATP*. It is with this notion that it is also possible to gain important insight into precisely how this thinking of creative 'evolution' differs from that found in autopoiesis. Autopoiesis, which stresses the active role played by an organism in its evolution, has distinct philosophical antecedents, including, among others, Leibniz, Kant, Hegel, and Heidegger and visible in their approaches to the question of the organism. In its present-day manifestation of complexity theory it is seen by many to provide an important corrective to the reductionism and determinism of the neo-Darwinian paradigm which has dominated biology in the postwar period. As we have seen, what makes Deleuze's biophilosophy so complex is that it freely draws on resources from these differing traditions and so has aspects it shares with complexity theory and other aspects which owe a debt to the revolution of neo-Darwinism (notably population thinking). The rapport between the work of Deleuze and Guattari and autopoiesis needs to be construed as having both sympathetic and critical aspects. To appreciate this we need to be clear what is intended by their stress on the 'machinic' character of evolution.

Deleuze and Guattari contend that the existence of 'transversal communication' between different lineages serves to 'scramble genealogical trees' (Deleuze and Guattari 1980: 18; 1988: 11). Genealogies exist, therefore, but they have to be viewed within a wider cartography of evolution that involves different modes of becoming. A genealogical series does not constitute the ground of evolution but always issues from a machinic phylum. The machinic phylum gives rise to specific animal forms and species but at the same time, on account of the transversal movement of material forces, genealogical lineages are constantly punctuated by novel becomings. These are modes of becoming which cut across arborescent schemas that move from the least to the more differentiated and function not via filiations but rather through the creation of novel alliances. A line of becoming is not defined in terms of connectable points, or by the points which compose it, since it has only a 'middle'. This middle does not play the role of an average but rather serves as the means by which life enjoys 'the absolute speed of movement'. Thus, a 'becoming is neither one nor two, nor the relation of the two; it is the in-between, the border or line of flight' that runs perpendicular to both and which enables an assemblage to function in a way that 'sweeps away selective pressures' [*la pression sélective*] (360; 293–4). We have noted that this conception of a 'becoming' would not convince the orthodox Darwinian in its claim regarding selective pressures. What it does do, however, is to draw our attention to the complex character of selection, namely, that it cannot be treated in reified terms independent of the activity of organisms that are implicated in ethological assemblages.

This turn towards the machinic assemblage is important for the idea of a 'creative evolution' because it broadens the conception we have of living systems as open systems. This can be brought out by making the contrast with autopoiesis. The key innovation effected by autopoiesis is to grant to living systems a dynamical capacity for change through self-maintenance, which means, at least in part, that such systems must function as open systems (only relatively so, as we shall see). The functioning of the autopoietic organism or machine is not reducible to its particular genetic structure or composition. In other words, what are important are not the components of the system but the dynamic relations between them. Autopoietic entities engender and specify their own organization and limits/boundaries, functioning as unitary, individuated, and closed to relations of input and output. Such entities are understood as being 'organizationally closed', which does not mean that they do not interact with an environment but rather that such interaction is always informed and determined by the organization of the particular autopoietic entity. An autopoietic organism evolves by engaging in an endless turnover of components under conditions of continuous perturbations and compensation of these perturbations. Any interference with their operation *outside* their domain of compensations will result in disintegration (see Maturana and Varela 1980: 73ff.). We see here that the theory of autopoiesis equates change that does not conform to the internal and self-directed organization of the entity in question with destruction, dissolution, and abolition.

For a machinic thinking autopoiesis fails to appreciate the extent to which all living systems and their boundaries are caught up in machinic assemblages that involve modes of transversal becoming. Although autopoiesis grants a high degree of autonomy to a living system it ultimately posits systems that are entropically and informationally closed. In defining what constitutes the system as 'open' by placing the stress on operational closure, which can only conserve the boundaries of the organism, it blocks off access to an appreciation of the dynamical and processual character of machinic evolution and is led to present a stark choice between either entropy or maximum performance. In placing the emphasis on living systems as guided solely by concerns with survival and self-maintenance, even though these are to be understood as endogenously driven and monitored, the theory of autopoiesis too much resembles the theory it seeks to supersede, namely orthodox Darwinism with its focus on discrete units of selection.

Machinic evolution refers to the synthesis of heterogeneities and involves the formation of a 'consistency'. A machinic assemblage connects and convolutes the disparate in terms of potential fields and virtual elements, and crosses techno-ontological thresholds without fidelity to relations of genus or species. With the notion of a machinic assemblage it is important to appreciate that the emphasis is not simply on the bonds that are created between the different components but rather on the virtual dynamics which constitute it. The departure from autopoiesis lies in the emphasis placed on its machinic character (Guattari 1992: 56; 1995: 35). If there is autopoiesis then it has to be conceived as operating on the machinic level. This is to think machinic autopoiesis *as* machinic heterogenesis and to introduce into the autopoietic model the necessary disequilibrium and far-from equilibrium conditions required for a truly creative model of evolution, in which evolution does not simply involve self-reproduction through the dissipation of outside forces and nullification of dimensions of alterity. The relation between machinic heterogenesis and autopoiesis should become clearer once we have comprehended the moves undertaken in Deleuze and Guattari's conception of an ethology of assemblages.

From an ethology of behaviour to an ethology of assemblages

It is interesting to note that one of the biologists now being rediscovered by those seeking a more embodied, contextualist, and dynamical approach to intelligence is von Uexküll (1864–1944), precisely the figure whose reworking of evolution in terms of ethology Deleuze was attracted to (see also Heidegger 1995 and Merleau-Ponty 1994). In new work in the philosophy of mind the emphasis is on distributed intelligence with different cognitions being explained in terms of different material consistencies rather than in terms of an appeal to representationalist conceptions of agency. 'The mind is a leaky organ', one notable contributor to the philosophy of mind has recently written, constantly escaping its alleged natural confines and mingling shamelessly with body and world (Clark 1997: 53). These developments call for a major reconfiguration of ethology since behaviour can no longer be localized in individuals conceived as preformed homunculi, but has to be treated epigenetically as a function of complex material systems which cut across individuals (assemblages) and which traverse phyletic lineages and organismic boundaries (rhizomes). This requires the articulation of a *distributed* conception of agency. The challenge is to show that 'nature' consists of a field of multiplicities, assemblages of heterogeneous components (human, animal, viral, molecular, etc.), in which 'creative evolution' can be shown to involve blocks of becoming.

Here 'ethology' loses its classical focus on 'behaviour' and becomes concerned with the directional movement of assemblages.[19] Ethology can be utilized as a 'privileged *molar* dimension' (my emphasis) on account of the fact that it shows how the most diverse components, from the biochemical to the social, 'crystallize in assemblages [*agencements*][20] that respect neither the distinction between orders nor the hierarchy of forms' (Deleuze and Guattari 1980: 415; 1988: 336; compare 630/504). Acentred and complex adaptive systems are held together by 'transversals', which themselves are special kinds of components that play the role of specialized vectors of deterritorialization. The contention is that it is not linear causal relations that hold an assemblage together but its most deterritorialized component, which is able to ensure that the population of intensities and forces which inform it remains a multiplicity (the example they give is that of the refrain) (414; 336):

> One launches forth, hazards an improvisation. But to improvise is to join with the World, or meld with it. One ventures from home on the thread of a tune. Along sonorous, gestural, motor lines that mark the customary path of a child and graft themselves onto or begin to bud 'lines of drift' with different loops, knots, speeds, movements, gestures, sonorities.
>
> (383; 312)

Before proceeding with an inquiry into the nature of this move to an ethology of assemblages it is first necessary to understand the work being done by the notion of deterritorialization.

As Eugene Holland has shown, the terms deterritorialization and reterritorialization, derived initially from Lacanian psychoanalysis, serve to rearticulate Freud's libido and Marx's notion of labour-power, referring, in the first mode, to the freeing up of the libido from its investment in pre-established objects (the mother's breast, the Oedipal triangle) and, in the second, to the freeing of labour-power from particular modes of production (such as the seigneurial plot of land, the assembly line, etc.) (Holland 1991: 57–8). The libidinal is expanded to include an array of human and nonhuman forms of invested energy, including physical, perceptual, cognitive, and productive articulations of energy. In key respects Deleuze and Guattari's reconfiguration of libidinal flows or desire comes close to Nietzsche's articulation of the will to power as the major principle of a 'historical method' which seeks to show how nonlinear change is possible due to all things, from physiological organs to social customs and art forms, enjoying a functional indeterminacy (form is fluid, 'meaning' even more so) (see Nietzsche 1994: Essay 2, Section 12). Deleuze and Guattari are not positing, it should be noted, an abstract opposition between 'territory' and 'deterritorialization', simply because they maintain that any given territory or enclosure of a thing enjoys vectors of deterritorialization and is, in fact, constituted by them *as a territory* (informing the becoming of what it is).

It is the failure to appreciate this point – namely, the fact that Deleuze and Guattari take the question of territory seriously, say in the case of the animal and the function of a refrain – that might explain why the critique Baudrillard has advanced against Deleuze is difficult to sustain (Baudrillard 1994: 140–1). Nothing is more errant and more nomadic in appearance than animals, Baudrillard notes, and yet their law is that of the territory (139). He then makes the challenging criticism that with the privileging of desire, of animality, and of the rhizome Deleuze is simply projecting the schema of deterritorialization that constitutes the economic system of capital as an ideal nomadism. But in making this criticism Baudrillard too readily equates the deterritorialization of Deleuze's ethological rhizomatics with the deterritorialization of capital. He is keen to reclaim the territory for the animal, and resist the rhizomatic move, because he believes that to depict animal behaviour as naturally nomadic is to become complicit with capital's disregard for animals and their territories. He writes, powerfully:

Nature, liberty, desire, etc., do not even express a dream the oppo-site of capital, they directly translate the progress or the ravages of this culture, they even anticipate it, because they dream of total de-territorialization where the system never opposes anything but what is relative: the demand of 'liberty' is never anything but going fur-ther than the system, but in the same direction.

(Baudrillard 1994: 141)

Deterritorialization, however, is doing rather different work in Deleuze and Guattari's rhizomatics. In *A Thousand Plateaus* deterritorialization is said to become 'absolute' when it enhances 'lines of flight' to the power of a germinal or vital line of life and draws a plane of consistency (the nondialectical synthesis of heterogeneous and disparate elements). Elsewhere it is said to name a becoming that reaches a continuum of intensities that are valuable in themselves: that is, a world of intensities where 'all forms come undone' (Deleuze and Guattari 1975: 24; 1986: 13). In other words, the 'absolute' denotes not a quantity (something that exceeds all relative quantities), but a certain mode of movement that is held to be neither transcendent nor undifferentiated. In addition, its absolute character is said not to be dependent on the fastness of a speed. Deleuze and Guattari argue that it can be reached through relative slowness and delay, and they provide neoteny or retarded development as an example. With the notion of deterritorialization now clarified let us return to Deleuze and Guattari's highly innovative reworking of ethology.

The tradition of ethology, as it developed in the twentieth century in the work of its major figures such as Lorenz and Tinbergen, has focused attention on the causes of 'instinct' (or 'innate behaviour') over 'learning processes' (see Tinbergen 1969: 2–3). The notion of 'reflex' serves to capture this emphasis on the innate capacities of organisms, denoting a simple and stereotyped response that centres on the central nervous system and involving only a part of the organism, being a response to localized sensory stimuli (see Thorpe 1979: 89). For Lorenz, who sought to explain instinctive behaviour not with the notion of an internal drive but with the concept of a 'reaction-specific energy', the instincts – or what he also called 'movement forms' – are modes of behaviour that are just as constant and fixed as anatomical structures (see McFarland 1996: 390). The selective responses to external stimuli are explained in terms of an in-built mechanism which can account for the recognition of sign stimuli called the 'innate releasing mechanism' or IRM (see Tinbergen 1969: 42ff., who offers IRM as a 'free translation' of the German term derived from Uexküll and Lorenz of 'das angeborene auslösende Schema). The response released by the IRM is stereotypical, representing a part of the animal's innate repertoire of fixed-action patterns. For Lorenz and Tinbergen the movements of certain motor patterns, which are seen as functioning in terms of a rigid innateness, possess significant taxonomic value for the classification of a species, a genus, and even a whole phylum. For them instinctive actions are distinguishable in their form from all receptor processes; they are not only distinguishable from 'experience' in the broadest sense of the word, but also from the stimuli that affect the organism during their operation.

More recent advances in ethology have shown that the fixity of behaviour patterns result not simply from the rigidity and precision with which an animal performs an action, but derives more from the control of the intensity of the action, involving its timing and its speed (Thorpe 1979: 127; see also Merleau-Ponty 1988: 163–4). Once attention shifts to the mutual interaction of animal and environment, the question as to whether behaviour is predominantly innate or learned becomes largely otiose, leading only to erroneous conclusions and fruitless arguments. For contemporary ethologists the more interesting question concerns whether a given item of behaviour is environmentally stable or environmentally labile. Moreover, contemporary ethologists have developed a much richer conception of animal adaptation, arguing, for example, that processes involving selectivity in responsiveness need not be purely innate but are also influenced by 'learning' (see McFarland 1996: 392). Current research in ethology seeks to demonstrate the variability and plasticity of reflexes, with one showing that so-called 'fixed action patterns' vary in duration and in the sequence and number of components from one occasion to the next. Moreover, they are regarded as subject to modification during performance on the basis of feedback from their targets (see the work of G. Barlow discussed in Taylor Parker and Baars 1990: 76ff.).

Every territory encompasses or cuts across the territories of other species.[21] This gives expression to what Deleuze and Guattari call, following the work of Uexküll, 'a melodic, polyphonic and contrapuntal conception of nature' (Deleuze and Guattari 1991: 175; 1994: 185). Examples of the musical character of complex evolution include birdsong, the spider's web, the shell of the mollusc which upon its death becomes the habitat of the hermit crab,[22] and the tick (on the tick compare Heidegger 1995: 263–4 and Deleuze 1981: 167; 1988: 124–5). This is to replace a teleological conception of nature with a melodic one in which the distinction between art and nature (natural technique) is revealed as an arbitrary one. It is the relationships of 'counterpoint', such as that of the shell of the dead mollusc and the hermit crab, which joins planes together and forms compounds of sensations and blocs, which then can be seen to be the principal influence on 'becomings' (Deleuze and Guattari 1991: 175; 1994: 185). The 'territory' of the animal implies not simply the transformation of organic functions, such as sexuality, procreation, feeding, and so on, but equally the emergence of 'sensibilia' which 'cease to be merely functional and become expressive features, making possible a transformation of functions' (174; 183). It is this emergence of 'pure sensory qualities' that Deleuze and Guattari label 'art'. The animal 'becoming' as art means nothing less than a *creative* mode of evolution is at play in which there is an 'outpouring of features, colours, and sounds' (174; 184).

In contrast to prevailing histories of ethology, therefore, which construe Uexküll's work as belonging to the classical tradition and with its stress on instinct, Deleuze and Guattari identify a more 'musical' conception of ethology in his work. The most important suggestion that they extrapolate from his work, and which they then deploy to novel and far-reaching effect, taking it well beyond the domain Uexküll was analysing, is the idea that becomings-animal involve not only the selection of adaptive traits but also the play of physico-chemical intensities and zones of proximity that cut across phyletic lineages, so involving a musical 'becoming' of life. Deleuze and Guattari adopt his notion of 'double articulation' but extend it across spectrum of life from the energetic and physico-chemical to the geological and technical strata. The problem in all cases of articulation is that of 'the organism', that is, 'how to "make" the body an organism' (Deleuze and Guattari 1980: 55–6; 1988: 41). However, Deleuze and Guattari make a key move which takes this reconfiguration of ethology away from a focus on 'behaviour' to a concentration on 'assemblages'. It is an error, they insist, to get locked into fruitless debates about nature and nurture or instinct and learning. Rather, the principal problematic in need of attention is that of 'consistency', that is, of how the various components of a territorial assemblage, as well as those of different assemblages, hold together. In considering an answer to this question they do not accept the response of twentieth-century ethologists such as Tinbergen, simply because they regard it as still caught up, in spite of the shift from bios to ethos, in a linear, hierarchized, centralized, arborescent schema (403ff.; 327ff.). On the model of ethology developed by Tinbergen, for example, the focus is on the central nervous system in which a functional centre is seduced automatically into operation, releasing an appetitive behaviour in search of specific stimuli. Other centres then act to release new appetitive behaviours, finally leading to the activation of other centres of behaviour such as fighting, nesting, and courtship. The problem with this model is that it is constructed by a series of binaries, such as inhibition-release and innate-acquired, and unable to approach the question of consistency on the level of a rhizomatic assemblage. In spite of ethology's advance over ethnology, therefore – the 'advance' consists of not dividing an indivisible terrain into forms of kingship, politics, economics, etc. – it runs the risk of reintroducing 'souls' and 'centres' of activity simply by orienting behaviour along the axis of spurious binaries.

For Deleuze and Guattari behaviour cannot be modelled in terms of linear or hierarchical relations between centres of activation; rather, it becomes necessary to think of 'packets of relations' that are 'steered by molecules', in which the coordination of behaviour may be *either* positive *or* negative but never in accordance with a direct, linear relation (1980: 403ff.; 1988: 327ff.). On this model of a 'whole behavioural-biological "machinics" [*machinique biologique-comportementale*]' the problem of consistency is treated as a problem of molecular engineering and not one of molar formation (the latter being the problem of how to build an organism) (405; 328). This is to situate the problem of the 'innate-acquired' in the more dynamic context of a rhizome in which the natal gets decoded and the acquired is subject to territorialization. This, however, is to think beyond 'behaviour' and move in the direction of the 'assemblage', in which in any system local operations are coordinated and a global result is actualized ('synchronized') 'without a central agency' (26; 17).

A useful way to think of this is in terms of an 'autocatalytic set'.[23] In chemistry a catalyst denotes a molecule which serves to speed up or slow down another chemical reaction, inhibiting or promoting a reaction. It was the Swiss chemist Baron Jakob Berzelius who in 1836 gave the name 'catalysis' to the type of chemical reaction that takes place in the presence of certain bodies, which he compared to heat (see Jacob 1974: 98). Its innovation was to point a way out of a difficulty that the discoveries of Lavoisier and Laplace had created concerning whether the dynamics of oxidation and of chemical reactions in general were the same in living and inanimate matter. Berzelius showed that catalytic force can be used to explain reactions between chemicals where there is no affinity but only a presence or contact (for further insight see Osborn 1918: 286–7). Autocatalysis refers to the feedback process when the product of a reaction catalyzes its own production, in which the process of production and combination involves nonlinearities: that is, it takes place in more complicated ways than simple addition (Coveney and Highfield 1995: 157). The notion of an autocatalytic set plays a crucial role in Stuart Kauffman's speculations on the origins of life, where it is utilized in relation to catalytic polymers and the capacities of such sets to evolve independently of a genome (Kauffman 1993: 285ff.). For Kauffman the notion of an autocatalytic set offers the most plausible model for understanding the very 'origins' of life.[24]

On a systems model, therefore, what matters most are not the components of a system but the dynamic interactions between them. As Jantsch notes:

> An organism is not defined by the sum of the properties of its cells. In chemical reaction systems, certain molecules which do not participate in the reaction may act as catalysts and thereby influence the overall dynamic system in a decisive way.
>
> (Jantsch 1980: 24; see also 103ff.)

In biology specific proteins, the enzymes, play a catalytic role, serving to encourage a system to follow a new reaction path. The key point is articulated by Kauffman: 'in a collectively autocatalytic set, there is no central directing agency. There is no separate genome, no DNA' (Kauffman 1995: 275). Rather, there is a 'molecular economy' made up of ecologies of competition, mutualism, symbiosis, and host–parasite relations. Similarly, on Deleuze and Guattari's model there is no centre which is formally controlling behaviour; that is, there is not a top-down subject or agent of evolution (whether a nervous centre or an homunculus), but only a distributed intelligence that encompasses brains, bodies, molecules and worlds, all of which exist in terms of overlapping territories. It is on this point that they can be seen to depart from the figuration of autocatalysis in the theory of autopoiesis or self-organization.[25] Autocatalysis is not to be located simply or solely in autopoietic terms within the bounds of an organism, but rather situated rhizomatically in terms of an assemblage understood as a multiplicity. The 'refrain' [*la ritournelle*] itself, for example, can best be conceived as a crystal of 'space-time' which both acts upon the sound and light that surrounds it, extracting vibrations, decompositions, and transformations, *and* fulfils the role of a catalyst, serving not simply to regulate the speed of the exchanges and reactions in its surroundings, but, more importantly, assuring that indirect interactions between elements that are devoid of 'so-called natural affinity' take place: 'The refrain is therefore of the crystal or protein type' (Deleuze and Guattari 1980: 430; 1988: 348). As Guattari notes, refrains or ritornellos function in rhythmic and plastic forms, prosodic segments, signatures and proper names, as well as transversally between different substances (Guattari 1996: 162).

Deleuze and Guattari show the extent to which the invention of the human machine is the product of an assemblage made up of a number of different components and events, such as retarded development (neoteny), a synergistic deterritorialization of organs (such as the hand and the foot),[26] and a correlative deterritorialization of the milieu (notably deforestation) (all such deterritorializations then give rise to compensatory reterritorializations). The latter proves particularly crucial since the transition from the forest to the steppes exerts a tremendous selective pressure of deterritorialization on the body and technology.[27] The properties of the human being, such as technology, language, free hand and supple larynx, etc., are products of a new *distribution* (Deleuze and Guattari 1980: 79; 1988: 60). It is only zoocentrism which leads us to maintaining that it is the emergence of human beings which marks the absolute 'point' of origin of this new distribution. On the model of the rhizome 'evolution' can no longer be characterized by 'points' of origin but is to be approached in terms of the distributed technics of an assemblage.

Recent work in cognitive science and mainstream AI has begun to move in the direction of an ethology of assemblages in which behaviour only makes sense when viewed as a component in an assemblage. As Andy Clark puts it, the 'Rational Deliberator' needs to give way to the 'Adaptive Responder', with brain, body, world, and artifact locked together in the highly complex network of evolutionary, intelligent activity (Clark 1997: 33). Taking his inspiration from Uexküll and Merleau-Ponty, Clark notes that behaviour, brain activity, and the organism's being in the world, take place via what he calls 'the complex cooperative dance of a variety of internal and external sources of variance' (1997: 175). His main point is to insist that the philosophy of mind must give up its disembodied, atemporal, and intellectualist vision of the mind, and, along with it, the image of the mind as a controller of embodied action. An individualist, isolationist, and intellectualist approach has served to hinder and damage the development of innovative cross-disciplinary research in both the biological and cognitive sciences. He notes, for example, that in the field of biology the majority of biologists have selected as their focus of study the individual organism as the locus of adaptive structure and behaviour.

The task, says Clark, is to understand, through laboratory work in robotics for example, how adaptive behaviours are assembled in ways that both respond to local contexts and reveal intrinsic dynamics. This means that the brain is as much in the world as it is in the head. Human reasoners, he argues, are '*distributed* cognitive engines' which draw on external resources and aids (technics) to perform specific computational tasks (1997: 68–9). For Clark it is erroneous, therefore, to allow individual brains to take all the credit for the flow of thoughts and adaptive responses since there is always a rich collaboration between brain and world. As he writes in a crucial passage, 'although specific thoughts remain tied to individual brains, the flow of reason and the informational transformations it involves seem to criss-cross brain and world....This flow counts for more than do the snapshots provided by single thoughts or experiences'. In short, the engine of reason is 'bounded neither by skin nor skull' (69). Clark rightly notes that this new model of living systems raises new questions and problems for biology, most notably concerning how we are to understand adaptive success once it is appreciated that that success accrues not to individual brains but to multiple 'brain–body coalitions in ecologically realistic environments' (98). As we have seen, while Deleuze and Guattari clearly acknowledge the shaping influence of the pressures exerted by selection, they wish at the same time to emphasize the importance of novel becomings which considerably alter the character of these pressures. Selective pressures must be seen as working within specific ecological contexts and rich ethological dimensions, while 'Selection' cannot be treated as some quasi-theological agent of biology that functions automatically and independently of the brains, bodies, technics, and physico-chemical intensities which make up the world (see Deleuze and Guattari 1980: 400; 1988: 325, on the role played by 'interassemblages' in modifying selective pressures).

Clark makes an innovative move when he construes the rapport between mind and world in terms of the mediation provided by technics. The fact that only now are researchers like Clark taking this dimension of evolution-ary activity seriously shows the extent to which cognitive science has been retarded by emerging from out of the background of a methodological individualism and intellectualism. In the case of the human mind, Clark argues that the external structures provided by social institutions and technical aids serve to both complement individual cognitive profiles and diffuse human reasoning across wider and wider social and material networks, and whose collective computations exhibit their own special dynamic properties. As he pithily and incisively puts it: 'We use intelligence to structure our environment so that we can succeed with *less* intelligence' (1997: 180). However, we would stop short if we thought it was simply a matter of organisms and life-systems deploying technics to mediate their environment, since such technics also serve to both modify the character of an environment and transform the being who manipulates them.

Becomings – animal of the human

After having noted that 'not all Life is confined to the organic strata', and that the more intense and powerful life remains anorganic, Deleuze and Guattari then contend: 'There are also nonhuman Becomings of human beings that overspill the anthropomorphic strata in all directions' (1980: 628; 1988: 503). It is important to note that these molecular and so-called 'nonhuman' becomings remain articulated as becomings peculiar and specific to the 'human'. What exactly are these becomings? Do they rest on an untenable extension of an ethological rhizomatics?

The ethological approach seeks to define a body not in terms of organs and functions, and as characteristics of species and genus, but rather in terms of 'affects' (which are not mere feelings or affections, but harmonies of tone, colour, etc.). Affects do not bring about the transformation of one body into another, but rather something passes from one to the other. Affects communicate on the level of becomings, which involve 'neither an imitation nor an experienced sympathy....It is not a resemblance, although there is resemblance. But it is only a produced resemblance' (Deleuze and Guattari 1991: 164; 1994: 173). This is sensation conceived as a zone of indiscernibility or indetermination in which 'things', such as beasts and people, reach the point that precedes any 'natural differentiation', achieving a dissolution of forms. Deleuze and Guattari insist that mimicry is a bad concept in biology since it does not explain what is genuinely novel or creative in becomings. Mimicry is dependent on a binary logic to describe phenomena that involve neither simple imitation nor mere resemblance. Uexküll is important to the articulation of an ethology because his account of animal worlds can be used to disclose the active and passive affects 'of which the animal is capable in the individuated assemblage of which it is a part' (Deleuze and Guattari 1980: 314; 1988: 257; this is from the section of the plateau entitled 'Becoming-Intense, Becoming-Animal, Becoming-Imperceptible...' where Deleuze and Guattari explicitly relate Uexküll to the older tradition of ethological ethics, including Spinoza). For example, the three affects of the tick only assume generic characteristics from the standpoint of physiology, where the focus is on organs and their functions, but not from that of ethology, where what is important is the movement of longitude (relations) and latitude (degrees), and the rhythm of speeds and slownesses (see Spinoza 1955: 96; Spinoza's attention to motion and rest is really a way of thinking about systems of energy).[28]

Deleuze and Guattari construe the exchange of affects on the level of human becomings as not involving some dubious and highly problematic return to origins, 'as if beneath civilization we would rediscover, in terms of resemblance, the persistence of a bestial or primitive humanity'. Rather, they claim that it is 'within our civilization's temperate surroundings that equatorial or glacial zones, which avoid the differentiation of genus, sex, orders, and kingdoms, currently function and prosper...what is animal, vegetable, mineral, or human in us is now indistinct' (Deleuze and Guattari 1991: 164; 1994: 174). An ethological ethics of the human is, in large part, directed against psychoanalysis and its reduction of human desire to an Oedipal model. Becomings-animal are not to be conceived on the order of representation but are solely affective, so that, for example, a horse for Little Hans, the 1909 case study carried out by Freud, does not represent a member of a species but is an element within a machinic assemblage, such as 'draft horse-omnibus-street' (see Freud 1990b: 169ff.). In Freud's narrative the boy Hans is depicted as a 'little Oedipus' and his case includes not only the phobia of horses but also, unsurprisingly, giraffe phantasies – long necks = big 'widdlers'. (See especially 208–9 for Freud's diagrams of the horse–cart assemblages and treatment of the different kinds of horse at play in Hans's 'phantasies'.) Deleuze and Guattari insist that on the level of a becoming it is not at all a question of merely subjective reveries. Psychoanalysis is rebuked for its inability to 'see the reality of becoming-animal, that it is affect in itself' (1980: 317; 1988: 259). It is unable to locate any lines of flight because its focus is on animal-becomings as representatives of drives and of Oedipal influences. On an ethological model the composition of speeds and affects serves to bring into play new and different modes of individuation. This mode of becoming is described in provocative terms as a style of *'unnatural participation'*. It is in the betrayal of nature, in the invention of nuptials against nature, that genuinely 'creative' modes of evolution can be articulated (Deleuze 1977: 56; 1987: 44). A becoming-molecular involves a betrayal of the so-called natural 'molar' order of species and genus:

> You become animal only molecularly. You do not become a barking molar dog, but by barking, if it is done with enough feeling, with enough necessity and composition, you emit a molecular dog. Man does not become wolf, or vampire, as if he changed molar species; the vampire and werewolf are becomings of man, in other words, proximities between molecules in composition, relations of movement and rest, speed and slowness between emitted particles.
>
> (Deleuze 1977: 337; 1987: 275)

Representation is attacked by Deleuze and Guattari for concealing a culturalism and a moralism that upholds the unique irreducibility of the human form and order. They construe becomings-animal in terms of 'zones of proximity' denoting the co-presence of particles operating within a zone that is both topological and quantal. Thus, they argue, there 'is a reality of becoming-animal, even though one does not in reality become animal' (1980:

335; 1988: 273). To become-animal in any literal sense would be to fall into the trap of representation, limiting the body to the order of organs and functions. In participating in a becoming-animal it is not a question of constructing archetypes but rather of building new levels of existence, such as 'zones of liberated intensities' where forms come undone and are freed from the signifiers that formalize their content and expression. In involution the kingdom of animals is transformed into a body without organs defined by zones of intensity and proximity (vibrations and movements). Here the animal and the child cease to become 'subjects' and are better understood as 'events' that are caught up in complex assemblages which come into being without fidelity to notions of species or genera (320; 262). The becoming of nature in animal-worlds and human-worlds is inseparable from milieus, temperatures, climates, and times (wind, season, and hour), and these phenomena, too, constitute an essential part of the assemblage. Deleuze and Guattari construe relations of space and time not as the predicates of the thing but rather as the dimensions of assemblages or multiplicities. The animal, for example, has to invent, by merging with its environment or by entering into symbiotic complexes, the 'becoming of the evening', say, for example to stalk at five o'clock in the evening.

Deleuze and Guattari acknowledge the philosophical writings of Duns Scotus (*c.* 1266–1308) as the source for the word 'haecceity', where it names 'individuating difference' (Duns Scotus 1987: 4, 166). The word can also be found deployed in Leibniz's 'Discourse on Metaphysics' of 1686, where it names individual substances that can only be known in their *complete* virtual form by God (Leibniz 1973: 19).[29] Its deployment by Deleuze and Guattari is highly novel since it has no reference to either subject or substance; on the contrary, it endeavours to deprive both of these notions of their efficacy in order to grant primacy to a mode of individuation that is not that of a definite person, determined subject, or a formal substance. This mode is the *haecceitas* which concerns longitudinal relations of movement and rest between molecules/particles and latitudinal capacities of affect and affectedness. These absolute individuals or haecceities refer to degrees and intensities that combine with other degrees and intensities, and which, although irreducible to a subject, nevertheless compose an individual (only the 'shimmeriness' of the protoplasm is real, 'the shape is a dead crust...."A sort of disseminated consciousness, that's all there is of me. I feel as if my body were living empty, as I were in the other things – clouds and water" ', Lawrence 1994: 183, 232). For example, a degree of heat is an individuated warmth that is distinct from the so-called 'subject' that constitutes it. Similarly, times and lengths of day are not actually extensions but rather degrees or modulations. Subjects exist only as populations within a cartography of these longitudes of speed and latitudes of affect. On this model spatiotemporal relations need to be treated not as predicates of a thing but rather as dimensions of multiplicities. The plane of nature can be said to enjoy a unity precisely on account of the fact that it can be mapped out in terms of individuated multiplicities. This plane is populated by 'subjectless individuations' that constitute collective assemblages and which 'evolve' through speeds and powers/affects. 'Haecceity' is, therefore, the domain of the 'proper name' which speaks not of a subject but rather the agent of an infinitive. No specific, particular individual is designated by a proper name; on the contrary, the name refers to the operation of a 'depersonalization' (a de-oedipalization), in which an individual opens itself up to the multiplicities pervading and invading it (Deleuze and Guattari 1980: 51; 1988: 37).

Deleuze and Guattari insist that it is never a question of positing a simple opposition between molecular becomings and molar stratification, but of seeking to demonstrate that every molar formation has a molecular unconscious (a multiplicity or a population) which both marks its tendency to decompose and haunts its operation and organization. Moreover, becomings (molecular, animal, inhuman) always involve a 'molar extension, a human hyperconcentration, or prepares the way for them' (Deleuze and Guattari 1980: 48; 1988: 34). The critique of psychoanalysis emerges out of this insight. Psychoanalysis is to be critiqued for reducing becomings to the *one* complex, the complex of molar determination (Oedipus, castration). The point is not to deny that an assemblage of desire that involves a becoming-nonhuman of the human is devoid of an Oedipal apparatus; on the contrary, Deleuze and Guattari argue that there are *many* Oedipal statements entailed in such a becoming (50; 36). We are involved in a social formation, and we are caught up in Oedipal relations. Freud's psychoanalysis is critiqued not only because it reduces becomings solely to articulations of Oedipal desire, but rather for employing this enunciation in order to delude patients with the belief that through therapy they will be able to speak finally in their own name as unitary organisms: 'No sooner does Freud discover the greatest art of the unconscious, this art of molecular multiplicities, than we find him tirelessly at work bringing back molar unities...*the* father, *the* penis, *the* vagina' (39–40; 27). We have only to think of the case of the Wolf Man, who has to learn to accept that there are not several wolves but only one, and that this is his father; or take the case of the little boy Arpad with his 'poultry perversion' playing the game of the child as the small chicken and daddy as the big cock. The events in the poultry-yard do not simply gratify the young lad's sexual curiosity, but teach him a lesson in '*human* family-life' (Freud 1990: 191). Then there is the case of the dog phobia examined by who else but 'Dr Wulff', in which fear of the father is displaced, as are all animal identifications for Freud, on to dogs: ' "Doggie, I'll be good!" – that is, "I won't masturbate" ' (188). Indeed, it would be interesting to examine the extent to which Freud's entire analysis of totemism in 'Totem and Taboo' reduces the rhizomatic animal becomings contained within, and constrained by, totemic practices to an Oedipal filiative model. [30]

A productive reading of Freud would be one that claimed him for population thinking. 'Freud was Darwinian, neo-Darwinian', Deleuze and Guattari write, 'when he said that in the unconscious everything was a problem of populations' (1972: 333; 1984: 280). It is, however, a matter of two types of populations, of large aggregates and micromultiplicities. Both are invested in a collective field of social reality, for 'even a lone particle has an associated wave as a flow that defines the coexisting space of *its* presences'. 'Transverse multiplicities' serve to convey desire as a molecular phenomenon, 'as partial objects and flows, as opposed to aggregates and persons' (ibid.). In the unconscious there exist only populations and micromachines that function on a plane of inextricably linked forces.

A principal task of schizoanalysis or rhizomatics for Deleuze and Guattari is to expose the transcendental illusion of mass phenomena and molar aggregates by giving primacy to the molecular multiplicities which do not function in accordance with the superior laws of race or blood, genus or species. This is the 'domain of nondifference between the microphysical and the biological, there being as many living beings in the machine as there are machines in the living' (1972: 340; 1984: 286). At this molecular level of the formation and deformation of machines even the laws of thermodynamics will not be obeyed simply because 'the chain of assembly begins in a domain where by definition there are as yet no statistical laws' (ibid.) Deleuze and Guattari here are making a crucially important citation from Raymond Ruyer's *La genèse des formes vivantes* of 1958. Ruyer's point is that, no matter how large or complicated, an organism remains 'microscopic'. The only genesis of a machine worth talking about scientifically is that of a *schizogenesis*, that is, to focus on machines that evolve by way of breaks and flows, molecular becomings which induce action at a distance – transversal communications, inclusive disjunctions, and polyvocal conjunctions – and which produce and reproduce only 'selections, detachments, and remainders' (341; 287). No matter how structured and stratified a machine becomes, say through the imposition of a single object on a molecular mass, or the production of an organism as a single subject, these molar manifestations and statistical determinations can nevertheless still be treated as the 'same' machines as the desiring-machines: that is, as organic, technological, and social machines treated in their mass phenomenon. These are the same machines under determinate conditions (one being the selective pressures exerted by natural selection): 'There is only desire and environments, fields, forms, of herd instinct'. However, it is necessary to appreciate that the real engineering of desire only takes place at the submicroscopic level of the desiring-machines. This is simply because all 'molar functionalism' is false 'since the organic or social machines are not formed in the same way they function, and the technological machines are not assembled in the same way they are used' (342; 288). This means that large formations or aggregates function in a molar fashion but owe their production to a molecular assembling. They are false, or illusions, therefore, because they separate their own process of production from their distinct product. These molar formations come to signify and represent things and, in the process, divorce what is produced, such as reified persons and things, from their molecular conditions of production. It is for these reasons that Deleuze and Guattari enlist the support of Jacques Monod's mapping of the field of biology in terms of a microscopic cybernetics in which allosteric proteins,[31] for example, are seen as specialized products of a kind of molecular engineering: they enable interactions, positive or negative, between compounds without chemical affinity, so subordinating any reaction that comes about to the intervention of compounds which are chemically foreign and indifferent to this reaction (see Monod 1971: 62ff.). It is, therefore, the very 'gratuitousness' of such cellular systems which gives molecular evolution an almost 'limitless field for exploration and experiment' (Deleuze and Guattari 1972: 343; 1984: 288–9).

This is not to deny that for them the formation of molarities is without significance and real effect. Molar formations effect a unification and totalization of the molecular forces, whether this unity be a biological species or a socius. It is only in relation to this new order of unity that the partial objects of a molecular disorder now appear as a lack: 'In this way all desire will be fused to lack' (Deleuze and Guattari 1972: 409; 1984: 342). In psychoanalysis, for example, one is no longer dealing with the positive dispersion in a molecular multiplicity but solely with 'large vacuoles' that are the subjects of a global determination, such as those of neurosis and castration types. It is this welding of desire to lack which gives desire ends, goals, and intentions, whether defined and articulated personally or collectively. With this conquest of the molecular by the molar, desire is no longer being conceived in the 'real order of production', where it behaves as a molecular phenomenon devoid of goal or intention. Such statistical accumulation by molar aggregates is not, however, to be modelled along the lines of chance, as the result of a purely random result. Rather, it is the 'fruit of a selection' which exerts its nonrandom pressures on the elements of chance. On account of his insight into the entropic and molar character of natural selection Deleuze and Guattari rightly credit Nietzsche with inaugurating a 'fundamental intuition' that proves decisive for modern thought. Nietzsche suggests that large aggregates and numbers 'do not exist prior to a selective pressure that might elicit singular lines from them'; rather, such aggregates are the product of this selective pressure which crushes and regularizes the singularities of microevolution (410; 342–3; for Nietzsche 'contra Darwin', see Nietzsche 1968: Sections 684–5). Statistics, therefore, is never functional but always structural, serving to extract from molecular breaks and flows regularity of function and purpose of potentially chaotic behaviour. It concerns chains of phenomena which selection has already placed in a condition of partial dependence (such as the Markov chains) (see also Deleuze 1986: 125; 1988: 117). For further insight into these chains see Prigogine and Stengers (1985: 236–8, 273–6). In a genetic code, for example, the forms of gregariousness are never indifferent but 'refer back to the qualified forms that produce them by creative selection' (Deleuze and Guattari 1972: 410; 1984: 343). Molar and gregarious aggregates must never be viewed as the 'ground' of, but always as the *result* of, a process of selection.

On the advantages and disadvantages of Deleuze's reading of Uexküll

Let us now attend to the problems of this configuration of nonhuman becomings of the human. It may be helpful to recapitulate the essential claims that are being made. In the essay of 1981 entitled 'Spinoza and Us' Deleuze speaks of an ethological ethics which is to be mapped out in terms of the plane of immanence that distributes affects and which transcends distinctions between natural and artificial. This ethics involves the experimentation of affect in which a living thing is to be approached not in terms of its specific form and determinate functions but in terms of its capacities for affect and for being affected (Deleuze 1981: 167–8; 1988: 124; Deleuze and Guattari 1980: 314; 1988: 257). The ethological approach, Deleuze insists, is no less valid for human beings than for animals. On this plane of immanence or consistency ethics eschews the need for transcendent organization, leading to the formation of new extensive relations as well as the constitution of more intensive capacities and powers (Deleuze 1981: 168–9; 1988: 125–6). Although this ethics concerns bodies it needs to be noted that these are bodies of affect, bodies composed of 'an infinite number of particles', which are to be defined in two ways: first, kinetically in terms of relations of speed (motion and rest), and, second, dynamically in terms of their capacities for affecting and being affected. Ethics on this plane is strictly ethological since it does not concern forms and functions but rather relations of velocity between particles only of an unformed and incorporeal material. Thus: 'There is no longer a subject, but only individuating affective states of an anonymous force' (172; 128). The becoming-molecular of the barking dog in the case of the human is an example of the nonorganic life in which sonorous intensities of affect bring into play the novel dimension of a creative becoming.

The move towards a creative ethology is in danger of transforming the world into the mere *effect* of such becomings, which comes perilously close to philosophical idealism. This is the result, I would contend, of Deleuze and Guattari's treatment of the world as if it was entirely a creation of particular haecceities and individualities/singularities. Moreover, in many of the examples given of ethological assemblages that involve supposedly nonhuman becomings of the human the crucial component in the assemblage is more often than not the human one. It is, in fact, the technicized human that provides the unifying and privileged point of consistency in such an assemblage. This suggests that Deleuze and Guattari do not pay adequate or sufficient attention to the specific context of 'nonhuman' becomings of the human, which suggests that there is a failure in their work to address the specific character of human evolution *qua* a techno-organismic, not merely an affective, animal, and that the conception of technics in *ATP* is being utilized too universally in terms of a general ethology. Becoming a molecular barking dog in the case of the human might have a specific role to play in the context of drawing a line of flight in relation to social stratification and Oedipal molarization, but then the becoming is only intelligible in this highly specific context.

Failure to deal adequately with this context also reveals an alarming naivete on Deleuze and Guattari's part in relation to the question of the animal. They insist that becomings-animal operate beneath the molar level, that is, beneath the division of life into distinct kingdoms, etc., and involve an emission of particles and distribution of intensities and affects. They offer their position, therefore, as being decidedly unsentimental, insisting that anyone who loves cats or dogs is a sentimental fool, arguing that what is sentimental is the individuation of animals as pets (1980: 294; 1988: 240). One could contend, however, that it is equally sentimental to construct the animal as an assemblage, such as in the example given of Little Hans, with the affective distribution 'horse-omnibus-street'. The difference between the animal as pet and animal as assemblage is that there is a difference only in the kind of sentimental education at play in the two cases. Moreover, when Deleuze talks of the differences between types of horses in the essay 'Spinoza and Us', such as a plough horse, a draft horse, and a racehorse, as differences of affect, arguing that the difference between any of these is 'greater' than that between an ox and a plough horse, he fails to note that what unites and makes these differences between types of horse possible is the fact that they are domesticated animals. The differences between them are the product of breeding (there is no pure realm of affects in these examples).

As much as an ethological ethics opens up the human to the nonhuman, therefore, it is also blind to the specific ethology of the animal, as well as its place of occupation within the field of human technics. This is evident, I believe, in the way they articulate a response to animal exploitation and suffering, which does not seem to require assuming any social or political action, but rather is to be met by a becoming-animal on our part so that 'the animal also becomes something else'. They then add: 'The agony of a rat or the slaughter of a calf remains present in thought not through pity but as the zone of exchange between man and animal in which something of one passes into the other' (1991: 104; 1994: 109). The attempts of the human to destratify itself as an organism – which is the specific 'ethical' task given to it in *ATP* – and to accomplish this through the medium of becomings-animal, not only leave the social stratification of the animal exactly as it finds it but actually appears to require it for is own destratification and deterritorialization. I fail to see how these examples take us beyond the problem of human narcissism and solipsism.

In both 'Memories of Spinozist II' in *ATP* and 'Spinoza and Us', an essay written in 1981, Deleuze links together Spinoza and Uexküll in terms of an ethology of affects in which the emphasis is placed on the way in which animals construct the 'outside' in terms of an interiorization involving milieus and affects.[32] In Uexküll, however, one does not find a simple opposition between the animal conceived as 'species' and the animal conceived in terms of its 'affects'. Rather, Uexküll can be seen to be going back to an older tradition of thought in which a 'species' denotes not so much a biological entity in the way we have come to understand it since Darwin, where an organism or form of life is to be conceived in terms of relations of descent, but is to be defined in terms of a semiotics of affect. As a Latin term the word 'species' comes from the Greek designation of the outward appearance or form of a thing as if it was an immediate object of vision. It thus indicates something that is to be an object of knowledge in the specific sense that it can be grasped by the mind as an object of apprehension (see Deely 1994: 126ff.). In the Latin context the term is caught up in issues of epistemology regarding *specification* or the ability to *specify* form, an issue which is still present in Darwin's *Origin of Species*, although he nowhere feels the need to offer a precise definition of it, since he holds that every naturalist knows vaguely what it means. The term 'species' is 'one arbitrarily given for the sake of convenience to a set of individuals closely resembling each other' (see Darwin 1985: 108; compare Aristotle 1985: 99–101). In Latin scholastics we see the source of Uexküll's usage of the term, since it is linked with the idea that a physical entity is a 'cognitive organism' able to respond to its surroundings by 'becoming aware of the source of the modifications as *object*' (Deely 1994: 128). 'Species', therefore, are '*specifiers* or *specifications* of the milieu' (ibid.). In Uexküll's theory of *Umwelten* each species inhabits a species-specific objective world in which what constitutes an animal as a species are the affects through which it is able to make a 'world' for itself. (In his *Theoretical Biology* Uexküll treats Darwinism as a confused model of what evolution actually entails, calling it a religion rather than a science, and he attacks it for not permitting as a legitimate question of biological inquiry the notion of a 'plan of Nature' (1926: 264–5)).

J. von Uexküll founded the Institute of Umwelt Research at Hamburg University in 1926. His approach to the 'invisible worlds' of both animals and humans was inspired by Kant, seeking to explore in novel ways the 'phenomenal world' of the animal (its 'self-world'), while 'nature' itself is invoked as the great noumenon which lies 'eternally beyond the reach of knowledge' (Uexküll 1992: 390; see also Uexküll 1926: xv–xvi for a clear statement of his debt to Kant. Chapters 1 and 2 of this work are on 'Space' and 'Time'). He saw himself as taking up the challenge of Kant's Copernican revolution in thought, construing the task of biology as one of expanding Kant's revolution along two lines: first, analysing the role of the body, especially its perceptual organs and central nervous system, and, second, studying the relationship of other subjects – animals and plants – to their own distinctive worlds [*Merkwelten*]. One of the most innovative aspects of Uexküll's work lies in the insistence that the machines, devices, and technologies of animal and of human life are to be viewed as 'perceptual tools' and 'effector tools' that form a constitutive feature of the 'worlds' of living things. However, he does not accept the theory of those mechanists who claim that animals function as 'mere machines', since this is to neglect the dynamic and formative aspects of animal modes of existence, that is, the fact, that there is 'acting' and 'perceiving' taking place.

There is an important aspect to Uexküll's theory, detectable in the above, that does not properly figure in Deleuze's reading, namely, that the question of 'becoming' is construed as intrinsic to the activity of the animal-*Umwelt* in question. In other words, the peculiar 'animality' of the animal is not something simply 'given', and there needs to be posited an *animal becoming* as well as a *becoming-animal*.[33] Now, the challenge presented by Deleuze's reworking of ethology in terms of affects and transversality is to show just how complicated the 'becoming' of the animal is. But this does not mean that the question of the animal becoming what it is ought to be disregarded or rendered otiose.

Part of the difficulty with Deleuze and Guattari's approach stems from a failure sufficiently to acknowledge and make clear the specific character of becomings-animal of the human, such as the cultural contexts in which they take place and which can make them intelligible. If one looks at their essay on Kafka and minor literature, for example, it is quite clear that their reading of the figuration of such becomings is being carried out in the context of a *politics* of desire. They show, for example, that the 'becomings-animal' that populate Kafka's fiction operate in the context of a negotiation with values that are at once economic, juridical, bureaucratic, and technological.

The criticisms I am making do not deny that 'evolution' proceeds rhizomatically or that it can be characterized by modes of transversal communication, but rather are intended to show that the various 'becomings' that characterize 'evolution', and serve to make it nongenealogical and nonfiliative, cannot be treated as if they were all the same, so that, for example, we could move, simply but far too quickly, from talking about the transversal movement of the 'C' virus that is connected to both baboon DNA and the DNA of certain domestic cats, so involving a 'becoming-baboon in the cat', to talking about the becoming molecular-dog of a human being, as if these were of an equivalent order (1980: 17; 1988: 10). Of course, it could be argued that it is on account of the way in which Deleuze extends and applies ethology that he is able to demonstrate the possibility of nonhuman and overhuman becomings of the human. Equally, one could argue that this attempt to disclose nonhuman becomings of the human results in a 'violent' humanization of animal-worlds, as well as producing an idealistic account of nature and the cosmos.

These criticisms do not aim to wholly discredit the work of Deleuze and Guattari in its ethical and political dimensions. In the next section I want to show that when viewed in the context of a specifically human-social desire the attempt to construct deterritorialized lines of flight beyond the model of an Oedipal domestication has powerful aspects to it, aspects which can also be seen to be 'political' in the specific sense of contesting established molar constructions of sexual and gendered identity.

Becomings beyond memory

Weismann in A Thousand Plateaus *and a reading of Hardy's* Tess

It is in the context of the specifically ethical question concerning how one is to make oneself a body without organs that Deleuze and Guattari refashion in quite dramatic terms Weismann's germ-plasm in *ATP*. The body without organs, they bluntly state, is 'the egg'. This egg is to be conceived as a milieu of intensity and as a spatium that is not an extensity. This is because it is characterized as an intensive field of division, augmentation, foldings, migrations, and zones of proximity, which denotes not a field of undifferentiation but pure potentiality and pure virtuality: 'Zero intensity as a principle of production' (1980: 202; 1988: 165).

Deleuze and Guattari speak in this plateau on making oneself as a body without organs of a 'fundamental convergence' between embryology and mythology and between the biological egg and the cosmic or psychic egg. It is on this level of convergence that they will endeavour to address the *ethical* question of this egg as it concerns the becoming of the human organism. The body without organs conceived as the egg of zero intensity is said to be not at all regressive but always contemporary since it is always carried with one as one's own milieu of experimentation and as one's associated milieu. The self, in other words, is always becoming *with* this body without organs, a becoming in which the body exists beyond the opposition of fragmentation and totalization. It is clear that the 'organism' in question is the organism that has been formed by processes of organization and subjectification. So far as the *human* organism is concerned this is the 'self' or individual person [*Moi*] that has been made the 'subject' of a social and historical formation, and organized as something unified in the molar sense.

Deleuze and Guattari rework the neo-Darwinian legacy of Weismann in the plateau entitled 'How Do You Make Yourself a Body without Organs?' by refusing two things: first, the positing of an abstract opposition between the strata and the body without organs, and, second, positing the body without organs as existing prior to the organism, in which it would get figured as some kind of primitive state of innocence to which the organism tries, with great futility, to return. The body without organs, therefore, is always adjacent to the organism and always under construction. The crucial move made by Deleuze and Guattari is now to stage the relationship between the body without organs and the organism in terms of a reworking of the evolution of the child into the adult, of the egg into an organism, so that it becomes a creative involution, where involution does not entail embarking on a course of regression therapy:

> If it is tied to childhood, it is not in the sense of that the adult re-gresses to the child and the child to the Mother, but in the sense that the child, like the Dogon twin who takes a piece of the placenta with him, tears from the organic form of the Mother an intense and destratified matter that...constitutes his or her perpetual break with the past, his or her present experience, experimentation.
> (Deleuze and Guattari 1980: 203; 1988: 164)[34]

The body without organs is not, therefore, so much a memory of child-hood, that is, a memory which belongs to the child and which we can own as children, but rather a 'childhood block' which allows things to pass through it and to intensively distribute. This is a map of life populated by 'comparative densities and intensities' that articulates the contemporaneity of the adult and the child, in which one is not simply before the other. The body without organs on this model has become the site of 'intense germen' [germen intense] where there are no longer clear and distinct ideas about who is the child and who is the adult. 'This is what Freud failed to understand about Weismann', argue Deleuze and Guattari: that the child is the 'germinal contemporary' of its parents, not simply a descendant of its poor relations endowed with a weak life-force (ibid.). The body without organs cannot be taken up by a possessive individualism; it is, rather, the site of alterity and, as such, does not 'belong' to anyone, it 'is never yours or mine. It is always a body. It is no more projective than it is regressive. It is an involution, but always a contemporary, creative involution' (ibid.).

This reworking of Weismann's theory of the germ-plasm is laid out in the specific context of the ethical question that is posed in relation to the becoming of the organism. It reworks the notion of germinal life by seeking to show that the individual organism does not cease in the course of its lifetime to be constituted by the 'intense germen' that can be configured as a 'body without organs'. In other words, *the germinal life refers not to points of origin but solely to modes of creative becoming.* These becomings are always implicated in a machinic heterogenesis and rhizomatic assemblages.

As an illustration of what is at stake in the move away from straightforward genealogical memories to a creative becoming, and from the perspective of a concern with a possible 'nonhuman' becoming of the human, I offer a reading of Hardy's novel *Tess of the D'Urbervilles*. This is a novel about genealogy, in which the heroine's personal history goes right back to the Norman Conquest with the family genotype being presented 'as an active force within the surface "present" of the novel' (Morton 1984: 176) (Morton suggests that Hardy may have directly drawn the notion of the genotype from Weismann). Tess's lover, Angel Clare, breaks down when he learns that she is not a 'new-sprung child of nature' but rather 'the belated seedling of an effete aristocracy' (Hardy 1981: 257) (which eventually turns out to be spurious and an invention). *Tess* can be read as a pessimistic novel about genetic determinism and the impossibility of any overcoming of the problem of heredity. There is only the pain of life, no crack through which that pain might be transfigured. So strong is Hardy's rejection of Lamarckism and adherence to Darwinian doctrine, that is, to the absolute immutability of the germ-plasm as it passes relentlessly and remorselessly through the bodies of individuals generation after generation, that the social misfits and unfits like Tess and Jude must necessarily be extirpated. In the novel bodies writhe feverishly 'under the oppressiveness of an emotion thrust on them by cruel Nature's law', an emotion which they do not desire: 'The differences which distinguished them as individuals were abstracted by this passion, and each was but portion of one organism called sex' (169). From the point of view of conventional society the infatuation of bodies, including their excesses, is futile. Morton speaks of Hardy's commitment to a new form of tragedy, born of the new biology, in which the older concepts of fate and *hamartia* were to be replaced with those of instinct and inheritance. The heroes and heroines of Hardy's novels receive a hard education in life, being compelled to face up to the fact that they inhabit an inhuman world without human significance or value (Morton 1984: 201). Tess, for example, is shackled by her ancestry, caught up wholly in nature's diabolic plan (202). The only reality is the immutable flow of the germ plasm: 'what's the use of learning that I am one of a long row only – finding out that there is set down in some old book somebody just like me, and to know that I shall only act her part....The best is not to remember that your nature and your past doings have been just like thousands' and thousands' ' (Hardy 1981: 147). To all humankind Tess is nothing more than 'a passing thought' (109). Why must it be that this 'beautiful feminine tissue, sensitive as gossamer' is 'doomed' to receive in the course of time such a 'coarse pattern'? (89). Here the fatalism of Tess's people, people she can never escape from, reverberates throughout the novel: 'It was to be' (90). So all-powerful is Hardy's vision of the finality of germ-plasmic life that there is nothing else to recall other than the memory of one's fatal inheritance. Only memory, no becoming. The six Durbeyfield children are 'passengers in the Durbeyfield ship' who had never been asked if they wanted life on any terms, existing as helpless creatures caught up in 'Nature's plan' (33). Tess is so enmeshed in the miserable, meaningless fate of evolutionary life, which passes by in mute procession, that even the 'occasional heave of the wind' is the sign of 'some immense, sad soul, coterminous with the universe in space, and with history in time' (42). In the midst of the over-population of her mother's family, and its miserable condition, she is made to feel quite 'Malthusian' (48). Surplus is guilt and must be punished, and evolution treats its productions as little more than 'passing-strange destinies' (54). There will be a day that comes in the future, though she does not 'know the place in month, week, season, or year', when her people will be in a position to look back on her life and say that was 'the day poor Tess Durbeyfield died', and there will be 'nothing singular in their minds in the statement' (117).

It is interesting to note that Lawrence approached the study of Hardy from the perspective of the extravagant life which in the course of evolution has always exceeded the effort at self-preservation (Lawrence 1985: 7). Everything that comes into being from out of the void is created with a surplus. It is this excess of life which constitutes for Lawrence primary production, and, as such, it always exceeds *reproduction*: 'The excess is the thing itself at its maximum of being. If it had stopped short of this excess, it would not have been at all' (11).[35] In Hardy's fiction Lawrence locates the tragedy of 'always the same'. In novel after novel the overriding theme is the ultimate triumph of the species over the individual, the community over the new individual being: escape and you will die either from isolation or from the direct revenge of the community, or, at worst, from both (21). The constant revelation in Hardy's novels is of 'a great background, vital and vivid, which matters more than the people who move upon it' (28). But for Lawrence there is no way that the individuals in Hardy's world are ever allowed to access this dark, incomprehensible background world; it is always outwitting them, and driving them pitilessly into an inevitable fate of merciless and meaningless suffering. Here Lawrence notes the weakness of modern tragedy, in contrast to that of either Sophocles or Shakespeare, since in its world transgression against the social code can only bring nihilistic destruction, 'as though the social code worked our irrevocable fate' (30).

Lawrence locates a deep misogyny within Hardy's novels. The women in them, he argues, are not 'Female in any real sense' but merely 'passive subjects to the male' (1985: 95). The activity of humanity during its time on earth can be understood, he argues, in terms of two great efforts: 'to appreciate the Law and the effort to overcome the Law in Love' (123). His charge against Hardy is that he shows love as only ever in conflict with the law, with death guaranteed as the only result (126). Thus, for Hardy the 'Female shall not exist. Where it appears, it is a criminal tendency, to be stamped out' (95). Tess, on this reading, is caught up in an entropic genealogy, endowed with powers 'neither to alter nor to change nor to divert' (ibid.). She is a woman without purpose, nothing more than an acquiescent complement to the male: 'The female in her has become inert' (ibid.).[36]

In contrast to the Weismannian pessimism informing Hardy's novels, Lawrence will appeal to the biology of excess in which life is always 'brimming over', which is to say that it positively lives off new life and new generation, existing without measure and where the supply always exceeds the 'fit' (1985: 31). There is no 'true line of life', as in the continuity of the germ-plasm, only a sporting life that is elastic and discontinuous (49). Hardy's problem is that his characters, such as Jude, are endowed with a weak life-flow, unable to break free of the old adhesion, the germ-plasm sticks to their bodies, ensuring they cannot detach themselves from the common, from the mass, which bore them. Their unfortunate ends are always 'begged in beginning' (50). Again, no becoming, only an accursed share of genealogical life, an inescapable and damned fate. The shallowest life – that of public opinion – must always win (51). The eternal wheel rolls on, crushing anyone who dares to break free of its chain, while the great river of life ensures that time devours all her children. This is a bleak Darwinism that persists down to our own present time; witness, for example, Dawkins and his account of the 'river out of eden', which is equally a river without a future destiny – the 'unbroken lines of successful ancestors' (Dawkins 1995: 2), meaning that the ones who didn't quite fit fell to the wayside or were exterminated in the good old fashion of natural selection: it's all in the genes; there's only inheritance; life is not what you make it. 'It is not success that makes good genes. It is good genes that make success, and nothing an individual does during its lifetime has any effect whatever upon its genes' (3). This river of DNA is never contaminated by new life (4). 'We – and that means all living things – are survival machines programmed to propagate the digital database that did the programming' (19). Moreover, 'DNA neither knows nor cares. DNA just is. And we dance to its music' (133). It's all written in the genes then, and 'we' are the vehicles through which DNA flows on and on – and Tess must die at the *right* time.

But perhaps we don't need to read *Tess*, the becoming of a 'pure woman', solely in terms of a Darwinian pessimism or chronically as a straightforward exercise in genealogy. There are moments in the novel, for example, when Tess 'becomes' a rhizome, forging connections with other populations around her, when the bodies without organs begin to dance in harmony (Hardy 1981: 80), when Nature becomes almost Dionysian, mingling harmoniously 'with the spirit of wine' (84), and when she merges with the night, experiencing the loss of her solitude after dark in the woods: 'It is then that the plight of being alive becomes attenuated to its least possible dimensions. She had no fear of the shadows; her sole idea seemed to be to shun mankind'. And then on the hills and dales her 'flexuous and stealthy figure' becomes an integral part of the landscape, with natural processes around her intensifying to the point where they become a part of her own story, the interior having become a selected exterior and the exterior a projected interior (102). In moments such as these Tess comes to know that while she may have been forced to break an accepted social law, she has not broken any law known to the environment in which she exists as an 'anomaly' (103). She comes to learn that while she lives as a stranger and an alien in her homeland, the land of memory (of family, race, tribe, church and clan), it is no strange land that she herself inhabits, the land of a new becoming (106). So that while to all humankind she may be only a passing thought, to herself she is 'an experience, a passion, a structure of sensations' (109). The alien can only ever be defined negatively when it is placed in relation to the land of home. She feels the 'pulse of hopeful life' warm within her, looking for a 'nook which had no memories', while waiting for a new departure a fine spring arrives and 'the stir of germination was almost audible in the buds', moving her 'as it moved the wild animals' (117). Throughout the novel Hardy will depict Tess as a wild animal who resists all attempts to domesticate her, whether by family or lovers, 'disconnecting herself by littles from her eventful past at every step, obliterating her identity' (303). Her 'useless ancestors' lie entombed, while she recognizes, as she becomes equal to the event, that it is impossible that any event should leave upon her 'an impression that was not in time capable of transmutation' (121–2). People now begin to 'differentiate themselves in a chemical process', disintegrating 'into a number of varied fellow-creatures – beings of many minds, beings infinite in difference…winds in their different tempers, trees, waters and mists, shades and silences, and the voices of inanimate things' (138–9). At 'non-human hours' she becomes another kind of memory, not a memory of the human but of the overhuman, a memory of the future: 'The universe itself only came into being for Tess on the particular day in the particular year in which she was born' (177). True to the bitter end, Hardy will render ambiguous the meaning of Tess's commitment of the terrible deed (her murder of Alec d'Urberville): is it simply the aberration of a pure woman, an act she has been led to by some 'obscure strain' in her blood? Or might it be the becoming of a new woman, not simply a question of blood but of overcoming, the event of that death-instinct in which death turns against itself and transmutes all the instincts (419)? (Compare Lawrence on the meaning of this murder, 1985: 96–7.)

To suggest it is the latter would be to read *Tess* under the inspiration of Deleuze and Guattari's conception of the great novelist as an artist who 'invents unknown or unrecognized affects and brings them to light as the becoming of his characters', and where instinct is inseparable from a becoming-animal and becoming-imperceptible (Deleuze and Guattari 1991: 165; 1994: 174–5). Deleuze notes that in the case of Hardy's fiction the characters portrayed are not simply persons or subjects but constituted as blocs of variable sensations. Even though as a writer Hardy displays in his novels an extraordinary and 'strange [*curieux*] respect for the individual', this respect needs to be conceived not only in terms of a form of recognition (the self desiring to be recognized for what it is as an autonomous being), but rather as involving modes of individuation that function without reference to a determinate identity (which is why Tess is an anomaly):

> he saw himself and saw others as so many 'unique chances'...from which one combination or another had been drawn....Hardy invokes a sort of Greek destiny for the empiricist experimental world. Individuals, packets of sensations, run over the heath like a line of flight or a line of deterritorialization of the earth.
>
> (Deleuze 1977: 51; 1987: 39–40)

Writing, Deleuze and Guattari maintain, is nothing other than a becoming, not the story of a representation, but the production of new lines of life: 'blazing life lines...asignifying, asubjective, and faceless [*sans-visage*]' (Deleuze and Guattari 1980: 230; 1988: 187). The writer is a sorcerer who dismantles the self and who 'experiences the animal as the only population before which they are responsible' (294; 240). Love is dismantled so that we may become capable of a greater loving:

> To have dismantled one's self in order finally to be alone and meet the true double at the other end of the line. A clandestine passenger on a motionless voyage. To become like everybody else; but this, precisely, is a becoming only for one who knows how to be nobody, to no longer be anybody. To paint oneself gray on gray.
>
> (1980: 242; 1988: 197)

However, to strip this kind of becoming-imperceptible and becoming-animal of historical contextualization in the case of works of literature, such as Hardy's *Tess*, is both naive and apolitical. Deleuze's creative becomings cannot be offered as universalist 'truths' divorced from particular historical forces, but have to be made to work in specific contexts, involving, for example, a politics of gender and of race. The 'people to come' that *Tess* invokes is a new molecular population that necessarily requires a social revolution in relation to desire. As Deleuze and Guattari themselves note, the molecular 'lines of flight' make a return to the molar organizations in order to reshuffle their segments and rearrange the binary distribution of sexes and classes (Deleuze and Guattari 1980: 264–5; 1988: 216–17).

In *What is Philosophy?* Deleuze and Guattari insist, in fact, that creative fabulation has nothing to do with memory. Here memory is being understood in precise terms, however, as the 'memento' [*un souvenir* in French] (see also Deleuze 1985: 269; 1989: 207). A conception of *creative* memory needs to sever it from its function as a psychological faculty of recollections [*souvenirs*], so that it becomes conceived as the 'membrane' that allows for a correspondence between 'sheets of past and layers of reality', making relative insides and outsides communicate as interiors and exteriors. This conception of memory as a 'membrane' involves '*un double devenir*' (288; 221 – a good example where such a memory can be seen at play is Resnais's film *Providence*). A work of art, it is proposed, offers itself as a *monument*, not in the sense of something that commemorates the past, so stifling the present and the arrival of the future *as* futural, but as a 'bloc of present sensations' which evolve in terms of their own preservation and that are able to 'provide the event with the compound that celebrates it' (Deleuze and Guattari 1991: 158; 1994: 167). 'We write not with childhood memories but through blocs of childhood that are the becoming-child of the present' (158; 168). The task of the artist is one of giving existence to the event (of life and death) by going beyond the 'perceptual states and affective transitions of the lived' (ibid.). It is in this respect that the artist and novelist can be described as seers or 'becomers' who free perceptions and affections for a more intense, nonorganic and germinal life: 'It is always a question of freeing life wherever it is imprisoned, or of tempting it into an uncertain combat' (ibid.).

In the writings of Lawrence, for example, the emphasis is on the liberation of sensations and affects from their imprisonment in bodies that are overdetermined by molar identity-formation. Deleuze points out that in Lawrence there is always a double determination of characters through, on the one hand, the sentiments of an organic person, but, on the other hand, and more importantly, the traversing of the inorganic power of a vital body. This is the 'body without organs' that names an affective and intense body, a body of zones, gradients, and of poles such as the Sun and the Moon, Venus and Saturn (see Lawrence 1995: 77ff.). Lawrence presents us with this intense vitalism of a body-becoming that defies the organs and that defeats rigid organization (Deleuze 1993: 164). Or, as Lawrence expressed himself in a powerful essay on Poe to be found in his *Studies in Classic American Literature*: 'It is the souls of men that subtly impregnate stones, houses, mountains, continents, and give these their subtlest form. People only become subject to stones after having lost their integral souls'. The mystery of dissolution, as Lawrence called it, involves learning that 'There is a long way we can travel, after the death-break: after that point where the soul in intense suffering breaks, breaks away from its organic hold like a leaf that falls'. There are subtle realities far beyond 'phallic knowledge' and beyond the scope of 'phallic investigation' (Lawrence 1995c: 253–4).

What, then, are Deleuze's 'Memories of a Bergsonian'? What *kind* of memories are being referred to? The memories to be left behind are clearly those of a 'souvenir' kind. This does not, however, preclude the possibility of the constitution of new kinds of memory. In fact, whenever memory is being attacked in Deleuze and Guattari it is always a specific formation and function that is in mind. When they declare that 'becoming is antimemory' it is the punctual organization of memory found in standard genealogies, in which memory is referred back to a specific point of origin that is placed under critique.[37] Arborescent schemas, whether in biology or computer science, posit hierarchical systems with centres of subjectification and central automata functioning in terms of an organized memory. Channels of transmission are pre-established, the system pre-exists the individual, and any deviation from the norms of the system is to be regarded as an aberration (we may note that the Freudian schema of memory conforms to this type of system).

The critique of origin(s) also explains why rhizomatics cannot be readily associated with recent moves towards hybridized conceptions of identity within feminist and postcolonial theory that seek to go beyond a politics of autochthonous identity. Deleuze and Guattari contend that hybrids simply require a connection of points and do not facilitate a passing between them. A point always remains wedded to a point of origin. On a rhizomatic model, however, the becomings mapped out involve movements by which the line frees itself from the origin and points are rendered indiscernible. In order to locate another becoming (another memory), perhaps buried within the sediments of historical life and overcoded by hegemonic genealogies, or perhaps involving future becomings that are imperceptible to us, it is necessary to activate a rhizomatic and molecular conception of natality, such as, for example, the birth of a new earth no longer governed by laws of species (race) and genus:

> an immanent justice, the earth, and its non-chronological order in so far as each of us is directly from it, and not from parents: auto-chthony. It is in the earth that we die and atone for our birth....The earth as the primordial time of the autochthonous.
>
> (Deleuze 1985: 151; 1989: 115)

It is thus possible to speak of an 'architecture' of memory which does not appeal to a centre point, a fixed point of origin or a pre-established end. The rhizome uproots genealogy, revealing the autochthonous as the always becoming-heterogeneous of the earth ('The category of the root, the origin, is a category of dominion', Adorno 1966: 156; 1973: 155). The time of the ages of the earth is aeonic. There is a 'world-memory', a continuum of life or duration characterized by strange metamorphoses that cannot be restricted to a single character, a single family or a single group.

The most important impulse behind Deleuze's reconfiguration of memory, and its allied attempt to think germinal life in molecular terms, is, ultimately, a *political* one. In a condition where a people and the conditions for the expression of its revolutionary desire are missing the writer and cinema author perform a catalytic function, able through their solitude and marginality to articulate potential forces of change and produce utterances that are like the germs or 'seeds of a people to come' [*les germes du peuple à venir*] (Deleuze 1985: 288; 1989: 221). The cinema author, Deleuze notes, 'finds himself before a people which, from the point of view of culture, is doubly colonized', finding itself colonized by both stories that come from another place and by their own myths that have become like impersonal entities now at the service of the colonizer and the imperial presence. Under these conditions the author cannot allow himself to become the ethnologist of a people or invent a fiction that is solely of private interest. The art of storytelling, whether in literature or film, should assume neither the form of an impersonal myth nor that of a personal fiction, but rather take the form of a 'speech-act' through which characters are made to cross the boundary between that which separates the private from the political, so producing 'collective utterances' that perform the catalytic function of addressing the injustices of time and expressing 'the intolerable' in a way that gives a renewed vitality to the movement (the event) of time (286, 289; 219, 222). It is in this respect that all art is political, not addressing a people but contributing to the invention of one (283; 217). For the revolutionary artist it is always a question of populations: where statistics concern individual phenomena, 'antistatistical individuality operates only in relation to molecular populations' (Deleuze and Guattari 1980: 413; 1988: 335). 'The artist discards romantic figures, relinquishes both the forces of the earth and the forces of the people. The combat, if combat there is, has moved....The established powers have placed us in the situation of combat at once atomic and cosmic, galactic' (426; 345). The poet is one who 'lets loose molecular populations in the hope that this will sow the seeds of, or even engender, the people to come, that these populations will pass into a people to come, open a cosmos' (ibid.).

Deleuze's concern with the 'overhuman' becoming of the human should not be seen to be an isolated case within recent so-called 'continental philosophy'. It finds resonances, for example, with the work of Derrida, who has contested the tradition of metaphysics in terms of its attachment to a conception of life based on filiative models, a tradition that is at once biological, political, technological, etc. By contesting established molar conceptions in biology and philosophy the aim for Deleuze is to open up thought to the anomalous and the novel, so liberating such becomings from the stigma of the abnormal and the deviant. The anomalous resists schemas of classification and cannot be said to be either an individual or a species. It has to do with the movement of multiplicities and of populations. It thus differs from the abnormal since it cannot simply be defined negatively in relation to specific and generic characteristics. For Derrida it is because the question of genealogy cannot be reduced to the 'one' that it is possible to speak of a *politics* of memory (which is, in effect, a memory of the future) (compare Derrida 1997: 105 on the stakes of a 'deconstruction *of the* genealogical schema'). The preoccupation with rethinking number also runs through much of Derrida's writing and is restaged as a 'major' concern in the text *Politics of Friendship* (1994, trans. 1997: 1–25; 299), in which such minoritarian questions as the following are posed: 'what is a small number?', 'what is a democracy of singularities?', 'how does one calculate but differently?'. In addition he has spoken of the possibility of new non-communitarian politics inspired by the 'promise' of the spectral 'bond between singularities' emerging within late-capital, which 'exceeds all cultures, all languages, it even exceeds the concept of humanity...our dissatisfaction requires, at the same time...rethinking the limits between the human and the animal, the human and the natural, the human and the technical' (1994b: 49). Derrida argues that the concept of politics very rarely announces itself without an attachment of the State to the family (of man, always, perhaps), without, that is, a schema of filiation (1997: viii). Thinking beyond the law of number – against great numbers, where the greatest numbers include the strongest and weakest at same time (71) – is a thinking that also involves going beyond the law of genus and species/race (the politics of filiation), and calls for an altogether different language (299).

The emphasis on thinking the political in terms of new molecular identi-ties, such as non-denumerable multiplicities, and in relation to the estab-lished molar politics of denumeration and its apparatuses of capture, suggests that it is not simply a question of demonstrating the possibility of a political application of an ethological rhizomatics, but also of appreciating the political import of its novel ethological moment. In his 1968 study of Spinoza, Deleuze himself incisively notes that as a new thinking of the living body ethology offers 'a new conception of the embodied individual, of species, and of genera' (1968: 236; 1992: 257). He argues that we should not neglect the 'biological significance' of this new conception. Its chief importance, however, is said to be '*juridical* and *ethical*'. He suggests that once we pose the problem of rights at the level of heterogenous bodies then we necessarily transform the whole philosophy of right(s).

Conclusion: philosophy as absolute deterritorialization

In *What is Philosophy?* Deleuze and Guattari assign a specifically political task to philosophy conceived as a thinking of 'absolute deterritorialization'. This consists in the constitution of a 'new earth' and a 'new people'. The essential contrast I want to address in the conclusion to the chapter is the one presented in *What is Philosophy?* between the declared 'absolute' deterritorialization of philosophy and the merely 'relative' one effected by capital. The nature of this contrast is far from obvious.

The radical character of this conception of philosophy is evident in the opening sentences of the chapter on 'Geophilosophy', where it is argued that subject and object offer a 'poor approximation of thought'. Instead Deleuze and Guattari propose a model in which thinking involves the relationship 'of territory and the earth'. What is primary in this relationship, however, are the movements of deterritorialization that are constantly carried out on the earth. What makes deterritorialization 'relative', whether the deterritorialization in question be social, physical, or psychological, is the fact that the relationship of the earth with the movements of territory that take shape on it are historically determined. Deterritorialization is said to be 'absolute' when the earth is made to pass into a 'pure plane of immanence' that is not constrained by particular historical determinations – whether the relationships involved be social, geological, or even astronomical – and is able to enter into the immanence of a thought of Being and of Nature that contains 'infinite diagrammatic movements' (1991: 85; 1994: 88). This involves assigning a specific *critical* function to philosophy in relation to capital. We shall return to this shortly; for the moment let us note some further aspects to the contrast between absolute and relative modes of deterritorialization. First, deterritorialization on the plane of immanence is said not to preclude a reterritorialization – of forms, functions, and affects – but rather posits it in terms of 'the creation of a future new earth'. Second, and crucially, an absolute deterritorialization is only intelligible and operative in relation to 'certain still-to-be determined relationships with relative deterritorializations' (ibid.). The 'absolute', therefore, does not simply indicate an absolute that has been posited as simply outside history or beyond historical and social determination, but one that aims to transform our cognitive mapping of history and society.

Deleuze and Guattari insist that the connection between modern philosophy and capitalism is never simply an ideological one. Modern philosophy, in the forms of all the great thinkers from Kant to Husserl, has never ceased to push historical and social determination to a point of infinity where it might pass into something else. Philosophy is not to be construed as a mere 'agreeable commerce of the mind', as a commodity with its own unique exchange value, or as the disinterested sociability peculiar to 'Western democratic conversation'. It has a particular relation to movements of territories and populations. A rapport is identified between this conception of philosophy and Adorno's negative dialectics in terms of the essential untimeliness of the critical moment. The praxis of thought arrives at the *nonpropositional* form of the concept since it has no desire to compromise with the present through modes of communication, exchange, consensus, and opinion which common sense could all too readily understand and assimilate. Philosophy moves in the element of para-doxa not simply because it wants to associate itself with the least plausible opinion or because it advances contradictory opinions, but because it deploys a standard language to articulate something which belongs neither to the order of opinion nor that of the proposition. Philosophy remains untimely, out of joint with its time, in order to invoke a 'new earth' and a 'new people'. It is in this respect that philosophy can be said to be 'utopian'. It is utopia that links philosophy with its own time and epoch and, ultimately, becomes political in the sense that it 'takes the criticism of its own time to its highest point' (Deleuze and Guattari 1991: 95; 1994: 99).

The critical question to be considered concerns how philosophy can become political through constructions of absolute deterritorialization without becoming the impotence of a merely narcissistic critique or method of intensity. This, I would suggest, is the principal risk contained, maybe necessarily, within Deleuze's thinking of the event, and one that now needs to be examined. Living for the event makes life assume *for us* the character of the greatest weight. The only relationship with the present Deleuze and Guattari acknowledge is that of shame: 'We do not feel ourselves outside of our time but continue to undergo shameful compromises with it' (1991: 103; 1994: 108). But what does this shame reveal about ourselves and about philosophy today? Why is it offered by Deleuze and Guattari as the natural, almost perennial, condition for us, for philosophy, to be in? Is it in order to provoke us into remaining faithful to the germinal life of the event? Or does it perhaps suggest that we can never live up to the event? Could it be that within this logic of the infinite, of the absolute plane of immanence, we can still detect some possible shadows of God upon the face of the earth? Does the plane of immanence liberate life or might it enslave it, and us, to the unattainable? *'What is the seal of liberation? –* No longer being ashamed in front of oneself' (Nietzsche 1974: aphorism 275).

For Adorno what is ineffable about utopia is that which defies subsumption under a principle of identity. It is for this reason that 'philosophical critique transcends philosophy'. What is distinctive about the conception of philosophy advanced by Deleuze and Guattari is the extent to which it resists – in demanding *resistance to the present –* this need to transcend philosophy. In this respect we can see the extent to which Deleuze never departed from his 'Bergsonism' in terms of remaining faithful to the aim of thinking beyond the human condition and to do this through the praxis of philosophy. Indeed, in *What is Philosophy?* the word 'utopia' is deployed to designate the conjunction of philosophy (of the concept) with the 'present milieu' (a conjunction that is said to equal 'political philosophy'). The virtual potentiality of the concept is an emancipatory one in that it does not restrict philosophy to imitating, or being faithful to, what is given to it. The event of philosophy has, in addition, no relationship with a linear and progressivist conception of history. Philosophy is related to 'the lived' only in terms of its concept of the event (incorporeal, virtual, germinal, etc.). In thinking this event philosophy 'surveys the whole of the lived' and every state of affairs. It is a kind of speculative vitalism in which the event of the future, the always 'to-come', is to free the present state of affairs from its subjugation by entropic forces. A crucial distinction is made between 'becoming' and 'history'. Becoming involves experimentation – the creation of the new and the remarkable – but history provides merely the set of 'negative conditions' that 'makes possible the experimentation of something that escapes history' (Deleuze and Guattari 1991: 106; 1994: 111).

The danger of construing philosophical activity in this way, as the 'autopoietic' creation of concepts, is that of bestowing on it a spurious independence (the Fichteanism of a supposed 'self-positing' [*auto-position*] (Deleuze and Guattari 1991: 16; 1994: 11). In refusing a historical determination of philosophy – whose task is defined as a 'Nietzschean' one of inventing modes of existence and possibilities of and for life – Deleuze courts the risk of rendering the event of thought and the tasks of philosophy not simply indeterminate but without connection to anything other than philosophy's own desire as it floats abstractly on a plane of immanence uninformed by historical praxis and the historically specific predicaments of modern thought (this plane is said to be without substance and subject, entirely immanent to itself) (1991: 47; 1994: 45; see also Deleuze 1997: 3–9) (for Fichte, it should be noted, the correct conception of Critical Philosophy is that it be entirely immanent, by which he means autopoietic; see Fichte 1994: xxiv). The problems attendant on the concept, however, stem from the complex relation between the event and historical actuality and experience. This leads us to ask whether there remains on Deleuze's model the domination of the philosophical concept. This matter is not easy to decide upon for reasons to do with Deleuze's conception of philosophical practice and its 'superior' empiricism. I shall come back to this concern later in this section.

In *What is Philosophy?* Deleuze and Guattari are insistent that the event and the state of affairs denote two completely different orders of existence. This distinction lies behind their claim that no specific character can be given to the desire of revolution, since it is only the 'enthusiasm' that informs it which counts (obviously they are following Kant here; and compare the presentation of revolution and figuration of enthusiasm as a 'sign' of history in Lyotard 1989: 401–10, where a similar distinction is made between 'data' and 'signs'). They are interested in 'revolution' *not* in terms of its existence within a specific social field, which is held to be merely 'relative', but rather as a 'concept' and an 'event'. What matters, therefore, is the enthusiasm with which we posit revolution on an 'absolute' plane of immanence, which is compared to a 'presentation of the infinite in the here and now'. Thus, the crucial issue at play for them in the expression of enthusiasm is not that of the separation of the spectator from the actor but the distinction between the historical factors (the state of affairs) and the event that lies within the deed itself and can be extracted from it. This means that revolution refers strictly to itself and so enjoys a 'self-positing' and an 'absolute survey', just as the enthusiasm can be said to be entirely 'immanent' in which not even the 'disappointments of reason' can destroy the germinal existence of this self-positing event of revolution (Deleuze and Guattari 1991: 96–7; 1994: 100–1). In other words, a revolution does not derive its 'meaning' (or, in Lyotard's terms, 'sign') from history, where a dating might fix its time and place in terms of a serial continuum, but rather, when thought as the event, it is that which gives to history a *becoming*. But conceived as such, any unmediated opposition between the state of affairs and the event has to be regarded as abstract since it is only in terms of their rapport that both enjoy a logic of sense. The 'becomings' that punctuate equilibrial history are necessarily ambiguous and paradoxical. Positing revolution as a plane of immanence, or as absolute deterritorialization, is meaningless and hopeless *unless* its features can be connected up 'with what is real here and now in the struggle against capitalism, relaunching new struggles whenever the earlier one is betrayed' (96; 100; and on capital, depicted as the 'impostor-subject', compare again Lyotard 1989: 410).

Philosophy has become in Deleuze and Guattari's last work, strangely, uncannily, the 'home' of the revolution of desire, confirming Adorno's opening insight in *Negative Dialectics* that philosophy, once deemed obsolete, only lives on 'because its moment of realization [*Verwirklichung*] was missed' (Adorno 1966: 13; 1973: 3; translation modified). This is a 'moment' of philosophy, however – its failed historical realization and continued germinal life – that both Deleuze and Adorno affirm. The problem in Deleuze and Guattari's re-presentation of what philosophy is, however, does not so much concern philosophy's germinal existence, but rather the particular and peculiar configuration it assumes on their autopoietic model. It is this conception of philosophical activity as autopoietic that requires our sympathetic and critical treatment.

In insisting that the concept is neither given nor (pre-)formed, Deleuze and Guattari are placing the stress on the creation of a concept as a self-positing activity, the positing of itself that reveals 'an autopoietic characteristic by which it is recognized' (Deleuze and Guattari 1991: 16; 1994: 11). Positing is a creating, in which a concept enjoys a heterogenesis, able to exist in a state of survey by freely traversing its components: 'It is immediately co-present to all its components or variations, at no distance from them, passing back and forth through them' (26; 20–21).[38] Such a concept is an assemblage – 'like the configurations of a machine' (39; 36) – existing in a constellation of concepts in communication with the plane of immanence that allows concepts to move transversally. The concept of a 'bird', for example, will be found neither in its genus nor its species but only 'in the composition of its postures, colours, and songs'; it will lie in an indiscernibility that is not so much synesthetic as syneidetic. The plane of immanence, it should be noted, is not itself a concept which we can think but rather the acentred 'image' thought gives itself in order to exist as a free, autopoietic activity. The plane is said to be 'prephilosophical' in this regard, not in the sense of something pre-existent but as the 'absolute ground' of philosophy that cannot exist outside of it but only function as its fundamental presupposition, and as the 'power of a One-All like a moving desert that concepts come to populate' (43; 41). (The notion of 'ground' here is offered in the sense of philosophy's earth or deterritorialized zone, and not as a programme, design, or end.) Of course concepts have a relation to things, but these are self-generated problems that are judged to be immanent to the internal logic of philosophical pedagogy (21–2; 16). It is this internal logic of philosophical becoming which accounts for the emphasis placed on autopoietic closure in the realm of concept-creation: 'And this is really what the creation of concepts means: to connect internal, inseparable components to the point of closure or saturation so that we can no longer add or withdraw a component without changing the nature of the concept' (87; 90).

This is, admittedly, a complex notion of autopoiesis, since it wants to allow for dynamic and novel change. However, this change appears to be taking place solely in terms of the internal logic of a philosophical becoming with philosophy creating concepts in response to its own staging of problems. We have seen how on the level of a mapping of biology Deleuze and Guattari's conception of rhizomatics serves to attack the idea of an agent of 'evolution' conceived as the autopoietic subject of its own becoming (autopoiesis naming here any system which both produces and maintains its identity in the face of perturbations, shocks, entropic forces, etc.). It is thus curious that in their last work they should so heavily rely on an autopoietic model in order to develop a conception of what philosophy is. They are prepared to go as far as to maintain that a concept is 'anenergetic', positing an opposition between 'energy' and 'intensity' in which energy is conceived merely as the way in which intensity gets nullified in any extensive state of affairs. There appears to be little scope for a feedback process here, however, between the event (intensity) and the state of affairs, or between the pedagogy of the concept and the pedagogy of historical experience. But this lack of a feedback process might be a necessary feature of their autopoietic construction of the becoming of philosophy. And yet, as we have already noted, the articulation of a *political* project of philosophy rests on a commitment to the struggles taking place 'here and now' against capitalism and a concern with the joys and sufferings of human beings in what are said to be their 'constantly resumed' struggles.

For Adorno 'Bergsonism' in creating a novel type of cognition in order to grasp the nonconceptual washes away 'the dialectical salt…in an undifferentiated tide of life' (1966: 18; 1973: 8). By too readily disposing of the rigid general concept Bergson ran the risk effecting a 'cult of irrational immediacy' (ibid.), so offering 'a cult of freedom in the midst of unfreedom', that is the disjunction between the freedom of immanence and the facticity of historical reality remains uncomprehended and only concretely overcome in illusory terms: 'The aporetical concepts of philosophy are marks of what is objectively, not just cogitatively, unresolved' (154; 153). 'Absolutizing' duration rendering becoming 'pure' runs the danger of recoiling into the same 'timelessness' that is being chided in traditional Platonic and Aristotelian metaphysics. It could be argued, however, that much of Deleuze's reconfiguration of 'Bergsonism' – consisting in a vital empiricism, rhizomatics, machinism, and so on – is designed to refute such a cavalier and preemptive reading of Bergson. Adorno locates idealism as the peculiar predicament of philosophy within modernity: 'Necessity compels philosophy to operate with concepts, but this necessity must not be turned into the virtue of their priority' (21; 11). It could be contended that so long as we find ourselves in a philosophical condition – and such a condition seems an inescapable and necessary fate – then philosophy can only exist as the germinal life and as the virtual event. For Adorno philosophy was to be conceived as both the most serious thing of all and not at all that serious. However, what is for him a specific historical necessity is, on Deleuze's model, turned into the affirmation of an absolute deterritorialization.

Let me now bring this chapter to a close by making some concluding remarks on the predicaments confronting a philosophy of the concept. What is important for Adorno is to avoid any fetishism of the concept: 'Disenchantment with the concept is the antidote of philosophy'. Only such disenchantment can prevent the concept from becoming 'an absolute to itself'. The task, however, is to 'transcend' the concept by means of it. Although one admires his critique of any attempt to produce a bloated monster out of philosophy, bloated to the extent that its concept would become identical with the world (idealism), Adorno's solution to the predicaments of modernity is no more adequate or satisfying. An important passage in *Negative Dialectics* to take note of in this regard is the section on 'Indirectness by Objectivity', in which Adorno claims that within negative dialectics the 'transmission' of essence and phenomenon, of concept and thing, is no longer presented as 'the subjective moment in the object' (Adorno 1966: 169; 1973: 170). The source of transmission is no longer the *'subjective mechanism of their pre-formation and comprehension'* (my emphasis) but rather 'the objectivity heteronomous to the subject, the objectivity behind that which the subject can experience' (ibid.). Adorno notes that if 'judgement' (of facticity, of objectivity, of materiality) remains subjective in a historical milieu then the subject will necessarily parrot or ape the *consensus omnium*. In order to go over to the object and survey it free of imperial designs or claims the subject would have 'to resist the average value of such objectivity and to free itself as a subject'.

However, in seeking to think a different relation between subject and object Adorno is in danger of producing an uncreative model of evolution, in which the subject's relation to, and rapport with, alterity is encased in stasis, incapable of undergoing the experience of difference and heterogeneity, and instead of confirming its universal mastery over things it dutifully respects its cognitive and contemplative distance from them. 'The reconciled condition', Adorno says, 'would not be the philosophical imperialism of annexing the alien', but rather lie in the fact that the alien remain 'distant and different, beyond the heterogeneous and beyond that which is one's own' (1966: 190; 1973: 191). Philosophy's 'substance' would then consist in 'the diversity of objects that impinge upon it and of the objects it seeks', and this would be a diversity not compelled by any schema: 'to those objects, philosophy would truly give itself rather than use them as a mirror in which to reread itself' (23; 13). The problem with these statements and formulations concerning subject and object is that they remain within the ambit and limits of modern cognitivism.

For Eric Alliez there are decisive ontological stakes at issue in Deleuze's creation of a Bergsonism since, his contention goes, it provides the first major systematic break with the movement of philosophy from Kant to Husserl, offering a new ontological monism and materialism. Bergsonism can be used to provide immanence with an entirely new sense and sensibility, instituting being as an 'object of pure affirmation' (Alliez 1997: 85). As Deleuze and Guattari themselves claim, a mode of existence and a possibility of life are to be evaluated independently of any transcendent principle (God) or value (beyond good and evil). This means for them that the only 'criterion' available for a valuation of values is 'the tenor of existence, the intensification of life' (1991: 72; 1994: 74). The aesthetic turn in modernity, which is pronounced in Deleuze and Guattari's conception of philosophy with its attention to immanence, is, according to Alliez, a resistance against capital and its capture of the sensible. The aesthetic is 'the necessary form of "first philosophy" as heterogenesis of the becoming world'. Moreover, 'The whole idealist tradition has been constructed against the risk of the sensible, against the vital possibility of a "new Earth" and a "new people" who would refuse the intermediation of Capital by showing themselves against its will to ensure the absolute subsumption of Being in Method, of some Object in the unique Subject' (Alliez 1997: 85).

Beginning with Marx, historical materialism has contended, however, that the articulation of categories of subject and object is not just a problem of metaphysics to do with abstract and ahistorical issues of epistemology and ontology, but bound up with a specific mode of production and the juridical illusions it gives rise to. Adorno is insistent, for example, that Kant's synthetic unit(y) of apperception must be comprehended in relation to its material conditions of existence in bourgeois society (Adorno and Horkheimer 1979: 87). Moreover, modernity's obsession with the new might be a reflection of the fact that the new is no longer possible under prevailing conditions of historical and material existence. The only affirmation to be had in aesthetics is that of our own impotence in the face of the domination of the not-I (capitalist rationalization and instrumentalization). The difficulty of life is a historical predicament, not an ontological or metaphysical condition. For sure, within the Marxian tradition one can locate the fantasy of complete historical immanence, the presupposition of a subject–object identity formation in which history is no longer alien but has become self-determined and self-constituted, which would amount to true human history (the same model can be located in Nietzsche, as when he speaks of bringing to an end the gruesome nonsense and accident of history; for Nietzsche, however, this would be the inauguration of *over*human history).

Adorno's negative dialectics retain the critique of metaphysics as a critique of capital without entertaining the fantasy of historical auto-production or realization. For Deleuze, if thought allows itself the notion of a subject then immanence must always be immanent to a phallically conceived, foundational subject. This is not so for Adorno, for whom a transformative political praxis, as well as a new conception of the earth and the cosmos, requires the emancipation of the subject from the transcendental illusions of capital. This requires the overcoming of social alienation and political impotence through a rethinking of the subject and its contested dialectical becoming in history. The critique of capital is informed by a recognition of the need to conceive of new models of self-expression and collective subjectivity by first of all recognizing the actuality and potency of historical determinations of subjectivity. Jameson has cogently articulated the predicament in the following terms (his passage also shows the tremendous limitations of this way of thinking):

> Now the philosophical and anthropological evocation of the will to domination inherent in the identical concept gives way to a more vivid sense of the constraints of the economic system (commodity production, money, labour-power secretly inherent in all manifestations of identity itself); meanwhile, this infrastructure of the concept then also makes it clear why its effects...cannot simply be thought away by the thinking of a better thought, by new forms of philosophizing and more adequate (or even more Utopian) concepts. History already thinks the thinking subject and is inscribed in the forms through which it must necessarily think.
>
> (Jameson 1990: 24)

What can be usefully extracted from Adorno's navigation of the antinomies of modern thought and applied in the case of Deleuze, not in terms of an abstract criticism but simply in terms of staging a problem, is the recognition that a self-contained system of the concept which aspires to the production of the infinite (Adorno has Fichte in mind), is in danger of merely imitating the central antinomy of bourgeois society. This is a society which in order to preserve its identity, which is nothing more than a lack of identity, is forced to constantly expand and advance its frontiers. It is a society hooked on change and addicted to the new, but it is also a society in which the production of the new merely enacts a compulsive repetition, one that always confirms the power of its own devouring, insatiable appetite. In short, there is development but no becoming in this society since any 'becoming' that takes place simply confirms what is already known and experienced. There is only the illusion of change and transformation. In short, one might claim that this society never encounters the event, as the heterogeneous, as the future, as the genuinely alien or unfamiliar. There is only the universal boredom of the new as repetition of the same, or what Baudrillard calls the nihilism of everything.[39]

Clearly, 'absolute deterritorialization' is not equated by Deleuze and Guattari with such an appetite or with such a nihilism. The difficulties with their conception of philosophy, however, might be said to stem from the immense power credited to the concept and its lack of mediation by social and historical forces. For Deleuze, however, there is not a problem here, simply because the task of philosophy is not to bind us to an antinomical condition and to pay obeisance to any alleged determination by history (such an enslavement of thought is evident in Jameson's reading of Adorno), but rather to demonstrate, through the creation of concepts and the mapping of diagrammatic features, that it is always possible to go further than we thought since deterritorialization is the fundamental 'truth' of our condition. Such a conception of philosophy does not take its mediation from history but rather seeks to open history up to radical transformation and transmutation through novel alliances and becomings. Subjectivity is always machinic and rhizomatic. Deleuze's attachment to a vital empiricism, in which the emphasis is placed on the virtual dimension of any creative evolution or becoming, means, in part, that the tasks of thought can never be restricted to a comprehension of the antinomies of the present. The tasks of thought are far more risky and experimental, while the demands and exigencies of ethics are that much greater.

CONCLUSION
Fold and superfold

Thought demands 'only' movement that can be carried to infinity. What thought claims by right, what it *selects*, is infinite movement or the movement of the infinite. It is this that constitutes the image of thought.

(Deleuze and Guattari 1991: 40; 1994: 37, my emphasis)

In the conclusion to this study I want to turn attention to the return of the question of the fold as it emerges in Deleuze's book on Leibniz, and then, finally, contend with the claim that his philosophical project rests on a 'disavowal' of the specific 'complexity' of the human being in terms of its historicity and finitude.

Deleuze's return to the fold

The complex 'image' of thought provided by the rhizome expresses nature or germinal life as the open system *par excellence*. But it is important not to neglect the role played by the existence of limits and the constitution of boundaries in any viable conception of 'creative evolution'. Closure is not contra 'evolution' but one of its conditions of possibility. Deleuze fully recognizes this in his book on Leibniz, arguing that closure is not, strictly speaking, and contra Heidegger, the condition of being *in* the world but rather of being *for* the world' (Deleuze 1988: 36; 1993: 26). In constructing 'evolution' in terms of 'nature as music', Deleuze draws on several sources, including Uexküll, Ruyer, as well as Spinoza and Bergson, both of whom he returns to again and again in his writings. It is in *What is Philosophy?* that it becomes clear that for Deleuze there is no fundamental opposition between the naturalisms of Spinoza, Leibniz, and Bergson, or between an attention to monadic closure and to the univocal plane of immanence. In *What is Philosophy?*, for example, Deleuze refers to Bergson's conception of living beings as *musical beings* who compensate for their 'individuating closure' by forging an openness to the world through the creation of specific modes and forms of 'being', such as modulation, repetition, transposition, juxtaposition, and so on (Deleuze and Guattari 1991: 180; 1994: 190). This leads to a conception of the 'nonorganic life',[1] which is to be construed not in opposition to organismic life but as always accompanying it and informing the openness of living systems to the world.

The salient features of Deleuze's later conception of germinal life and 'nonorganic life' can be summarized as follows:

(a) All life involves individuating closure and this can be established in a number of ways: membrane, skin, a territory, all of which serve to bring into communication an interior and an exterior. There is no absolute inside or outside, no absolute life or death.

(b) The territory serves to transform organic functions such as sexuality, procreation, aggression, feeding, and so on. The territory in turn implies the existence of pure sensory qualities which do not operate merely functionally but evolve as features of expression that enjoy their own deterriorialized existence and actually make possible the transformation of functions. This, notes Deleuze, constitutes the 'philosophical concept of territory'. Deleuze's position is, ultimately, different from that of Merleau-Ponty, since it does not construe becomings-animal on the model of empathy but solely on the level of affect. Drawing on the work of Uexküll, Merleau-Ponty argues: 'It emerged that all zoology assumes from our side a methodical *Einfühlung* into animal behaviour, with the participation of the animal in our perceptive life and the participation of our perceptive life in animality' (Merleau-Ponty 1988: 165). Merleau-Ponty utilizes the idea of animal participation to critique what he perceives to be the 'philosophical artificialism' of neo-Darwinism in which an organism is taken to be possess a spurious self-enclosed unity, an essentialism in which it either is what it is or is not. This, argues Merleau-Ponty, is to neglect the most remarkable feature of living homeostases, that of 'invariance through fluctuation' (ibid.). The law of the organism is not one of all or nothing; rather, what we encounter are forms of life subject to 'dynamic, unstable equilibria in which every rearrangement resumes already latent activities and transfigures them by decentering them' (ibid.). Moreover, relations between species and species, and between species and 'man', cannot be conceived in terms of natural hierarchies; rather, there are only differences in quality, within which transcendence of one species by another is 'lateral' and not 'frontal', with 'all sorts of anticipations and reminiscences'.

(c) For Deleuze to think this process transversally and affectively is to introduce a musical conception into the philosophy of nature in which we no longer know what is art and what is natural technique (Deleuze and Guattari 1991: 176; 1994: 185). Nature has to be conceived in terms of a differentiation between 'House and Universe', or between the *Heim-*

lich and the *Unheimlich*, which really amounts to a distinction between the creation of territory and the movements of deterritorialization.

(d) This 'musical' conception of nature does not place in the foreground notions of adaptation or survival as found in Darwinian thinking. Rather, the emphasis is placed on the creation of a 'nonorganic life' neglected by Darwinism and other biophilosophies. Ronald Bogue has brought out the character of this 'life' especially well in connection to a phenomenon we examined in the previous chapter, the refrain, which as a differential rhythm that passes between milieus and territories 'has a life of its own, a nonorganic life that functions only as a creative line of flight, an autonomous, deterritorializing transverse vector of invention' (Bogue 1997: 479). In other words, what is 'functional' cannot be separated from what is 'expressive', and which allows for a transformation of functions to do with survival, procreation, etc. The 'house' of an animal does not simply provide shelter from cosmic forces but rather filters and selects them. It is these forces that produce in nature, and *for us*, zones of indiscernibility in which 'animal, plant, and molecular becomings correspond to cosmic or cosmogenetic forces' (Deleuze and Guattari 1991: 173; 1994: 183).

(e) Such molecular becomings of animal, plant, and other forms of life, can be rendered political in the case of the becoming of the human in that the creation of new sensations, percepts, and affect drawn from the intensities of nonorganic life, whether through revolution or the monument of art, installs new bonds between people and creates, in fact, the vision and riddle of a 'people to come'.

Contemporary thinking on creative evolution and ethology has much to learn from repeating an encounter with crucial trajectories of modern thought from Spinoza to Simondon. Spinoza's significance lies in his understanding of the 'fitness' of an organism in terms of its pathic and affective capacities. His definition of fitness as the capacity 'to do or suffer many things', expresses, as Hans Jonas has pointed out, the dual character of the organism (see Spinoza 1955: Part II, Prop. XIII, note, 92–3). Jonas states it succinctly and helpfully as involving the organism's '*autonomy* for itself, and its *openness* for the world', or spontaneity and receptivity (Jonas 1973: 278). For Jonas, Spinoza's emphasis on both passive power and active power as necessary to the organism's 'perfection', expresses the nature of organic life itself, in which its closure as a functional whole within an individual organism implies at the same time a correlative openness towards the world, since individuation is the condition of its communication and integration with the virtual whole.

It is precisely such a conception of the organism that Deleuze pursues in his book on the fold: 'The world must be placed in the subject in order that the subject can be for the world' (Deleuze 1988: 36; 1993: 26). The world as chaos is a pure abstraction, Deleuze notes in this work, since it is inseparable from a 'screen', such as an elastic membrane or an electromagnetic field, that makes something emerge from it. Here he reconfigures notions of the Many and the One, noting that if chaos was not harnessed by the cosmicized forces at play in both nonorganic life (individuations, haecceities, singularities, etc.) and the activity of organismic life it would remain a 'purely disjunctive diversity'. The One, however, is not to be conceived as a pregiven unity but as an emergent potentiality within a dynamic field. It is this kind of insight that leads him to putting a highly Bergsonian gloss on Leibniz's well-known and, from Voltaire onwards, frequently ridiculed doctrine of this world as the best of all possible worlds: 'The best of all worlds is not the one that reproduces the eternal, but the one in which new creations are produced, the one endowed with a capacity for innovation or creativity' (107–8; 79). He describes this as amounting to a 'teleological conversion of philosophy', but this conversion is to be understood in the Bergsonian sense of an open-ended and virtual finality. A few pages later in the text he returns to the Bergsonian vision of creative evolution he was keen to unfold in his 1966 text: 'The play of the world has changed in a unique way, because now it has become the play that diverges. Beings are pushed apart, kept open through divergent series and incompossible totalities that pull them outside, instead of being closed upon the compossible and convergent world that they express from within' (111; 81). For Deleuze, therefore, no 'great line of difference' separates the organic life from the inorganic life but rather 'crosses the one and the other' in terms of distinguishing what is 'individual' from what is a 'collective or mass phenomenon', or what is an 'absolute form' (potentiality and plasticity of forces) from what are 'massive figures or structures, molarities' (139; 104). Primacy is given once again to the 'primal forces' of individuation and individual beings, conceived in Leibniz's terms of sufficient reasons and indiscernibles and, in a way that once again closely resembles Nietzsche's critique of Darwinism with its reduction of the activity of life, involving the primacy of the 'spontaneous, form-shaping forces' that provide life with new directions and interpretations, to a question of adaptation.

The manner in which a reconciliation between Leibniz's stress on internal finality and Bergson's stress on external finality can be effected is contained, in fact, in Deleuze's *Bergsonism* of 1966. It emerges out of his treatment of Bergson's critique of both mechanism and finalism, the former assuming that everything can be calculated in terms of a state and the latter holding that everything is determinable in terms of a programme. In both models 'time' is reduced to a screen that merely holds the eternal from us, showing us successively 'in the course of time' (my expression) what a God or a superhuman intelligence would see in a glance. This illusion is inevitable, Deleuze argues, as long as we spatialize time in which space is granted a supplementary dimension. To agree with Bergson that time is not merely a fourth dimension of space is to insist that time enjoys a positivity which is identical to a ' "hesitation" of things' and to the 'creation of a world'. In positing a 'whole' to duration no conception of presence is being invoked since such a whole is no more, and no less, than virtual and capable of perpetual re-invention.

It is at this point in his reading that Deleuze brings internal and external finality together. If a living being can be compared to the 'whole' of the universe, this is not in terms of an analogy between two perfectly closed totalities such as a microcosm and a macrocosm. The finality of a living system is only possible on the condition that the whole onto which it opens is itself open. Hence Deleuze's pithy formulation: 'it is not the whole that closes like an organism, it is the organism that opens out onto a whole, like this virtual whole' (Deleuze 1966: 109; 1988: 105).

It is also in the Leibniz book of 1988 that Deleuze returns to Bergson's question of the human animal as a question concerning the *over*human. In *Bergsonism* Deleuze posed the question of the human and overhuman (what may still become of the human?) in terms that were at once Spinozist and Nietzschean. We cannot think of the overhuman – of 'man...going beyond his own plane as his own condition, in order to finally express naturing Nature' – in terms of intelligence simply because human intelligence is too much in the grip of a utilitarian memory that serves the needs of the present. Rather, Deleuze argues, it must be located in the domain of 'creative emotion', which precedes all representation and generates new ideas. This creative emotion does not have for itself an object but creates itself as its own object, it is an 'essence' that 'spreads itself over various objects, animals, plants, and the whole of nature' (1966: 116; 1988: 110). It is, says Deleuze, like a piece of music that expresses love in that its power resides not in the love for a particular person but upon its own essence which, once created, can then be extended to particular objects and individual things. In other words, it is a kind of singularity. Here Deleuze is giving expression to a curious hybrid involving a kind of Nietzschean or Bergsonian Platonism. He contends that when music cries so does humanity and the whole of nature with it. The feelings that creative emotion gives expression to are not, therefore, merely inner feelings. It would be more accurate to say, in accordance with this Nietzschean cum Bergsonian Platonism, that we are introduced and initiated *into them*. The creative emotion is now depicted as a 'cosmic Memory' that 'actualizes all the levels at the same time' and that 'liberates man from the plane or the level that is proper to him', and in this liberation he becomes a 'creator'; that is, the human becomes what it is – the overhuman: 'We, however, desire to become those that we are: the ones who are new, unique, and incomparable' (Nietzsche 1974: Section 335). Deleuze closes the *Bergsonism* book with the Zarathustrian-inspired vision and riddle of an open society made up of creators 'where we pass from one genius to another, through the intermediary of disciples or spectator or hearers' (Deleuze 1966: 118; 1988: 111).

In the book on Leibniz the question of the human and overhuman is posed in strikingly similar terms as a question of a soul which is forever tied to the body but which 'discovers a vertiginous animality...in the pleats of matter'. This is an 'organic or cerebral humanity...that allows it to rise up, and will make it ascend all over the folds' (1988: 17; 1993: 11). The task laid down to us is not only to become physicists of nature and physicians of culture but musicians of nature and artists of culture as well, to become with this vision and riddle.

Beyond the human?

It has been contended by commentators largely critical of Deleuze's project that it rests on a disavowal of the human, chiefly its historicity and finitude. The charge of disavowal is one to be taken seriously, though we may note that it would have had little truck with Deleuze since he would insist that historical and evolutionary questions about the human condition and the becoming of the human are inseparable from the transhuman ones posed by Spinoza and Nietzsche: 'what can a body do?' and, 'what may still become of the human?' We shall show shortly how these questions continue to inform and inspire Deleuze's thinking of the problem of 'human and beyond-human' in his 1986 book on Foucault.

The crucial issue is how the becoming of bodies and the becoming of the human and post-human are to be conceived and mapped out. Brian Massumi has argued that any theoretical move designed at 'ending' man ends up making human culture the meaning and measure of all things, and produces a kind of 'unfettered anthropomorphism'. He maintains that if the relation of the human to the nonhuman simply becomes a construct of human culture, or inertness, then the human/nonhuman distinction is devoid of meaning (Massumi 1996: 232). His attack is pertinently directed at a social constructivism that would naively posit the becomings of matter as little more than the fictions and constructs of our own practices and imaginings. But it is also applicable to contemporary accounts of cosmic evolutionism which equate our specific post-human mode of evolution with the alleged dynamics and directions of 'evolution' itself. Recent depictions of evolution as moving increasingly in the direction of a *post*human phase must be treated with suspicion since they have a tendency to biologize and hypostatize developments taking place in endocolonistic capitalism. In addition, they fail to relate the posthuman to the genealogical specificity of the human with regard to its technical evolution. For while it can be acknowledged that all evolutionary life is mediated by assemblages and technics, it is also necessary to consider whether there is something something specific and peculiar about the 'human' involvement with technics. This can only reside in the (contingent) character of its historical formation and deformation.

The second essay of Nietzsche's *On the Genealogy of Morality* raises a key question concerning the peculiar character of the human being: 'To breed an animal with the capacity to make promises: is this not the paradoxical task that nature has set itself in the case of the human? Is this not the real problem regarding the human?' The making of promises presupposes the disciplined cultivation of a faculty of memory and a sense of time by which we can make reality – and the human animal – something uniform, regular, calculable, and necessary. Nietzsche's emphasis on the capacity to make promises can, in fact, already be found articulated in Darwin's *Descent of Man*, where he speaks of 'moral qualities' and 'social instincts' as highly beneficial to the species and which have, 'in all probability', been acquired through natural selection. A moral being is simply one that has the capacity to compare its past and future actions. Conscience is nothing other than the 'resolve' to act differently in the future (Darwin 1993: 351–3). In contrast to this, however, Nietzsche stresses the role played by a system of mnemotechnics in the formation of the moral sense and conscience, that is, the becoming of the human as a creature of conscience and consciousness has taken place in terms of the fabrication of memory, language, sociability, etc., and all this has involved the mediation of a technical training, discipline, and punishment peculiar to the 'social straitjacket' and the violent methods and cruelties associated with the 'morality of custom'. Darwin comes close to this, although he has nothing to say on human technics, when he notes that human action is not dictated to solely by the struggle for existence. The moral qualities are advanced, directly or indirectly, more through the effects of habit, reasoning powers, instruction, religion, etc., than through natural selection (359–60). Selection is involved in providing the basis for the development of the moral sense, namely, in the selection of social instincts but little else.

But there remains a key difference between the two accounts. For Nietzsche this becoming of the human animal takes place in terms of an 'internalization' of instincts and drives, and it is this internalization which accounts for his distinctive nature *even in the domain of his technics and their experimentations*. Man not only experiments on other animals and nature in general, performing a scrambling of the codes of life, but is also the great *self*-experimenter (he invents a 'soul', creates a notion of 'evil', lives in accordance with ascetic ideals, suffers from the 'meaning' of his existence, and so on). Originally our inner world was stretched very thinly as though between two folds of skin. Once internalized, however, it rapidly expanded and extended itself. Eventually, the human found itself in a condition of enclosure where it rips itself apart and gnaws at itself like some caged animal. In its natural state the human condition is full of emptiness, and because its genetic make-up bestows so little this sick animal has to create for itself a torture-chamber, a hazardous wilderness, within.

To what extent, therefore, could we say that the human is implicated in a postbiological evolution as part of its very definition and one that significantly alters the character of natural selection in this case, almost to the point of it being inapplicable? What environment is the human adapting to if we accept that it is caught up in a fully 'artificial' evolution? Nietzsche's ideas find an echo in Lewis Mumford's approach to technics. For Mumford the human cannot be distinguished from the rest of nature simply in terms of its status as tool-making animal. 'In any adequate definition of technics', he writes, 'it should be plain that many insects, birds, and mammals had made far more radical innovations in the fabrication of containers', such as intricate nests and bowers, geometric beehives, urbanoid anthills, and so on, than our ancestors (Mumford 1967: 5). Tool-making only becomes significant for understanding human evolution when it is modified by linguistic symbols, aesthetic designs, and socially transmitted knowledge. Mumford is out to combat what he perceives to be a narrow 'technological rationalism' that goes back to Marx and stretches to the likes of de Chardin with his account of organized intelligence and the noospheric brain. For Mumford the 'aboriginal field' of human inventiveness lies not simply in the making of tools, but rather in the refashioning of bodily organs. Hence his aphoristic quip that reminds one of Nietzsche with his genealogy of morals: 'As compared with other anthropoids, one might refer without irony to man's superior irrationality' (10).

An apposite recent criticism of Deleuze that advances the charge of disavowal selects as its target *Anti-Oedipus* with its alleged 'metaphysics of energy'. The criticism is that in *Anti-Oedipus* Deleuze and Guattari lose sight of the inscription of the body as a site of meaning and memory – the analysis they trace via Nietzsche, Kafka, Mauss, Lévi-Strauss, and others in part three of the book – and 'enter a metaphysics of energy', so losing sight of the genealogical specificity and complexity of the human (Beardsworth 1996: 133). Such a criticism may prove incisive when applied to certain of Deleuze's texts but in the case of *Anti-Oedipus* it is stupendously wrong. This is because it misses the importance of the crucial transition Deleuze and Guattari identify taking place between a pre-modern and pre-capitalist society governed by codes and a capitalist one that can only evolve in accordance with an axiomatic in which the social machine itself organizes the decoded flows of production (including those of science and technology) always in the service and promotion of the 'ends' of capital. Codes continue to exist under capital, but only on the condition that they assume a function commensurate with the personifications of capital. It is within capital that the 'surplus value of code', a surplus which exists in all social formations and which necessitates social control, gets transmuted into a 'surplus value of flux' in which all flows of code are subject to a universal deterritorialization and decoding within capital. In order to effect the constant revolutionizing of the technical means of production capital must perpetually introduce cleavages and breaks into the flows of production. An axiomatic here simply denotes the translation of a code, a flow, an innovation or invention, into formulae that conform to the logic of capital. Capitalism is based, therefore, on a social machine that both decodes flows and reterritorializes them in terms of a highly specific system of axioms. The system functions inventively since new axioms can always be added to previous ones (for further insight into the axiomatic character of capital see Jameson 1997: 398–400). The key quote, which renders the criticism problematic, both philosophically and politically, runs as follows:

> The axiomatic does not need to write in bare flesh, to mark bodies and organs, nor does it it need to fashion a memory for man. In contrast to codes, the axiomatic finds in its different aspects its own organs of execution, perception, and memorization. Memory has become a bad thing. Above all, there is no longer any need of belief, and the capitalist is merely striking a pose when he bemoans the fact that nowadays no one believes in anything any more. Language no longer signifies something that must be believed, it indicates rather what is going to be done...despite the abundance of identity cards, files, and other means of control, capitalism does not even need to write in books to make up for the vanished body markings. Those are only relics, archaisms with a current function. The person has become 'private' in reality, insofar as he derives from abstract quantities....It is these quantities that are marked, no longer the persons themselves.
>
> (Deleuze and Guattari 1972: 298–9; 1984: 250–1)

Deleuze and Guattari do, in fact, subscribe to Nietzsche's claim in the *Genealogy of Morality* that society is from the first not so much a milieu for exchange and circulation, but more a 'socius of inscription', in which the essential thing is the marking of bodies as subjects and objects. Through the self-formative labour of the morality of custom, including its violence, cruelty, and stupidity, the human 'ceases to be a biological organism' and becomes a quite peculiar and specific social animal, a full body with attached organs that are attracted, repelled, and miraculated in accordance with the requirements of the social machine (1972: 169; 1984: 144; in analysing the social machine Deleuze and Guattari are heavily influenced by Mumford and his notion of the 'megamachine'). This process takes place primarily through the creation of a memory of, and for, man, a collective memory of words and signs imposed upon the ancient biocosmic memory.

The claim being made about capital is that the axiomatic which enables it to function immanently as a social machine (the limits it encounters are entirely its own limits) means that it no longer needs to write in bare flesh or to create a memory for the human. Human beings have now been colonized by capitalism and subjectified as private persons. Moreover, as a social machine capital is entirely novel in producing technical machines as constant capital with the result that humans have been rendered adjacent to these technical machines. Under capital the human substance no longer represents a variable capital of subjection, but has become instead transformed into an element of machinic enslavement: 'whence the fact that inscription no longer bears directly, or at least in theory has no need of bearing directly, on humans' (298; 251). It may be, therefore, that what is informing the move beyond memory and beyond the human, and in the direction of rhizomatic modes of becoming (anti-genealogy, anti-memory) in a work like *A Thousand Plateaus*, are these specific historical insights into the tendencies of capital, and which become increasingly actualized under late-modern capital (Marx himself had noted how capital as it develops steps forth '*in all its cynicism*', 1975: 342). If we now live 'beyond' the human – 'God is dead, everything is permitted' is the sentiment which drives the autopoietic system of capital – then a principal task of the critical project becomes that of identifying modes of becoming which resist the entropic valorization process that motors capital and its capture of the forces of biocosmic life.

So, the charge of disavowal cannot be located with reference to *Anti-Oedipus*. Indeed, the charge levelled at Deleuze of retreating into a metaphysics of energy is blind to the fact that he is, in part, responding to Marx's tremendous insight that it is *capital itself* which develops as a metaphysics of energy: '*"Wesens" bedingt sein kann, also eine "kosmopolitische", allgemeine jede Schranke, jedes Band unwerfende Energie entwickelt, um sich als die "einzige" Politik, Allgemeinheit, Schranke und Band an die Stelle zu setzen*' ('It [political economy/private property] develops a *cosmopolitan*, universal energy which breaks through every limit and every bond and posits itself as the *only* policy, the *only* universality, the *only* limit and the *only* bond') (Marx 1982: 384; 1975: 342).

One of the aims of Deleuze and Guattari's work is to show that it is possible to produce a critique of autopoietic capital without lasping into a simple-minded or naive humanism, and this task constitutes an essential aspect of their critical reading of Marx. The key notion, in fact, through which Deleuze and Guattari seek to remain both faithful to Marx's analysis and extend it is that of 'machinic surplus value'. With ever-increasing automation the organic composition of capital, in which variable capital defines a regime that directly subjects the worker (human surplus value), gives way to a new kind of machinic enslavement owing to the progressive increase in the proportion of constant capital. This framework then extends to all regions of society. In *A Thousand Plateaus* Deleuze and Guattari insist that they are not contradicting the theory of surplus value, but rather following Marx in recognizing that surplus value 'ceases to be localizable' within capitalism, where the circulation of capital comes to challenge the distinction between variable and constant capital.[2] So while it remains the case that all labour involves surplus labour, it cannot be said that surplus labour still requires labour. In capitalism surplus labour comes to operate less and less through the striation of space-time that corresponds to the 'physicosocial' concept of work. Human alienation through surplus labour comes to be replaced by a generalized 'machinic enslavement', extending to children, the unemployed, the retired, etc., and involving the media, entertainment, lifestyles, new urban models, and so on, with circulating capital recreating itself in terms of a 'smooth space' 'in which the destiny of human beings is recast'. The key distinction is no longer that between constant and variable capital, or even fixed and circulating capital, but between 'striated' capital (effected by modern state apparatuses) and 'smooth' capital (effected by the multinationals and globalization) (Deleuze and Guattari 1980: 613–14; 1988: 491–2).

The innovation of the notion of machinic surplus value is an important one for both a recognition of nonhuman sources of production and for the insights it provides into the actual character of late-modern capital that can longer be said to evolve in terms of traditional models of human labour, if it ever could. The notion has been further reworked by Guattari, who utilizes it to stress the emergent character of various processual interactions:

> Doubtless our category of 'process of production' subsumes the Marxist one, but it goes largely beyond it....These processual components have therefore to include material forces, human labour, social relations as well as investments of desire. In the cases where the *ordering* of these components leads to an enrichment of their potentialities – where the whole exceeds the sum of the parts – these processual interactions shall be called diagrammatic – and we shall speak of machinic surplus-value.
>
> (Guattari 1996: 234)

It is important to appreciate that although the notion of surplus value gets extended and redefined by Deleuze and Guattari it still functions critically in relation to the self-valorizations of capital.

The rhizomatics that is utilized by Deleuze and Guattari in order to remap evolutionary processes through a focus on assemblages can also be usefully extended to the domain of late-modern capitalism and social practices. Machinic assemblages characterize any given social formation and new assemblages of desire are always possible because society can be considered to be not only definable by its molar centres of power but equally by the molecular lines of flight that traverse it: 'centres of power [*pouvoir*] are defined much more by what escapes them or by their impotence [*impuissance*] than by their zone of power [*puissance*]' (Deleuze and Guattari

1980: 265; 1988: 217). Capitalism finds itself unable to exhaust the machinic surplus and flows it generates in spite of its attempts at repression and overcoding through state control and regulation and its media infantilism. This is why Deleuze and Guattari are keen to contest the idea of the state conceived as a 'master image', since such an image amounts to a fictitious representation that grants established powers and molar identity formations more control over the the transversal movements of material reality under capital then they are capable of enjoying. So while capitalism may to a certain extent be the 'master' of surplus value and its distribution, it does not 'dominate the flows from which surplus value derives' (276; 226). This insight does not appeal to a natural desire or flux but rather draws attention to the fact that the articulation of machinic subjectivity within the movements of capitalist production is rhizomatic, coming from multiple directions and also exceeding the utilitarian and productionist logics of capital in unpredictable and incalculable directions. The social field, therefore, is to be understood dynamically not only in terms of its contradictions (between forces and relations of production) but, equally, in terms of the deterritorializations taking place at the level of social and cultural practices, scientific innovations, new technical developments, and so on. These movements do not reside outside the social field but constitute its actual cartography or rhizome (Deleuze 1997a: 197; Guattari 1996: 240):

> The power of the productive process of Integrated World Capitalism seems inexorable, and its social effects incapable of being turned back; but it overturns so many things, comes into conflict with so many ways of life and social valorizations, that it does not seem at all absurd to anticipate that the development of new collective responses – new structures of declaration, evaluation, and action...might finally succeed in bringing it down.
>
> (Guattari, with Alliez, 1996: 246)

On this 'machinic' model capital, conceived as a social process of production, is critiqued in terms of its entropic tendencies. Capitalism tends towards an 'entropy of significational equivalences', in which every existential and political singularity that exceeds the law of equivalence is devalued (Guattari 1996: 169). However, it is necessary to appreciate that a machinic heterogenesis cannot be captured completely by capital in entropic terms on account of the fact that the ontogenetic evolution of machines is not reducible to the linear causalities of capital. Only a *theology* of capital would argue otherwise. A micropolitics of late capital sets out to construct a 'transmachinic bridge' in which the 'smoothing of the ontological texture of machinic material and diagrammatic feedbacks' are conceived as dimensions of intensification beyond the linear causalities of capitalist development (Guattari 1992: 79; 1995: 52). It is important to insist that although it is the machinic phylum which can be shown to 'direct' the historical rhizome of capitalism there is no finalization of history on this model. The 'destiny' of the phylum in relation to capital gets constantly played out 'in an equal match between social segmentarity and the evolution of modes of economic valorization' (1996: 240).

In opposition to the metaphysics of energy established by capital Deleuze's work can be shown to provide an *ethics* of energy, involving a 'creative evolution' reconfigured as a 'creative ethology', and where this ethics is understood in its Spinozist sense of bodies and its Nietzschean dimension regarding the future of the human. We are perhaps now ready to re-position the question of the overhuman in Deleuze. His staging of the problem complicates things by situating it on the level of the 'superfold'. If living beings enjoy only a precarious form and are a site for the transmutation of forces, then might it not be possible to locate the 'outside' as the correct place in which to pose the problem of the overhuman? Which is to say, if the human is neither fixed in its form nor determined in its function, but open to an affective becoming with nonhuman or extrahuman assemblages, then with what other forces is it capable of evolving within a play of chance and necessity?

Deleuze insists that this is *not* to suggest that the question of the future is one of achieving infinity by conquering finitude, but rather it is to think about the possibility of an 'unlimited finity', in which within any play of forces a finite number of components is capable of yielding a practically unlimited diversity of combinations (Deleuze 1986: 140; 1988: 131).[3] This 'active mechanism' of the future, or of the Outside, is the '*Surpli*' (superfold), and can be located in the foldings that belong to the chains of the genetic code, the foldings of an ethological deterritorialization, the potential of silicon in cybernetic machines and information technology, and so on. On this reading the overhuman signals not the death or disappearance of the human and something more than a simple change of concept; in short, it signals the arrival of new form of life that is neither God nor man – 'and which, it is hoped, will not prove worse than its previous two forms' (Deleuze 1986: 141; 1988: 132).

Deleuze brings Foucault into contact with Spinoza and Nietzsche on the issue of the 'death of man' in which the task becomes one of locating in the human a set of forces that 'resist' this death (1986: 100; 1988: 93). It is not, therefore, Deleuze suggests, a question addressed of the 'human compound', but rather of the 'dark forces of finitude' that are not initially human but which, through a historical formation, have entered into relation with the forces that make up the human. As Nietzsche wrote in a startling passage:

> We are in the phase of modesty of consciousness…perhaps the entire evolution of spirit is a question of the body; it is the history of the development of a higher body that emerges into our sensibility.

The organic is rising to yet higher levels....In the long run, it is not a question of man at all: he is to be overcome.

(Nietzsche 1968: Section 676)

Does this vision and riddle of the overhuman amount to another humanization of the world, or does it speak of a becoming of the human with the world that amounts to a 'superior' expression of it? The charge of 'disavowal' proves to be an elusive one to hold against Deleuze and one that is difficult to place exactly. Of course, one might argue that Deleuze's 'philosophy of nature' and his quest for a 'superior human nature' rest on a flight from history and finitude. But, as already argued, to sustain this criticism one needs to engage with the 'ethics' of germinal life inspired by Bergson, Nietzsche, and Spinoza: How can one learn to live 'in' duration? What can a body do? What may still become of the human?

There is no doubt that the human continues to enjoy a privileged status in Deleuze's thinking, although this is the privilege of the *over*human. One difficulty with this view, it might be suggested, is that it appears to be conceivable only to the extent that a fundamental rapport and connection between the human and the infinite movement of the plane of immanence can be established (compare Spinoza 1955: 260–3, letter to Oldenburg November 20, 1665). In the 'Leibniz' book Deleuze portrays the human as the being of nature that has the possibility of traversing the various folds and fields of matter and bringing them into novel modes of communication. Similarly, at the end of the appendix on the superman to the book on Foucault Deleuze identifies the superfold *with* the superhuman. He speaks of this human as the 'man' who is 'in charge' of the animals (of capturing fragments of codes from diverse species), who is 'in charge' of the rocks and of inorganic matter (the new domain of silicon), and, finally, who is 'in charge' of the being of language (making the invisible visible). In fact, this conception of man refers back to the conception put forward in *Anti-Oedipus*, where man is conceived not as the 'king of creation', but rather as the 'being' who is intimately related to the 'profound life *of all forms or types [genres] of being*', and who is thus said to be 'responsible' [*chargé*] for 'even the stars and animal life' since he is the 'eternal custodian of the machines of the universe' (Deleuze and Guattari 1972: 10; 1984: 4, my emphasis). The aim should not be to abstractly negate the value and validity of Deleuze's insight, but rather to inquire after the 'ethics' of this remarkable conception of the human and in relation to those forces and formative powers of life, the novel alliances and creative becomings, which it is held to be 'in charge' of. It is, in short, to ask after the nature of the responsibility Deleuze speaks of and the character of his *selection*. Any reading of Deleuze that aspires to be worthy of the event of his thought must return, again and again, to this question of the great weight and burden of the event of a germinal life. Nothing less than our conceptions of the finite and the infinite are at stake, and the best and the worst they bring in their wake.

In *Sense and Non-Sense* Merleau-Ponty describes Bergsonism as a 'philosophy of immanence' and reproaches it for describing the world only in terms of its most general structures and features (duration, openness to the future). His major criticism is to claim that Bergsonism lacks a 'picture of human history' that might provide its intuitions with some content (Merleau-Ponty 1964b: 97). The same could perhaps be said of Deleuze's own Bergsonian-inspired philosophy. In extending this criticism of Bergsonism to Deleuze, however, we are in danger of producing a fetish of History. It is clear that for Deleuze history does not ex-ist apart from a thinking of the event and independent of our producing cartographies of it free of both points of origin and ends or goals given in advance, and free of the 'glorification' of an imperial Subject. History on this model is produced and invented through the pursuit of different series and different levels, and the crossing of thresholds. The attempt is no longer made to expose phenomena and statements in terms of their vertical or horizontal dimensions; rather, there is only the freedom to move transversally and create the 'mobile diagonal line'. The diagram, for example, enjoys a superior existence to the world taken as a brute fact, since it does not function in order to represent a 'persisting world' but rather aims to produce a 'new kind of reality' and a new model of truth. It is not implicated in a subject of history and neither does it pretend to survey history. Its task is to 'make history' by 'unmaking preceding realities and significations', and it aims to do this by producing 'unexpected conjunctions' and 'improbable continuums'. It thus 'doubles history with a sense of continual evolution' (Deleuze 1986: 43; 1988: 35).

For Deleuze, therefore, the task of going 'beyond' what history has made of us, to be carried out through the production of new lines of thought and life, does not at all rest on a negation, or disavowal, of history and politics, but rather on a fundamental reconfiguration of them. The aim of this reconfiguration is to open up history and politics to a 'creative evolution' by showing the vital possibilities of what one might call a rhizomatics of historical time, in which the diagram moves beyond the limits of a filiative history and politics and weaves a supple and transversal network of novel alliances that is always perpendicular to the vertical structure of established and official history. This is the very specific contribution Deleuzianism makes to the philosophical discourse of modernity and to our practices of becoming those who we are.

Perhaps the principal task confronting any serious reading of Deleuze is to examine and scrutinize the tremendous claims made by him on behalf of thought, philosophy, the concept, and the diagram. Deleuze conceives philosophy as involved in the creation of concepts and production of diagrams that enable us to think and map new possibilities of existence and, moreover, give to the forces of change and evolution the status of the event. In this reworked and revitalized Kantianism the world (Being) is accessed not through the vehicle of a transcendental or phenomenological subject, but with the aid of concepts and diagrams that make up a transcendental field without a subject and which opens up a realm of language-being, of thought-being, of nature-being, and so on. Philosophy on this model remains faithful to the human, to the forces of finitude that constitute and consume its corporeality, by allowing it the freedom of the incorporeal event, and by seeking to demonstrate the possibility of a thinking 'beyond' the human condition. Deleuze's quest for a 'superior empiricism' is always sustained throughout the course of his writing by the repeated invention of philosophy as a new transcendentalism (our knowledge does not have to conform to the objects given to us in experience, rather the reality of process must conform to the events of our diagrams and concepts). This study has attempted to show that this new transcendentalism is both the source of Deleuze's innovations as a philosopher and the site of the principal predicaments of his thinking in relation to the fateful experimentations with the matter of life it aspires to conduct.

We must perform our critical engagement with Deleuze not in terms of a simple condemnation or a mere repudiation, but in terms of the ongoing battle we have with the problems, predicaments, and pretensions of philosophy. It cannot simply be, however, a question of being for or against Deleuze; rather, the task should be one of implicating him in the critical and clinical questions that constitute the very fold of *our* being and our becoming those who we are.

NOTES

INTRODUCTION: REPEATING THE DIFFERENCE
OF DELEUZE

1 Yovel notes an essential paradox in Spinoza's 'ethics': any striving to overcome duration can only be expressed in terms of duration. This means that finite beings, who necessarily pass away and perish, can only attain eternity *within* duration, 'that is, within the span of this life as an event in the immanent world' (1992: 156). For further insight into the ethics of this 'event' in terms of Deleuze's reading of Nietzsche – and Nietzsche's reading of Spinoza – see the section entitled 'Staging the event' in Chapter 2 of the present study (p. 121). On the 'co-existence' of eternity and duration see the entries on these terms in Deleuze's *Spinoza: Practical Philosophy* (1981/1988).

2 Pierre Macherey has taken Deleuze to task in his reading of Spinoza on this question of 'joyful *passions*', suggesting that it may not bring Deleuze closer to Spinoza but rather is another example of the way in which he distorts him so as to produce an 'expressive' reading. Macherey contends that for Spinoza *all* passions are 'sad' ones, even those that appear to articulate joys (1996: 153). By definition passions belong to the realm of external causes in which we exist separated from our power of acting. He does appreciate, however, that everything turns on Deleuze's conception of the 'point of transmutation', as Deleuze calls it, which is to take place in the cultivation of joyful passions into active joys. Deleuze himself is insistent that Spinozism has to be approached in terms of a *practical* philosophy, which means, in large measure, effecting a double reading of the *Ethics*, once geometrically in relation to the definitions and propositions, and, again and again, 'volcanically' in relation to the broken chain of the scholia: 'The entire *Ethics* is a voyage in immanence: but immanence is the unconscious itself, and the conquest of the unconscious. Ethical *joy* is the correlate of speculative *affirmation*' (Deleuze 1981: 43; 1988: 29).

3 It should be noted that in his reading of Spinoza Deleuze, unlike many commentators, insists upon the importance of marking a difference in the terms *Affectio* (affection) and *Affectus* (affect). Whereas the latter refers to the power of something acting in the dimension of its continuous variation, the former refers to the *mixture* of bodies, that is, to their *inter*action, and so concerns the 'modified' body rather than the 'modifying' one. The ethical question for Deleuze revolves around the physical question of bodies and their powers of being affected, so that what distinguishes two animals, say a lion and a snake, are not their species and genera characteristics, but rather the fact that they are not capable of the same affections. For further insight into the the link and the 'difference in nature' between the two, see the entry on 'Affections, Affects' in

Deleuze's *Spinoza: Practical Philosophy* (1981/1988) and Spinoza 1955: Book III, Definition 3. In Deleuze's later work the distinction is reworked and presented in somewhat different terms with the stress placed on the domain of 'affect'. This now involves a nonevolutionist 'becoming': that is, a becoming which brings into play novel complexes beyond a genealogical order of filiation and descent. This mode of becoming does not involve the transformation of one thing into another, but rather the passing or communication of something from one thing to another, implying a 'zone of indetermination' where everything that exists – things, beasts, and persons – are brought into relation in ways that precede their 'natural differentiation'. This is said to be the real domain of affects. See, for example, Deleuze and Guattari 1991: 163; 1994: 173: 'The affect goes beyond affections no less than the percept goes beyond perceptions'. This is crucial for Deleuze since it means that an affect or a percept does not belong to any original place or site of affection or perception, and hence opens up the possibility of rhizomatic (non-genealogical and non-filiative) 'becomings'.

1 THE DIFFERENCE OF BERGSON: DURATION AND CREATIVE EVOLUTION

1 In this early essay Deleuze argues that this 'raising' to self-consciousness is what constitutes the 'significance' of the human. However, he insists that we must not exaggerate the function of this historical consciousness simply because duration taken by itself is already consciousness. The function of history is simply to reanimate a consciousness of matter that has a tendency to become numbed. History denotes the point where consciousness re-emerges out of its traversal of matter. While history is, therefore, only ever a 'matter of fact', there is an 'identity in principle' [*identité de droit*] between difference itself and consciousness of difference (1956: 95) (of course, it is 'memory' that is to serve as this principle). This demotion of history and historical consciousness runs throughout Deleuze's writings, and derives, I would contend, from the influence of Bergson. As to why 'Bergsonism' refuses to encounter history: the crucial insights are provided by Merleau-Ponty in his essay 'Bergson in the Making' (1964c: 187–90). One interesting observation Merleau-Ponty makes is that Bergson was too 'optimistic' about the individual and its 'power to regain sources', and too 'pessimistic' regarding social life, to ever accept 'a definition of history as a "justified scandal" ' (188). One has the feeling that the remark may be equally apposite in the case of Deleuze.

2 The Bergson–Einstein debate opens up a veritable can of worms, and I can only make a few reasonably informed remarks here. The most incisive commentators, such as Čapek, argue that Bergson's philosophy is very close to Einstein's thinking and there was no need for him to oppose or contradict his work (Čapek 1971: 238–57). Both Deleuze and Merleau-Ponty insist on reading Bergson's engagement with Einstein in strictly *philosophical* terms (Bergson was doing metaphysics; Einstein science). In an essay on 'Einstein and the Crisis of Reason' Merleau-Ponty wrote that *Duration and Simultaneity* will 'install itself yet more resolutely in the perceived world' (1964c: 185), which has not proved prophetic. In the 'afterword' to the English translation of his *Bergsonism*, Deleuze argues that Bergson's aim was not, in fact, to refute or correct Einstein, but rather to provide Relativity with the 'metaphysics' it lacked. While both theories assert the invariance of the speed of light, a number of issues are at stake, such as whether Einstein confuses the virtual and the actual, whether his theory produces another spatialization of time (while taking this way of conceiving time to new imaginative heights), which reduces time to the simultaneity and relativity of

instants, and whether, even more contentiously, the theory actually eliminates time altogether (on this last point see Stengers 1997: 40; for a full treatment of the issue see Čapek 1971, and for a helpful summary see Lacey 1989: 59–66).

For a critical slant on Bergson's response to Relativity and the 'paradox' of the twins, which involves the one accelerating into space in a rocket ageing less than the other who remains on Earth, see Ray 1991: 25–6, 44–5. For Ray, Bergson's commitment to a single time entails a 'Newtonian belief' in a single global time frame in which different kinds of clock tick at the same rate. In positing both a multiplicity of (imaginary) times and a single time Bergson is said to want his relativistic cake while eating it baked *à la Newton*. This, however, seems to miss entirely the point being made about the 'single time', which is not at all Newtonian. The key question is whether Bergson's (erroneous) view that the two clocks of the twins will turn out to be simultaneous – experiments have proved Bergson wrong and confirmed the calculations of Relativity (his error was to assume that acceleration is also relative when it cannot be) – deprives his thinking of its critical 'philosophical' force, since it is clear that the difference concerns the measurement of time and not the (lived) experience of time (duration conceived as the virtual co-existence of the degrees of a single time). On the model of Relativity time simply *is* the difference of measurement; that is, the theory concerns different intervals of time between the 'two same events' and it is these time differences between events that are *relative*. They do not involve any 'actual' duration (see Davies 1995: 60). For Bergson the slowing of clocks is only as real as the shrinking of objects by distance, and so their status is entirely phantasmal (Bergson 1965b: 163ff.). As Ray points out, the peculiar challenge of the special theory of Relativity is to contest the view that the travelling twin is little more than a phantom in the imagination of the physicist. For Bergson the paradox vanishes only when we make the distinction between a multiplicity of imaginary times and the single time. This is the point he makes in *Duration and Simultaneity* (1922) when he reflects on the debate, at which both he and Einstein were present, that took place during a meeting of the Philosophical Society at the Collège de France in April 1922. For advocates of Relativity, however, the paradox dissolves when it is recognized that events happen at spatially different locations and involve observers in different states of motion (Davies 1994: 65). Deleuze boldly defends Bergson's solution of the paradox, arguing that the twins paradox only works in the case of the arguments of physicists with the unacknowledged aid and interpolation of symbolical tricks (1966: 84–7; 1988: 83–5). I would contend that the two views are not incompatible and, ultimately, have to be seen to be doing quite different work.

3 This, as Deleuze points out, is the significance of the move made by Bergson between *Time and Free Will* and *Matter and Memory*, in which psychology is compelled to take the 'leap' into ontology and where Being is conceived as a virtual multiplicity (Deleuze 1966: 75–6; 1988: 76–7; see also Čapek 1971: 189–93). On time compare Kant, *Critique of Pure Reason* (1974/1978: A35–6/B52–3), where it is treated as a 'purely subjective condition' of human intuition, constituting a 'transcendental ideality'. When abstracted from its subjective conditions of possibility time becomes 'nothing'. Time for Kant is that in which 'all change of appearances has to be thought' even though it itself 'remains and does not change' (A182/B225). Deleuze seeks to bring about a novel alliance between Kant and Bergson on time in his *Cinéma 2* book on the 'Time-Image' (Deleuze 1985: 110–11; 1989: 82–3). On the difficulty of effecting such an alliance see Čapek 1971: 135–6.

4 Of recent commentators Etienne Balibar has perhaps gone furthest in interpreting Spinoza's substance as 'nothing *other* than individuals', in which

individuals are to be conceived as neither given matters nor perfect forms but as effects of processes of individuation and individualization. For Balibar, therefore, substance is, in actuality, the name used to designate the 'causal unity' of an infinite multiplicity of modes. This highly challenging and novel reading of Spinoza is in large part made possible by utilizing Simondon's theory of individuation, which is treated in Chapter 2 (see Balibar 1997: 8–10).

5 However, as Deleuze notes in his early 1956 essay on Bergson, resemblance is an important notion in biology that serves to reveal 'the identity of what differs from itself', proving that 'a same virtuality realises itself in the divergence of series', and showing the 'essence' that subsists in change at the same time that divergence shows that *change* operates as and 'in the essence' (where the tendency to change is by no means accidental but constitutes the essence of the *élan vital*) (Deleuze 1956: 93; 1997: 12). On the difference between the processes of actualization and realization see also Deleuze 1988: 141; 1993: 105. The question of resemblance is taken up again in *Difference and Repetition*, where Deleuze considers the following two propositions – 'only that which is alike differs' and 'only differences are alike' – and argues that whereas the former posits resemblance as a *condition* of difference, the latter understands resemblance, identity, analogy, and so on as *products* of a primary difference or a primary system of differences (Deleuze 1968: 153–4; 1994: 116–17). Deleuze treats the insight, utilized at length by Bergson, into the independent evolution of organs such as the eye across many different lineages as evidence of his argument that resemblance needs to be subordinated to a principle of differentiation and to difference.

6 It is on this point that Deleuze will insist upon distinguishing Bergson's conception of wholes from set theory: 'Sets [*ensembles*] are closed, and everything which is closed is artificially closed. Sets are always sets of parts. But a whole is not closed, it is open; and it has no parts except in a very special sense, since it cannot be divided without changing qualitatively at each stage of the division' (Deleuze 1983: 21; 1986: 10). In *Cinéma 1* Deleuze considers the plane of immanence as the 'infinite set' that serves to ensure that all systems remain 'open' systems both in terms of the internal parts of each system and between different systems. It is the plane of transversal movement. It is 'machinism'. More than this, this plane makes possible finite sets and closed systems, conditioning them as a facet of creative evolution, by allowing them to be 'cut' from its universe of perpetual movement through the exteriority of its parts (87–8; 59). There is a further consideration of set theory in Deleuze and Guattari 1991: 113–14; 1994: 120–1. See also Deleuze 1990: 80; 1995: 55: 'The whole ranges over all sets and is precisely what stops them becoming "wholly" closed'.

7 For a recent articulation in biology of the kind of problematic Bergson, and then later Deleuze, are trying to deal with see Johnson 1988, who writes: 'Evolution is the outcome of the ultimate ascendancy of the trend toward increasing diversity and acceleration of the energy flow, counteracted and retarded by the individual species attempting to proceed in the direction of greater homogeneity and deceleration of the energy flow' (1988: 81).

8 As Milič Čapek has argued, in an exemplary treatment of the question of Bergson and modern physics, it is odd that Bergson should appeal, of all physicists, to Boltzmann in order to lend support to this point since Boltzmann was of the mechanist school and committed to the idea of the reversibility of all mechanical processes and to the corpuscular-kinetic model of the universe that is under attack in Bergson (the model that makes change reducible merely to an appearance relative to our ignorance). Čapek suggests that Bergson simply overlooked the wider mechanistic context on which Boltzmann's views rested

and that he involuntarily 'bergsonizes' them. Čapek also makes the supremely important point that for Bergson the second law's positing of a gradual increase of entropy is *one* manifestation of the unidirectionality of time and irreversibility of becoming, it cannot be construed as providing an exhaustive characterization of them. For further insight see Čapek 1971, especially Appendix III 'Bergson's Thoughts on Entropy and Cosmogony', 368–95.

9 Constantin Boundas has ably defended Bergson's philosophy of duration against the 'charge' of 'another philosophy of presence' (see Boundas 1996: 100–1).

10 Deleuze appears to recognize this in one place in his texts. In his book on Foucault (1986), where we find the most extended discussion of Merleau-Ponty in his *oeuvre*, Deleuze argues that in both Heidegger and Merleau-Ponty we find a 'surpassing' of the phenomenological standpoint (intentionality, crudely stated) towards 'the fold of Being' (1986: 116–17; 1988: 110). For further insight into this reading of phenomenology see the conclusion to Chapter 2 (p. 129). Deleuze's appreciation of Merleau-Ponty here is based on a reading of *The Visible and Invisible*, but it is curious that he does not note anywhere, at least to my knowledge, the important work on the philosophy of nature and the engagement with biology. See also the interesting but brief remarks in Alliez 1995: 48 n. 1, 68, 74 n. 4, 75–6 n. 2, 76, 114.

11 Of course, one is aware that the eternal return is presented as an ethical doctrine, as well as a cosmological one, in the 1962 *Nietzsche et la philosophie*. However, it is not until the two works of the late 1960s that Deleuze configures the eternal return in relation to his biophilosophical concerns and an ethics of individuation.

2 DIFFERENCE AND REPETITION: THE GERMINAL LIFE OF THE EVENT

1 The idea that a creative praxis of thought requires an 'engendering' of thinking is something that Deleuze locates in both Bergson and Artaud. In opposing 'genitality' to both innateness and reminiscence both are seen to propose the principles of a 'transcendental empiricism' that aims to think those 'bitches of impossibility' (Deleuze 1968: 192; 1994: 147–8). See also Bergson 1962: 208; 1983: 207: 'It is not enough to determine, by careful analysis, the categories of thought; they must be engendered' (translation slightly modified).

2 Sartre construes the distinction as one between the 'I' [*Je*] conceived as the ego in the sense of a unity of action and the 'me' [*Moi*] as the ego in the sense of a unity of states and qualities. He argues that this is merely a functional and grammatical distinction between two aspects of one and the same reality (1972: 60). Sartre's aim in this work is to de-posit the 'transcendental I' in order to overcome the subject–object duality whose function in philosophy is held to be 'purely logical': 'The World has not created the *me*: the *me* has not created the World. These are two aspects for absolute, impersonal consciousness, and it is by virtue of this consciousness that they are connected' (105–6). In *What is Philosophy?* Deleuze and Guattari argue that Sartre's supposition of an *impersonal* transcendental field restores to immanence its rights (1991: 49; 1994: 47).

3 It needs to be appreciated that Deleuze derives the 'superior empiricism' of his transcendental empiricism from a refusal to distinguish the transcendental form of a faculty from its disjointed, transcendent exercise (1968: 186; 1994: 143). The objects of the transcendent are, for Deleuze, not simply outside of the world of human experience and understanding; on the contrary, they serve to mediate its actual orientation in the world. It is the innovation of Kant's conception of 'Ideas' which allows us, says Deleuze, to fully explore this world: 'An object

outside experience can be represented only in problematic form; this does not mean that Ideas have no real object, but that problems *qua* problems are the real objects of Ideas' (219; 169).

4 In his essays of 1993 gathered together as *Critique et Clinique* Deleuze will become even more unorthodox in his reading of the history of thought, claiming Spinoza to be the true founder of critique and then naming his four successors as: Nietzsche, Artaud, Kafka, and Lawrence (1993: 158). It is also in this collection, in an essay entitled 'Sur quatre formules poétiques qui pourraient résumer la philosophie kantienne', that Deleuze reads the superiority of the 'aesthetic' of the *Critique of Judgement* (an aesthetic of the beautiful and the sublime) in relation to the 'aesthetic' of the first Critique, where it is restricted to determining how the sensible comes to be related to an object in space and time. Now the sensible is taken to be valid in itself and unfolds in a 'pathos' that is beyond logic and beyond the order of time (40–9, especially 48). See also the shorter account given in Deleuze's preface to the English translation of his book on Kant (1983: vii–xiii, especially p. xii).

5 Deleuze deploys the notion of metastability to inform his theory of the Event in *The Logic of Sense*, speaking, for example, of 'potential energy' as 'the energy of the pure event'. The event is never an event of 'realization' (Deleuze 1969: 125; 1990: 103). The position alters somewhat in *What is Philosophy?* where the *concept* of the event is said to be 'anenergetic': 'energy is not intensity but rather the way in which the latter is deployed and nullified in an extensive state of affairs' (Deleuze and Guattari 1991: 26; 1994: 21).

6 It is at this point that Deleuze distances himself from Foucault. In an essay published in 1994 on 'Desire and Pleasure' Deleuze tells of the time he and Foucault last met, with Foucault saying he could not 'bear' the word 'desire', and with Deleuze replying that he found the word 'pleasure' almost intolerable (Deleuze 1997a: 189).

7 'the Freudian conception of the death-instinct, understood as a return to inanimate matter, remains inseparable from the positing of an ultimate term, the model of a material and brute repetition, and the conflictual dualism between life and death' (Deleuze 1968: 137; 1994: 103–4). Deleuze further contends that Freud's model of death suffers from being individualist, subjective, solipsistic and monadic. It concerns a 'difference', that of the ego or self (the One), which only deserves to perish. He will go on to insist that the unconscious is not governed by either degradation or contradiction, which leads him to proposing an entirely different conception of the workings of the unconscious: not involved in limitation and opposition and conflict but rather informed by 'questioning and problematising'. He is thus able to declare: 'The celebrated phrase "the unconscious knows no negative" must be taken literally' (143; 108).

8 On the nature of the passive synthesis compare Merleau-Ponty's *Phenomenology of Perception,* in which it constitutes the event of subjectivity: 'What is called passivity is not the acceptance by us of an alien reality…it is being encompassed, a being in a situation – prior to which we do not exist – which we are perpetually resuming and which is constitutive of us.…A passive synthesis is a contradiction in terms if the synthesis is a process of composition, and if the passivity consists in being the recipient of a multiplicity instead of its composer. What we meant by passive synthesis was that we make our way into multiplicity, but that we do not synthesize it' (1962: 427).

9 Deleuze offers an incisive discussion of the predicament Freud finds himself in, in which the repetition of the death-drive is modelled on the model of 'cancelled difference' while Eros is left free to generate new differences, in *Difference and Repetition* (1968: 144; 1994: 109, 317 n. 18).

10 In another reading of Freud on the death-drive through Lacan, Boothby is inclined to associate it more with the 'Real' and the 'ineffable exigency of the body' (1991: 20). By disassociating the notion from Freud's dubious biologism, and the conundrums it leads to, he actually produces a reading of the death-drive that comes close to the position of Deleuze. His argument is that for Lacan the traumatic force of the death-drive is to be related not to the biological organism but rather to the authoritarian unity of the imaginary ego. Boothby notes that in the essay on the pleasure principle Freud offers two conflicting and mutually exclusive accounts of what it means to posit the death-drive as being 'beyond' the pleasure principle (as both an intensification of tension and its reduction towards absolute minimum), and suggests that it is the unacknow-ledged, but decisive, shift in the essay from a psychological to a biological discourse that explains this confusing turnaround on Freud's part. Boothby is right to claim that the productive basis of the notion of the death-drive lies in the way it draws attention to the excessive discharge of vital energies that are compelled to struggle against the constraints of the unitary ego (77, 87). The death-drive names that 'force of destructiveness' that is directed not against the other but against the *self*. It is, therefore, to be construed not as a drive to murder but rather as a drive to 'suicide', and helps to explain certain enigmatic features of human existence such as bodily violation and dismemberment (11, 40). Of course, what is missing from this helpful and incisive reading of the death-drive is an account of the molar formation of the organism or the unitary ego. As a result, Boothby is forced to sever the link with biology completely, rather than re-think the character of Freud's biology (see Ansell Pearson *Viroid Life*, 1997: 'Dead or Alive', 57ff. for further insight).

11 Gillian Rose argues that death, like repetition, enjoys two meanings in Freud: on the one hand it is linked, negatively, with the compulsion to repeat, and amounts to a blocked memory and a passive relation to death; on the other hand it is linked, positively, to the risking of one's life, the 'daring death' that heralds a 'perfect memory', an 'active relation to death' and a positive repetition *forwards* (see Rose 1992: 104, 109). But this distinction between the two deaths fails to appreciate the extent to which for Freud the active desire for death, through the risking of life (one's own and that of others), is equally implicated in *regression* and the latent destructive tendencies of the instincts (enduring life by preparing oneself for death means, says Freud in the 1915 essay, taking 'the backward step'; Rose recognizes this in the case of the 'Beyond the Pleasure Principle' essay but restricts it to this piece). I think the reason why Rose is misguided in her reading of Freud on death is because she pays insufficient attention to what is at stake in Freud's 1915 essay.

12 Does the wound exist apart from its becoming *and* its memory? Is memory both the wound and the overcoming of the wound? As Deleuze notes, 'A scar is the sign not of a past wound but of "the present fact of having been wounded" ' (Deleuze 1969: 105; 1994: 77). On the memory of wounds see Lawrence, espe-cially 1995b: 261: 'But why the memory of the wounds and the death? Surely Christ rose with healed hands and feet, sound and strong and glad? Surely the passage of the cross and the tomb was forgotten? But no – always the memory of the wounds, always the smell of grave-cloths? A small thing was Resurrection, compared with the Cross and the death, in this cycle'. See also Nietzsche in the foreword to *Twilight of the Idols*: 'A maxim whose origin I withhold from learned curiosity has long been my motto: "*increscunt animi, virescit volnere virtus*" ("the spirit grows, strength is restored by wounding")'.

13 According to Deleuze, Nietzsche's perspectivism operates as an art of inclusive disjunctions, in which divergence is not a principle of exclusion and disjunction

is not a means of separation; rather, incompossibility 'is now a means of communication' (Deleuze 1969: 203; 1990: 174). Deleuze stresses that in the case of inclusive disjunctions it is not a question of making the disjunction a conjunction; rather, it remains a disjunction that continues to bear on divergence: 'But this divergence is affirmed in such a way that the *either...or* itself becomes a pure affirmation. Instead of a certain number of predicates being excluded from a thing in virtue of the identity of its concept, each "thing" opens itself up to the infinity of predicates through which it passes, as it loses its center, that is, its identity as concept or as self. The communication of events replaces the exclusion of predicates' (ibid.). The 'schizo' experience, for example, 'is and remains in disjunction', not reducing two contraries to an identity of the same, but affirming their 'distance as that which relates the two as different' (Deleuze and Guattari 1972: 96; 1984: 76–7). Perspectivism, on Deleuze's reading, is not a relativism, in which truth would vary according to the subject in question; rather, perspectives denote the conditions within which 'the truth of a variation appears to the subject' (Deleuze 1988: 27; 1993: 20).

14 One should consult the astute remark by François Zourabichvili in this regard: 'Yet nevertheless, there is no life, and therefore no health, without a minimal organism. Life is non-organic, but its relation to the organism is one of reciprocal presupposition: it builds a house in order to take flight from it, it is House-Cosmos, a house standing against the cosmos, tuned into the cosmos, haunting as much as inhabiting it' (1996: 198).

15 On beatitude see also Deleuze's book on Spinoza and expressionism (Deleuze 1968: 282ff.; 1990: 308ff.). Clement Rosset has argued that beatitude constitutes the central and constant theme of Nietzsche's thought – 'I would willingly say the *only* theme' (Rosset 1993: 26). Nietzsche's expression of a beat-philosophy informs the entire endeavour of 'gay science' (Nietzsche 1974: Sections 276–7), with its commitment, in the eternal engagement with the trials of life, to the 'art of cheerfulness' [*Heiterkeit*]. It would be instructive to determine the *difference* between types and theories of beatitude, whether Spinozist, Fichtean, Nietzschean, or Stoic. Clearly Nietzsche's challenge resides in the attempt to think theodicy without God, a challenge most evident in his reading of Leibniz (see Nietzsche 1968: Sections 411, 419, 1019) (this is a move which, according to Rosset, makes Nietzsche more Leibnizian – more cheerful – than Leibniz).

3 THE MEMORIES OF A BERGSONIAN: FROM CREATIVE EVOLUTION TO CREATIVE ETHOLOGY

1 Deleuze's little book on Spinoza's practical philosophy was first published in an earlier skeletal form in 1970. The version most readers are familiar with is the later 1981 edition (translation 1988) published in the wake of *ATP*, and containing three new essays including the crucial piece entitled 'Spinoza and Us'.

2 With its commitment to the idea of the fixity of species, preformationism plays an awkward role in Leibniz's thought, standing in contradiction to his attachment to the principle of continuity. Deleuze endeavours to complicate the character of Leibniz's preformationism (a doctrine of internal destiny) in *The Fold* (1988: 12ff.; 1993: 8ff.). Leibniz's logic of monads provides an early account of endogenously produced change ('evolution', unfolding). In number 11 of the *Monadology* it is stipulated that all natural changes of a monad must proceed from 'an internal principle'; in Section 12 it is noted in addition to this principle of change (a principle of individuation) that the 'material forces' of such change must enjoy an 'internal complexity'; this is a complexity, Section 13 goes on to add, that enfolds 'a multiplicity in unity'. Rescher notes, commenting on these

sections, the internal complexity is so differentiated, running on into endless variation, that it can be compared to an irrational number, the infinite decimal expansion of *pi* or the square root of 2. See Rescher's translation of the *Monadology*, and commentary on it, in Leibniz 1991: 18, 68–71. The importance of the irrational number (infinite variation) in Leibniz is commented upon by Deleuze (1988: 17ff., 65).

John Mullarkey has argued that Deleuze's late turn to Leibniz results in a major switch in philosophical orientation away from the materialism and monism of Spinoza's infinite substance (the plane of immanence) to a nonreductive materialism that recognizes that each individual enjoys a 'dominant unity on account of the enfolded intricacy of its material content' (Mullarkey 1997: 458). Although this remark is highly incisive, I offer a different reading of the treatment of Leibniz, showing how it is consonant with Deleuze's reading of Spinoza, in the conclusion to this present study. One should also note that Spinoza returns as the 'prince' of the philosophers in *What is Philosophy?* Revealing in this regard are Deleuze's remarks to the English translator of his book on Spinoza and expressionism (1968/1992), Martin Joughin, where Deleuze identifies in quite clear and precise terms the nature of his readings of Leibniz and Spinoza. He says that one of the most original aspects of his book, which is also a feature of the reading of Spinoza in the contemporary text *Difference and Repetition*, is the attention given to the composition of finite modes in which substance is made to turn on these modes. Deleuze confides that it was his reading of Leibniz which enabled him to go in the direction of the modes so as to articulate the 'expressive character of particular individuals'. Nevertheless, he reveals that he considers himself to be a Spinozist rather than a Leibnizian because only Spinoza's philosophy attends to the *immanence* of being, and, moreover, it is only by mapping both substance and modes onto a plane of immanence that the Leibnizian turn makes sense and comes fully to life. It is for this reason, therefore, that Spinoza remains for him the 'prince of philosophers'.

3 *A note on the molar/molecular distinction.* The nature of this distinction has been ably articulated by Brian Massumi, whose account I now draw on (1992: 54–5). The difference between the two is not one of size or scale, but rather concerns composition and organization. Molecular is not simply or necessarily being associated or equated with 'small' or molar with 'large'. It is not, therefore, simply a distinction between part and whole, organ and organism, or individual and society. Rather, there can be molarities of any magnitude. Massumi brings out the fundamental operative difference when he notes that molecular populations are composed in terms of local connections between particles, whereas in the case of a molar population, such as a society, an organism, or a person, the connections have become stabilized and homogenized through rigidification and reification. In a molarity a population of particles is subject to well-defined boundaries and gets stratified as an identifiable whole. In a population that enjoys novel modes of becoming, the particles are correlated but not in a rigidified manner and its boundaries are constantly fluctuating. Deleuze and Guattari stress that the two configurations coexist and cross over into one another. Perceptions and emotions, for example, are subject to a molar organization and a rigid segmentarity within the philosophy of mind and within psychoanalysis. But this does not preclude a universe where these perceptions and affects function unconsciously and molecularly, and where their expression cannot be so readily stratified into accepted norms and conventions of meaning and value.

4 Kampis's own text criticizes both biology and cognitive science for basing an account of 'creative phenomena' on mechanistic, computational, and

representational models. The models deployed by physics and mathematics, he argues, privilege invariant structures which can be made amenable to a formal treatment, but which, in actuality, constitute merely a special case. His 'Bergsonian' thesis is that nature is made up of self-modifying, complexity-increasing systems that are capable of producing 'the most curious phenomena, far beyond reach of ordinary mechanistic systems'. At the same time he wishes to show that one can approach such systems with a notion of creation that strips it of its mystical connotations (Kampis 1991: 2).

5 We might note that new experimental evidence of the late 1980s showed that unicellular organisms, such as bacteria, can modify their DNA in a directed, adaptive manner.

6 See also Kauffman 1995: 50–1: 'DNA replicates only as part of a complex, collectively autocatalytic network of reactions and enzymes in cells. No RNA molecules replicate themselves...life is the natural accomplishment of catalysts in sufficiently complex nonequilibrium chemical systems'. An extensive treatment of Kauffman's ideas can be found in Chapter 16 of Depew and Weber 1996: 429ff. On the feedback role played by enzymes in the autocatalysis of DNA replication see also Prigogine and Stengers 1985: 153–4. It is interesting to note that in their critique of the tendency among molecular biologists to reduce the production of global order to the actions of individual molecules, Prigogine and Stengers have recourse to Bergson's attack on naive vitalist conceptions of his day, in which evolution can only be conceived by referring it to some pre-existing goal. The same error is committed by molecular biologists who deploy metaphors of 'organizer', 'regulator', and 'genetic program' (174–5).

7 The evolution of the proofreading capabilities of DNA is based on sound engineering principles: the way in which to send a message through a noisy channel (one that is subject to mistakes) is to send the message more than once by deploying redundancy. The important feature of the double helix is that it is 'double', since the existence of the two chains means that the memory of genes is redundantly coded, making proofreading possible. The peculiar character of DNA's redundancy is that the genetic message is represented in complementary, not identical, copies (see Watson 1970: 163ff.).

8 On content and expression, and their mutual solidarity, see Hjelmslev 1969: 47–60, and Deleuze and Guattari's modification in 1980: 58–9; 1988: 43–4.

9 For further insight into invariance see Chapter 6 of Monod on 'Invariance and Perturbations' (1971: 99ff.). As Monod notes, the quest for invariants has always revealed a 'Platonic ambition' at the heart of science (101). See the critical, but also crass, analysis of Monod ('The theologian of molecular transcendence') in Baudrillard 1993: 59–61.

10 Mutations are typically grouped into four major types, including 'base substitution' when one nucleotide base in a DNA sequence gets replaced by another; a change in a DNA sequence produced by the insertion or deletion of a nucleotide base; and alterations to DNA sequences involving either a reconfiguration that results from the chopping out and re-insertion of part of the sequence in an inverted order or the duplication, or deletion, of a whole sequence. See also Eigen 1992: 25–30; Pollack 1994: 30–4.

11 All these phenomena, which effect an 'enlargement' of the genetic programme, are examined at length in Jacob 1974: 291ff. Jacob treats recombination as the example of a non-additive reassortment of the genetic programme, but he also gives the example of the production of 'added' genetic material that allows a cell to acquire new structures and execute new functions, such as is involved, for example, in the sexual differentiation in certain kinds of bacteria. Interestingly, he sees the nucleic-acid sequence contained in these supernumerary elements as

escaping the constraints of stability and normalization imposed by natural selection on account of their gratuitous ('dispensable') character. Such elements amount to a 'free addition for the cell', a sort of virtual 'reserve' of code. The lesson is that if one is to talk of stability (normality) one must equally speak in the same breath of *variability* (abnormality): 'at the level of the bacterial population…the nucleic-acid text appears to be perpetually disorganized by copying errors, by recombinatorial spoonerisms, by additions or omissions' (Jacob 1974: 292). This insight, however, needs to be linked up with Deleuze and Guattari's insight into deterritorialization regarding the code and the territory.

12 'Transversals' is a term found in the history of mathematics, such as Lazare Carnot's *Essai sur la theorie des transversales* of 1806. A more immediate precursor can be found in the work of René Thom, the pioneer of catastrophe theory, where transversality concerns the ways in which the smooth curves of analysis intersect or 'cut' each other. Deleuze and Guattari utilize Thom's work in *A Thousand Plateaus*.

13 I appreciate that this does raise complex and difficult questions in relation to suicide, a phenomenon Deleuze confronts in *The Logic of Sense*. Whichever way we turn everything seems dismal, hopeless, a lost cause and a losing battle. Surrender would seem to be the only honourable thing. Total *Gelassenheit* as opposed to total mobilization. Better the death than the health which we are given. But for Deleuze this is death conceived within the frame of Nietzsche's superior (because germinal) 'great health' (1969: 188; 1990: 161). Deleuze contends that killing oneself is, in fact, *not to go far enough*. Suicide is an act of too much generosity towards the self and not enough towards life. By risking 'everything' it, paradoxically, doesn't risk nearly enough. In Milan Kundera's novel *Immortality* (1991: 195–200) suicide is depicted not so much as a desire to 'vanish' as a desire to 'stay', so engraving the self on the memories of those who remain behind, forcing the body into the other and perhaps 'crushing' it.

In a piece entitled 'Inquest' Artaud poses the question whether suicide is a solution to the problem of life. For him this is not to pose a moral question, since the suicide he wishes to access is without any representational value. This is not the suicide that would lead us to the side 'beyond' death but rather the one that would allow us to 'retrace our steps on the yonder side of existence rather than the side of death'. One does not hunger for death but for this '*anterior state of suicide*' (Artaud 1965: 60). In thinking through the problem of suicide in this way Artaud's labours are directed towards a destratification of the body and the self. This is to work against the arbitrary inheritance of a certain 'natural selection'. Suicide becomes for Artaud a way of violently reconquering the self, of invading the organized body, freeing the self from the conditioned reflexes of the organs, so artificially selecting and designing the body and the self for a free existence; now life is 'no longer an absurd accident whereby I think what I am told to think' (56). In carrying out the work of this 'anterior suicide' one chooses the direction of one's faculties and tendencies, placing oneself between the 'beautiful and the hideous, the good and evil' (ibid.). For it is clear that 'life itself is no solution', simply because 'life has no kind of existence which is chosen, consented to, and self-determined', but is merely a 'mere series of hungers and adverse forces, of petty contradictions which succeed or miscarry according to the circumstances of an odious gamble' (ibid.). It is clear that for him the problem is one of combatting genealogy with a new kind of becoming. 'After all, we are only trees', he writes, and it is written in the nook and cranny of one's family tree that we should kill ourselves unfreely on some given day. In emancipating the body from its organization as an organism, therefore, one is choosing to turn one's destiny *inside out*, so smashing one's pre-destination (57). For insight into the eminently rational character of suicide (the character of Deleuze's own suicide), in cases where the burden of life becomes too much to bear, whether generated by age, sickness, or misfortune, see Hume's essay on the subject: 'If Suicide be supposed a crime, it is only cowardice can impel us to it. If it be no crime, both prudence and courage should engage us to rid ourselves at once of existence when it becomes a burden' (1965: 160). The rational suicide does not simply devalue or negate life; on the contrary, it contains its own affirmation of it.

14 In *ATP* Deleuze now has a 'minor' geometry which is able to produce a more interesting account of number. In *Le Bergsonisme* number is seen to possess only differences in degree (division without any change in kind) (1966: 34–5; 1988: 41). In *ATP* the concern is with an ordinal and nomadic number which distributes itself in a 'smooth space'; that is, which does not divide without changing its nature each time. A distinction is made between the 'numbered number' that pertains to striated space and the 'numbering number' peculiar to smooth space. This is the ordinal number, the number that distributes itself and does not divide without changing its 'nature' on each division, and the units of which represent distances and not magnitudes. The notion of the 'numbering number' is, in fact, already named as such by Deleuze in his 1956 essay on Bergson: 'We will see that one of Bergson's most curious ideas is that difference itself has a number, a virtual number, a sort of numbering number [*nombre nombrant*]' (Deleuze 1956: 83; 1997: 4). Here the notion is introduced in the context of a critique of a principle of utility, that is, of the way in which utilitarian calculations homogenize difference by spreading out the degrees contained within it until there is *only* the difference in degree (differences of proportion).

15 If we could think it 'pure duration', Bergson writes, would be 'pure heterogeneity', that is, a 'succession of qualitative changes, which melt into and permeate one another, without precise outlines, without any tendency to externalize themselves in relation to one another, without any affiliation with number' (Bergson 1960: 104). This is a 'continuous' qualitative multiplicity, an intensive magnitude 'if intensities can be called magnitudes', not subject to measurement or quantification (106).

16 Modern evolutionary theory, beginning with Lamarck in the early nineteenth century, who first proposed the name 'biology' in his *Zoological Philosophy* of 1802, has approached the phenomenon of life in terms of a genealogical 'order of nature' (Lamarck 1914: 6). Lamarck recognized that the division of life into classes, orders, families, and genera represents an artificial device constructed for the purpose of human convenience. He sought to replace this artificial classification with one that would be in correspondence with nature itself. Affinities [*rapports*] between animals were to be determined by the family resemblance between their most essential organs and systems of organs (such as the nervous system, the respiratory system, and so on). While controverting the theological doctrine of the fixity of species, modern evolutionary theory remains wedded to an evolutionist or perfectionist model with distinct kingdoms and phyla of life. For example, Lamarck depicted a *distribution générale* of animals, in which a linear series of evolution is arrived at by simply determining the position of each species with reference to other species, and then ascertaining the exact point of the series that must be allotted to it (whereas 'classification' is an arbitrary art, 'distribution' is a genuine science) (20ff.; 56ff.). In Lamarck's schema there is a straight line of development leading from the most simple to the highly complex, with evolution being depicted as a continuous process of increasing complexity from the homogeneous to the heterogeneous (a view that was to form a corner-

stone of Herbert Spencer's formula of evolution). Lamarck saw this process of increasing complexity as endogenous to the activity of life (life is synonymous for him with 'vital movements' working under the mysterious influence of an exciting cause) (47, 211ff.). It should be noted that for Lamarck this process is marked by anomalies: 'Progress in the complexity of organization exhibits anomalies here and there in the general series of animals, due to the influence of environment and of acquired habits' (Lamarck 1914: 70). His attention, however, is not focused on these anomalies but simply on the straight line of increasing complexity.

Darwin breaks with this linear and vertical model, depicting the evolution of life forms horizontally as a 'descent with modification'. The only known cause of the similarity of organic beings, Darwin argues, is that of 'propinquity of descent', the hidden bond that is only partially revealed to us by our classifications (Darwin 1985: 399). In his 'B' Notebook Darwin reflects that perhaps the tree of life would better be described as 'the coral of life', which would then serve to 'excessively' complicate the view of evolution as one of the 'constant succession of germs in progress' (Darwin 1987: 177).

17 For a comprehensive treatment of the Cuvier–Geoffroy relationship in its historical and intellectual context see the excellent study by Appel 1987, and for a succinct account of the essential terms of the confrontation see Depew and Weber 1996: 48–50.

18 Dawkins has noted the proximity of his theory of the 'selfish gene' to the neo-Darwinian turn of Weismann's doctrine of the continuity of the germ-plasm. See Dawkins 1989: 11.

19 In the second half of the nineteenth century the terms ethology and ecology were often mistaken for each other. In the latter third of the century the term 'oecology' was used by the German biologist Ernst Haeckel to refer to the study of the organism in its environment, while 'ethology' was largely understood in J. S. Mill's sense as denoting the study of the formation or building of character (whether individual, national, or collective). For further insight into the history of the term 'ethology' from the seventeenth to the nineteenth centuries see Thorpe 1979: 9ff., who dates the first *bona fide* work in ethology to a work by a French nobleman's gamekeeper, C. G. Leroy, first published in 1764 and translated into English 100 years later as *The Intelligence and Affectability of Animals from a Philosophic Point of View, with a few letters on Man* (London, Chapman & Hall). This French tradition becomes more firmly established in the nineteenth century with the work of Lamarck and his followers, notably Geoffroy Saint-Hilaire and his son Isidore.

20 The chief problematic connected with an 'assemblage' is that of consistency. Millett (1997: 64) has usefully pointed out that the translation of *agencement* as 'assemblage' ignores the fact that both terms exist in French, with *assemblage* denoting something totalized and unified in terms of a linear ordering, while *agencement* implies a more distributed conception of 'agency'.

21 There is an important dicussion on the 'territory' in *ATP* that I am not dealing with here, in which Deleuze and Guattari make some complicated distinctions and criticize the pioneering work in ethology of Konrad Lorenz for making aggressiveness the basis of the territory; contra Lorenz they insist that aggressive behaviour does not serve to explain territorialization since such behaviour derives from it (Deleuze and Guattari 1980: 386ff.; 1988: 315ff.). On aggression and modern ethology see also Barnett 1988: 56–77.

22 Unlike most crabs the hermit crab is not fully protected by a hard cuticle but has soft hinder segments. It inserts its tail into empty seashells such as those of a whelk, and moves from 'house' to 'house' as it continues to grow.

23 The term has been extended in the way deployed here in unpublished work by Michael Eardley, and by Jantsch (1980) who selects the 'rhizome' as an example of a new image of thought (of creative evolution), in which the emphasis is on non-equilibrium, heterogeneity of associations and relations, and the autocatalytic self-reinforcement of fluctuations, and which, as such, constantly cuts across linear historical time: the 'rhizome suspends historical time, not because it falls out of it, but because *it creates historical time*' (1980: 233–4; see also 304–5). On the specific catalytic function of proteins see Monod 1971: 47ff.

24 'Life crystallizes at a critical molecular diversity because catalytic closure itself crystallizes' (Kauffman 1995: 64; see also 274ff.). Catalytic closure, it should be noted, is not to be conceived in terms of a property of any single molecule but is an emergent property of the evolution of a system.

25 For insight into the relation between autocatalycity and autopoiesis see Roque 1985: 117–19, and Prigogine and Stengers 1985: 133–7. Csanyi has noted that while autopoiesis has done some important theoretical work, it is unsuitable for modelling 'real systems' because its emphasis on closed, stable systems provides no adequate understanding of virtual change, in which in an autogenetic cycle components can get closed out, new components can enter, certain components can mutate, and so on (Csanyi 1996: 152–3).

26 In the context of this evolution involving a new distribution it is somewhat short-sighted to define the hand simply in terms of its status as an 'organ'; rather, its evolution involves a coding, a dynamic structuration, and a dynamic formation, in which it gets extended in tools that are themselves active forms and formed matters, and in which tools of production produce artefacts that serve as new tools, and so on, establishing a positive feedback loop involving the assemblage of hand, tool, producer, and product, leading to 'breakages, communications and diffusions, nomadisms and sedentarities, multiple thresholds and speeds of relative deterritorialization in human populations' (Deleuze and Guattari 1980: 79; 1988: 61).

27 For solid Darwinian accounts of human origins which support Deleuze and Guattari's thesis see Foley 1997. Foley adopts an ethological approach but argues that we don't need to choose between biology (or genetics) and ethology (behaviour).

28 In an effort to overcome the reductionism of a certain genetics, in *Difference and Repetition* Deleuze appeals to the 'comparative speeds and slownesses' which inform the movement of the virtual and its actualization (1968: 239ff.; 1994: 185ff.). A complex of genes by itself will produce nothing independent of its reciprocal relations and determinations in the context of the embryo viewed as a 'virtual whole'. Brian Goodwin argues: 'Genes are certainly involved at all stages...but to understand their action we need to understand the dynamic context within which they work, which is the morphogenetic field' (1994: 147). However, it should be noted that even ultra-Darwinians like Richard Dawkins do not advocate genetic reductionism or determinism. On the contrary, Dawkins has insisted that the effect a gene has is *not* a simple property of the gene itself, meaning that genes cannot be construed as containing a blueprint for a body (Dawkins 1991: 296).

29 In his book on 'The Fold' Deleuze seeks to overturn the orthodox reception of Leibniz, and hence his placement in the history of modern thought, advancing a particularly novel interpretation of Leibniz on substance. Deleuze argues that Leibniz's view that every substance draws from within itself all its activity in the world only makes sense when it is read as suggesting that predication is not attribution, so that the concept denotes not a 'logical' being but a 'metaphysical' one (one that exists beyond itself, in excess of itself), neither a generality nor a

universality but an 'individual' that is defined by *predicates-as-events*. In other words, the predicate '*is the proposition itself*', an activity of movement and change, to be grasped not in terms of constant attributes that refer to a given subject but rather in terms of an individual conceived as a virtual becoming or event (becoming what one is). See especially Deleuze 1988: 'Sufficient Reason', 55ff.; 1993: 41ff.; and 88; 65–6 for how this reading of Leibniz makes possible a conception of intrinsic difference. And compare Nietzsche 1968 Section 552, p. 298: 'Duration, identity with itself, being are inherent neither in that which is called subject nor in that which is called object: they are complexes of events apparently durable in comparison with other complexes – e.g., through the difference in tempo of the event (rest–motion, firm–loose: opposites that do not exist in themselves and that only express variations in degree that from a certain perspective appear to be opposites. There are no opposites...)'.

30 It is because societies, primitive and other, reduce animal becomings to relations of totemic and symbolic correspondence that Deleuze and Guattari deem the politics of such becomings to be so ambiguous (1980: 303; 1988: 247–8).

31 Allosteric proteins are molecules whose conformations alter in response to environmental situations. They play a key role in the regulation of critical biochemical pathways, acting as a feedback monitoring device in cybernetic circuits both inside and outside cells. There is a thorough treatment of protein life in Chapters 3, 4, and 5 of Monod's *Chance and Necessity*, covering their catalytic, regulatory, and constructive functions (1971: 45ff.), while their importance has been recently argued for in Cairns-Smith (1996: 76–8). Monod sees the stereospecific interactions of protein regulation and construction playing a crucial role in what he calls the 'teleonomic performance' of a living being, where teleonomy refers to its existence as an 'oriented, coherent, and constructive activity'. He treats allosteric enzymes in Chapter 4, which is on 'microscopic cybernetics', and examines the phenomena of feedback inhibition and feedback activation – which also includes a discussion of the role played by certain 'dark' precursors in the activation of enzymes (62–4). His key insight, articulated in Chapter 5 on 'molecular ontogenesis', is that the process of 'spontaneous and autonomous morphogenesis' rests upon the stereospecific recognition properties of proteins; in other words, 'morphogenesis is primarily a microscopic process that manifests itself in macroscopic structures' (81). On proteins see also Eigen 1992: 72–4, and Jacob 1974: 304–6, where he poses the question of which comes first, proteins or nucleic acids, as a chicken and egg question, noting that without nucleic acids, proteins have no future, while without proteins, nucleic acids remain inert.

32 In fact, there are a number of important sources for Deleuze's conception of the rapport between the inside and the outside, or the interior and exterior, including, in addition to Uexküll, the work of Raymond Ruyer (who figures in many of Deleuze's texts right up to his last collaborative effort with Guattari, *What is Philosophy?*) and Simondon. In Ruyer, for example, the boundary between 'internal' and 'external' is construed not as a frontier between two totally different or independent realms, but rather as a 'junction' between two distinct but connected *circuits*. In addition, Ruyer wishes to grant to the organism what one might call an 'originary technicity', that is, for him the way in which a living system overcomes the limitations of an auto-subjectivity – the problem of entropy or a closed system – is through the development of 'organs' external to those of the organism. This is the function of an 'external circuit'. See Ruyer 1946: 44ff., 268ff. on 'internal and external circuits'; and for insight into Ruyer in the context of work in cybernetics see Tomlin 1955: 133ff. For further insight

into Deleuze's utilization of the likes of Ruyer and Uexküll see Bogue (1997), and forthcoming work by Paul Bains.

33 It was in terms of this kind of reading of Uexküll that Heidegger sought to critique the Darwinian model of selection that can only depict the organism in terms of a passive adaptation. Heidegger lays stress on an originary structure of animality, in which an animal organism is defined as one that is capable of a 'proper peculiarity', namely the possibility of 'retaining itself within itself' (Heidegger 1995: 235). On this reading the term 'organism' does not refer to this or that being, but rather designates *a particular and fundamental manner of being* (ibid.). Animality is not to be defined by recourse to either physiology or morphology simply because an animal becomes what it is on account of its peculiar manner of being, which consists in the fact that it finds itself caught up in a world of its making. It is this insight which explains why Heidegger finds the Darwinian model of adaptation so wanting: 'The organism is not something independent in its own right which then adapts itself. On the contrary, the organism adapts a particular environment *into* it in each case' (264). It also explains Heidegger's quasi-Lamarckism, in which the 'proper peculiarity' of animal life – as opposed to the neither alive nor dead world of the stone – resides in the capability to create organs: in other words, it is not the organs which make the animal, but the animal which engenders, as a being in its own world, the organs peculiar to it. Finally, such a view also serves to explain Heidegger's eschewal of the need to refer to any vital principle (effective force, soul, or consciousness) to account for the dynamic character of animality: the vital character of the animal – that it is to be regarded as a being capable of life – resides in its peculiar becoming in its world. It is in terms of movement, especially movement away from itself, that the animality of the animal is to be defined, a movement in which it not only sustains itself but retains itself as a specific unity and 'gives this unity to itself for the first time' (235). For Heidegger, therefore, an 'ecological' definition of the animality of the animal is one which articulates itself in terms of a specific conception of animal behaviour: 'Behaviour and its forms are not something which radiate outward and allow the animal to run ahead along certain paths. Rather *behaviour* is precisely an *intrinsic retention* and *intrinsic absorption*, although no reflection is involved' (238). It is on account of the fact that the animal is 'captivated' by its world that it is capable of life. However, for Heidegger it is also the closed, all-encompassing, character of this captivation – the fact that it is *so* captivating – which also explains why the animal can be defined as 'poor in the world' and as incapable of a 'historical' existence (compare Adorno, 'The animal's world is devoid of concept', in Adorno and Horkheimer 1979: 246; and Bataille 1992: 18: 'For the animal, nothing is given through time'). Captivation does not denote an enduring state that is present within the animal but defines the 'essential moment' of its animality, such as the specific character of the worm's movement of escape and the mole's movement of pursuit. It is these 'behaviours', which cannot be reduced to simple processes understood as a mere series of events, that are said to define the peculiar environing world of the animal.

34 The Dogon myth is already being read in terms of a 'mythical Weismannism' in *Anti-Oedipus*. See Deleuze and Guatarri 1972: 186; 1984: 158.

35 One could also enlist in support of this thinking beyond reproduction Bataille's conception of general economy. See Bataille 'Laws of General Economy', 1988: 27–41: 'The very principle of living matter requires that the chemical operations of life, which demand an expenditure of energy, be gainful, productive of surpluses' (27). But even an extreme thinker of excess like Bataille commits the error of confusing reproductive sexuality with erotic expenditure, equating the

excess from which reproduction springs with the excess of death: 'We must never forget that the multiplication of beings goes hand in hand with death' (Bataille 1987: 100–1; see also Bataille 1993: 84–6: 'there is no doubt that death is the youth of the world' – *but* within the continuing cycle of species reproduction). As Baudrillard points out 'reproduction as such has no excess' since it is always 'a matter of positive economy and a functional death', one from which the species eternally reproduces itself (Baudrillard 1993: 157).

36 One could read the story of Ursula Brangwen in Lawrence's *The Rainbow* as representing an alternative depiction of the becoming of woman to Hardy's *Tess*. See Lawrence 1995b, especially 263ff., where the narrator speaks of Ursula's passage from girlhood to womanhood as an emerging out of nothingness and undifferentiated mass (the germ-plasm) and assuming the responsibility of new life (263).

37 Just as Bergson's *Matter and Memory* elides the distinction between '*memoire*' and '*souvenir*', so do Deleuze and Guattari in *ATP*. Having declared a 'becoming' to be an '*anti-memoire*', they go on to state almost in the next line that memory [*Le souvenir*] is 'always a function of reterritorialization' (1980: 360).

38 This turn to the concept amounts to a significant departure from the Bergsonism of the 1960s. Bergsonian intuition, Deleuze had argued in his 1966 essay, is concerned with specifying the conditions of real experience, that is, experience that is neither general nor abstract. Conceived as such an enterprise philosophy consists in going beyond the 'human condition' and in the direction of the inhuman and overhuman but *not* in terms of 'going beyond experience toward concepts. For concepts only define, in the Kantian manner, the conditions of all possible experience in general. Here, on the other hand, it is a case of real experience in all its peculiarities. And if we must broaden it, or even go beyond it, this is only in order to find the articulations on which these peculiarities depend' (Deleuze 1966: 19; 1988: 28).

39 It is because capital brings about the reduction of everything to rules of equivalence, effecting a universal neutralization of value, that Baudrillard holds that the experience of nihilism – of the nothing in Nietzsche's sense as well as Bataille's (Bataille 1993: 429–30) – is no longer available to us. We simply have the nihilism of everything and anything, which is the nihilism of the system itself (Baudrillard 1994: 159–64).

CONCLUSION: FOLD AND SUPERFOLD

1 There are a number of sources from which Deleuze takes over the 'prodigious idea' of nonorganic or inorganic life. In addition to Bergson, whom we have focused on in this study, two other important sources are Marx and the historian Wilhelm Worringer on Gothic art. On the latter see Deleuze and Guattari 1980: 512, 622–4; 1988: 411, 498–99; and Worringer 1927, e.g. 41–2, 71–6. In the 'Economic and Philosophical Manuscripts of 1844' Marx employs the notion of inorganic life in terms of a historical account of the evolution of the human as a 'species-being'. The notion is used to fix distinctions between the human and the animal and the human and the machine. Species-life, which consists in living from inorganic nature, is not peculiar to the human but common to animal existence. However, Marx adheres to the view, common to the philosophical tradition (Hegel, Heidegger, and Bataille for whom only the human and its domain of consciousness can be regarded as transcending an 'immanent animality', see Bataille 1989: 'The Poetic Fallacy of Animality', 20–3, and 'The Animal is in the World like Water in Water', 23–5), which claims that while the animal is

a creature of need and *immediately* at one with its life, the human is distinguished by the fact that it makes its activity an object of will and consciousness. The production of life by the animal is caught up in the satisfaction of immediate needs and so, Marx claims, exists without art and artifice: 'Animals produce only according to the standards and needs of the species to which they belong, while man is capable of producing according to the standards of every species and of applying to each object its inherent standard; hence man also produces in accordance with the laws of beauty' (Marx 1975: 329). The challenge of rhizomatics would be to show (a) Marx assumes that we can readily determine and isolate the 'species-life' to which the animal 'belongs'; (b) that all creative 'evolution' and creative ethology involves art and artifice; and (c) that the question of technics only makes sense on the level of assemblages and that these function in terms of the heterogeneity of their components. Communism is posited by Marx as a fully developed 'naturalism' that equals 'humanism' and as a fully developed 'humanism' that *equals* 'naturalism' (348). In other words, nature only exists for man as *anthropological* nature (see the important remarks on the natural sciences and on Industry, 355ff.).

There is a superb critical reading of Marx on the question of nature, especially in relation to the 1844 manuscripts in Merleau-Ponty 1968: 274–5; 1988: 130ff. For Merleau-Ponty the 'return' to a concept of nature never involves a turning away from questions of history, spirit, and the human, but rather laying the ground of a 'solution' to these problems that aims not to be immaterialist: 'an ontology which leaves nature in silence shuts itself in the incorporeal and for this very reason gives a fantastic image of the human, spirit and history' (1988: 130). It is clear from the 'Working Notes' at the end of *Visible and Invisible* that Merleau-Ponty was working on a concept of nature that requires a new philosophy beyond the cleavage of God, man, and creatures (this is Spinoza's division, Merleau-Ponty notes), and which is not simply a philosophy of nature but an account of the 'man–animal *intertwining*' (1968: 274).

2 In his remarks on the 'automatic system of machinery' in the *Grundrisse* Marx notes how, as capital evolves as a self-functioning system, 'dead labour' (objectified labour) becomes the very condition of life and comes to determine the process of production itself. This is a direct result of the transmogrification of labour into the machine or, to be more precise, into a system of machinery that proceeds via an autonomous logic: 'The production process has ceased to be a labour process in the sense of a process dominated by labour as its governing unity. Labour appears, rather, merely as a conscious organ, scattered among the individual living workers at numerous points of the mechanical system' (Marx 1973: 693). The living unity of the system now resides in the 'living (active) machinery' which confronts the human worker 'as a mighty organism' (ibid.). This denotes for Marx the point at which the labour process has become itself a mere moment in the auto-realization of capital. Marx further insists that the development of the means of labour into automatic machinery, and through which objectified labour assumes the form of the force of production itself, is not 'an accidental moment of capital' but rather has to be grasped as 'the historical reshaping of the traditional, inherited means of labour into a form adequate to capital' (694).

3 Deleuze contends that this unlimited finity (the Superfold) is what Nietzsche was naming with the eternal return. But it is clear from the way in which Nietzsche construes the matter as one of finite centres of force being caught up in a play of finite combinations *within infinite time*, that, on this model at least, his thinking of time as eternal return is not at all compatible with 'Bergsonism' for which time has to be conceived in terms of inventive duration. Indeed, Nietzsche was

contra the idea that there was novelty, and what he called 'the power of infinite transformations', in time. No doubt it is this kind of position that led him to advance, in one of its experimental formulations, perhaps its most disabling one, the eternal return as a truly nihilistic thought. When Nietzsche writes about the 'great dice game of existence' in terms of passing through a *calculable* number of combinations, he is conceiving time spatially in terms of an empty container that gets filled up by the eternal return of the same (because finite) configurations of force within discrete cycles of repetition: 'In infinite time, every possible combination would at some time or another be realized; more, it would be realized an infinite number of times' (1968: 1066; 1987: Volume 13, 376). Nietzsche construes the recurrence (*Wiederkehr*) of combinations taking place in terms of the 'circular movement of an absolutely identical series', and argues, no doubt recognizing how close he is to such a position, that this conception is not mechanistic on account of the fact that mechanism would determine a *final state* to this process of repetition, as opposed to an *infinite* recurrence of identical cases. Nietzsche explicitly identifies the quest for a conception of 'creative evolution' – a quest he regards as fundamentally deluded – with Spinoza's *'deus sive natura'* in 1968: 1062; 1987: Volume 11: 556–7 (it is in this passage that he insists that the world lacks the capacity for eternal novelty). The arguments Nietzsche develops in such passages show the limitations of a straightforward cosmological application of the eternal return.

BIBLIOGRAPHY

This bibliography includes all the primary and secondary material referred to in the text, as well as additional material that I consulted in the course of its writing.

Works by Bergson

(1959/1991), *Matière et Mémoire*, Paris, Presses Universitaires de France; *Matter and Memory*, trans. N. M. Paul and W. S. Palmer, New York, Zone Books.

(1960), *Time and Free Will*, trans. F. L. Pogson, New York, Harper Torchbooks.

(1962/1983), *L'Évolution Créatrice*, Paris, PUF; trans. *Creative Evolution*, authorized trans. A. Mitchell, Lanham, MD, University Press of America.

(1963), *The Two Sources of Morality and Religion*, trans. R. Ashley Audra and C. Brereton, Indiana, University of Notre Dame Press.

(1965), *The Creative Mind*, trans. M. L. Andison, Totowa, NJ, Littlefield, Adams & Co.

(1965b), *Duration and Simultaneity (with Reference to Einstein's Theory)*, trans. L. Jacobson, Indianapolis, Bobbs-Merrill.

(1972), *Mélanges*, annotated by A. Robinet, Paris, PUF.

(1975), *Mind-Energy*, trans. H. Wildon Carr, Westport, CT, Greenwood Press.

Works on Bergson

Adamson, G. (1997), 'Seditious Duration: Bergson and Serres', *Antithesis* (special issue on 'Time and Memory') (8: 2), 239–57.

Burwick, F. and Douglass, P. (eds) (1992), *The Crisis in Modernism: Bergson and the Vitalist Controversy*, Cambridge, Cambridge University Press.

Capek, M. (1971), *Bergson and Modern Physics*, Dordrecht, D. Reidel.

——(1992), 'Microphysical Indeterminacy and Freedom: Bergson and Pierce', in F. Burwick and P. Douglass (eds), *The Crisis in Modernism: Bergson and the Vitalist Controversy*, Cambridge, Cambridge University Press, 171–89.

de Issekutz Wolsky, M. and Wolsky, A. A., 'Bergson's Vitalism in the Light of Modern Biology', in F. Burwick and P. Douglass (eds), *The Crisis in Modernism: Bergson and the Vitalist Controversy*, Cambridge, Cambridge University Press, 153–70.

James, W. (1977), *A Pluralistic Universe*, Cambridge, MA, Harvard University Press.

Lacey, A. R. (1993), *Bergson*, London, Routledge.

Lewis, W. (1927), *Time and Western Man*, London, Chatto & Windus.

Lovejoy, A. O. (1961), *The Reason, the Understanding, and Time*, Baltimore, The Johns Hopkins Press.

Merleau-Ponty, M. (1964), 'Bergson in the Making', *Signs* (1964c), trans. R. C. McCleary, Evanston, Northwestern University Press, 182–92.

——(1988), 'Bergson', in *In Praise of Philosophy and Other Essays*, trans. J. Wild, J. Edie, J. O' Neill, Evanston, Northwestern University Press, 9–33.

Moore, F. C. T. (1996), *Bergson: Thinking Backwards*, Cambridge, Cambridge University Press.

Mullarkey, J. (1995), 'Bergson's Method of Multiplicity', *Metaphilosophy* (26: 3), 230–59.

——(ed.) (1998, forthcoming), *The New Bergson*, Manchester, Manchester University Press.

Pilkington, A. E. (1976), *Bergson and His Influence: A Reassessment*, Cambridge, Cambridge University Press.

Rostrevor, G. (1921), *Bergson and Future Philosophy: An Essay on the Scope of Intelligence*, London, Macmillan.

Russell, B. (1914), *The Philosophy of Bergson*, London, Macmillan.

——(1961), 'Bergson', in *History of Western Philosophy*, London, George Allen & Unwin, 2nd edn), 756–66.

Serres, M. (1977), 'Boltzmann et Bergson', in Serres, *Hermes IV: La Distribution*, Paris, Éditions de Minuit, 127–42.

Solomon, J. (1911), *Bergson*, London, Constable & Co.

Works by Deleuze

(1956), 'Bergson: 1859–1941', in M. Merleau-Ponty (ed.), *Les philosophes célèbres*, Paris, Éditions Lucien Mazenod, 292–9.

(1956/1997), 'La conception de la différence chez Bergson', *Les Études Bergsoniennes*, 4, Paris, Presses Universitaires de France, 77–113; 'Bergson's conception of difference', unpublished translation by M. McMahon, 1–25.

——(1962/1983), *Nietzsche et la philosophie*, Paris, PUF; *Nietzsche and Philosophy*, trans. H. Tomlinson, London, Athlone Press.

(1963/1984), *La philosophie critique de Kant. Doctrine des facultés*, Paris, PUF; *Kant's Critical Philosophy. The Doctrine of the Faculties*, trans. H. Tomlinson and B. Habberjam, London, Athlone Press.

(1966/1988), *Le Bergsonisme*, Paris, PUF; *Bergsonism*, trans. H. Tomlinson and B. Habberjam, New York, Zone Books.

(1968/1992), *Spinoza et le problème de l'expression*, Paris, Les Éditions de Minuit; *Expressionism in Philosophy: Spinoza*, trans. M. Joughin, New York, Zone Books.

(1968/1994), *Différence et répétition*, Paris, PUF; *Difference and Repetition*, trans. P. Patton, London, Athlone Press.

(1969/1990), *Logique du sens*, Paris, Les Éditions de Minuit; *Logic of Sense*, trans. M. Lester with C. Stivale, London, Athlone Press.

(1977/1987), *Dialogues*, Paris, Flammarion; *Dialogues with Clare Parnet*, trans. H. Tomlinson and B. Habberjam, London, Athlone Press.

(1981/1988), *Spinoza: Philosophie pratique*, Paris, Les Éditions de Minuit; *Spinoza: Practical Philosophy*, trans. R. Hurley, San Francisco, City Light Books.

(1983/1986), *Cinéma 1. L'Image-mouvement*, Paris, Les Éditions de Minuit; *Cinema 1. The Movement-Image*, trans. H. Tomlinson and B. Habberjam, London, Athlone Press.

(1985/1989), *Cinéma 2. L'Image-temps*, Paris, Les Éditions de Minuit; *Cinema 2. The Time-Image*, trans. H. Tomlinson and R. Galeta, London, Athlone Press.

(1986/88), *Foucault*, Paris, Les Éditions de Minuit; *Foucault*, trans. S. Hand, London, Athlone Press.

(1988/1993), *Le Pli. Leibniz et le Baroque*, Paris, Les Éditions de Minuit; *The Fold. Leibniz and the Baroque*, trans. T. Conley, London, Athlone Press.

(1990/1995), *Pourparlers 1972–1990*, Paris, Les Éditions de Minuit; *Negotiations 1972–1990*, trans. M. Joughin, New York, Columbia University Press.

(1991), *Empiricism and Subjectivity. An Essay on Hume's Theory of Human Nature*, trans. with an Introduction by C. V. Boundas, New York, Columbia University Press.

(1993), *Critique et clinique*, Paris, Les Éditions de Minuit.

(1997a), 'Desire and Pleasure', trans. D. W. Smith, in A. I. Davidson (ed.), *Foucault and his Interlocutors*, Chicago, University of Chicago Press, 183–95.

(1997b), 'Immanence: A Life…', trans. N. Millett, *Theory, Culture, and Society* (14: 2), 3–9.

Works by Deleuze and Guattari

(1972/1984), *L'Anti-Oedipe*, Paris, Les Éditions de Minuit; *Anti-Oedipus*, trans. R. Hurley, M. Seem, and H. R. Lane, London, Athlone Press.

(1973/1995), 'Bilan programme pour machines désirantes', appendix to *L'Anti-Oedipe*, Paris, Les Éditions de Minuit, 1972/3, 463–87; 'Balance-Sheet Program for Desiring Machines', trans. R. Hurley in F. Guattari, *Chaosophy*, ed. S. Lotringer, New York, *Semiotext(e)*, 119–50.

(1975/1986), *Kafka. Pour une littérature mineure*, Paris, Les Éditions de Minuit; *Kafka. Toward a Minor Literature*, trans. D. Polan, Minneapolis, University of Minnesota Press.

(1980/1988), *Mille Plateaux*, Paris, PUF; *A Thousand Plateaus*, trans. B. Massumi, University of Minnesota Press.

(1991/1994), *Qu'est-ce que la philosophie?*, Paris, Les Éditions de Minuit; *What is Philosophy?*, trans. G. Burchell and H. Tomlinson, London, Verso.

Works on Deleuze

Agamben, G. (1998), 'L'immanence absolue', in E. Alliez, *Gilles Deleuze. Une vie philosophique*, Institut Synthélabo pour le progrès de la connaissance, Les Plessis-Robinson, 165–89.

Alliez, E. (1993), *La Signature du monde: Qu'est-ce que la philosophie de Deleuze et Guattari?*, Paris, Les Éditions du Cerf.

——(1997), 'Questionnaire on Deleuze', *Theory, Culture, and Society* 14(2), 81–9.

——(1998), 'Sur le bergsonisme de Deleuze', in E. Alliez, *Deleuze. Une vie philosophique*, Institut Synthélabo pour le progrès de la connaissance, Les Plessis-Robinson, 243–65.

Ansell Pearson, K. (ed.) (1997), *Deleuze and Philosophy. The Difference Engineer*, London, Routledge.

Badiou, A. (1994), 'Gilles Deleuze, The Fold: Leibniz and the Baroque', in C. V. Boundas and D. Olkowski, *Gilles Deleuze and the Theater of Philosophy*, London and New York, Routledge, 51–73.

——(1997), *Deleuze. 'La clameur de l'Être'*, Hachette.

Bogue, R. (1989), *Deleuze and Guattari*, London, Routledge.

——(1997), 'Art and Territory', in I. Buchanan (ed.), *A Deleuzian Century?*, special issue of *The South Atlantic Quarterly* (96:3), 465–83.

Boundas, C. V. (1996), 'Deleuze-Bergson; an Ontology of the Virtual', in P. Patton (ed.), *Deleuze: A Critical Reader*, Oxford, Basil Blackwell, 81–107.

Boundas, C. V. and Olkowski, D. (eds) (1994), *Deleuze and the Theater of Philosophy*, New York, Routledge.

Buchanan, I. (ed.) (1997), *A Deleuzian Century?*, special issue of *The South Atlantic Quarterly* (96: 3).

Caygill, H. (1997), 'The Topology of Selection: The Limits of Deleuze's Biophilosophy', in K. Ansell Pearson (ed.), *Deleuze and Philosophy: The Difference Engineer*, London, Routledge, 149–63.

Clemens, J. (1997), 'A Thousand Tiny Stupidities: Why I Hate Deleuze (and Guattari)', *Antithesis* (8: 2), 179–203.

Conley, T. (1997), 'From Multiplicities to Folds: On Style and Form in Deleuze', in I. Buchanan (ed.), *A Deleuzian Century?*, special issue of *The South Atlantic Quarterly* (96:2), 629–46.

de Landa, M. (1992), 'Nonorganic Life', in J. Crary and S. Kwinter, *Incorporations: Zone 6*, New York, Zone Books, 128–68.

dos Santos, L. G. (1998), 'Code primitif/code génétique: la consistance d'un voisinage', in E. Alliez (ed.), *Gilles Deleuze. Une vie philosophique*, Institut Synthélabo pour le progrès de la connaissance, Les Plessis-Robinson, 421–9.

Douglass, P. (1992), 'Deleuze's Bergson: Bergson redux', in F. Burwick and P. Douglass, *The Crisis in Modernism: Bergson and the Vitalist Controversy*, Cambridge, Cambridge University Press, 368–87.

——(1992b), 'Deleuze and the Endurance of Bergson', *Thought* (67), 47–61.

Foucault, M. (1977), 'Theatrum Philosophicum', in *Language, Counter-Memory, and Practice*, trans. D. F. Bouchard and S. Simon, Oxford, Basil Blackwell, 165–99.

Gatens, M. (1996), 'Through a Spinozist Lens: Ethology, Difference, Power', in P. Patton (ed.), *Deleuze. A Critical Reader*, Oxford, Basil Blackwell, 162–88.

Goodchild, P. (1994), *Gilles Deleuze and the Question of Philosophy*, London, Associated University Presses.

——(1996), *Deleuze and Guattari; An Introduction to the Politics of Desire*, London, Sage.

——(1997), 'Deleuzian Ethics', *Theory, Culture, and Society* (14: 2), 39–51.

Grosz, E. (1994), 'A Thousand Tiny Sexes: Feminism and Rhizomatics', in C. V. Boundas and D. Olkowski (eds), *Deleuze and the Theater of Philosophy*, New York, Routledge, 187–213.

Hallward, P. (1997), 'Gilles Deleuze and the Redemption from Interest', *Radical Philosophy* (81), 6–22.

Hardt, M. (1993), *Gilles Deleuze: An Apprenticeship in Philosophy*, London, UCL Press.

Heffernan, N. (1994), 'Oedipus Wrecks? Or, Whatever Happened to Deleuze and Guattari? Rereading *Capitalism and Schizophrenia*', in B. McGuirk (ed.), *Redirections in Critical Theory: Truth, Self, Action, History*, London, Routledge, 111–65.

Holland, E. W. (1991), 'Deterritorializing "Deterritorialization" – From the *Anti-Oedipus* to *A Thousand Plateaus*', *SubStance* (66), 55–65.

——(1997), 'Marx and Poststructuralist Philosophies of Difference', in I. Buchanan (ed.), *A Deleuzian Century?*, special issue of *The South Atlantic Quarterly* (96: 3), 525–43.

Jameson, F. (1997), ' Marxism and Dualism in Deleuze', in I. Buchanan (ed.), *A Deleuzian Century?*, special issue of *The South Atlantic Quarterly* (96:3), 393–417.

Lebrun, G. (1998), 'Le transcendental et son image', in E. Alliez (ed.), *Deleuze. Une vie philosophique*, Institut Synthélabo pour le progrès de la connaissance, Les Plessis-Robinson, 207–33.

Macherey, P. (1996), 'The Encounter with Spinoza', in P. Patton (ed.), *Deleuze: A Critical Reader*, Oxford, Basil Blackwell, 139–61.

Mackenzie, I. (1997), 'Creativity as Criticism: The Philosophical Constructivism of Deleuze and Guattari', *Radical Philosophy* (86), 7–19.

Malabou, C. (1996), 'Who's Afraid of Hegelian Wolves?', in P. Patton, *Deleuze: A Critical Reader*, Oxford, Basil Blackwell, 114–39.

Massumi, B. (1992), *A User's Guide to Capitalism and Schizophrenia*, Cambridge, MA, MIT Press.

——(1996), 'The Autonomy of Affect', in P. Patton (ed.), *Deleuze: A Critical Reader*, Oxford, Basil Blackwell, 217–40.

May, T. (1994), 'Difference and Unity in Gilles Deleuze', in C. V. Boundas and D. Olkowski (eds), *Deleuze and The Theater of Philosophy*, New York, Routledge, 33–51.

Mullarkey, J. (1997), 'Deleuze and Materialism: One or Several Matters', in I. Buchanan, *A Deleuzian Century?*, special issue of *The South Atlantic Quarterly* (96:3), 439–65.

Patton, P. (1994), 'Anti-Platonism and Art', in C. V. Boundas and D. Olkowski (eds), *Deleuze and the Theater of Philosophy*, New York, Routledge, 141–57.

——(1996), 'Introduction', P. Patton (ed.), *Deleuze: A Critical Reader*, Oxford, Basil Blackwell, 1–18.

Prado Jnr, B. (1998), 'Sur le "plan d'immanence" ', in E. Alliez, *Deleuze. Une vie philosophique*, Institut Synthélabo pour le progrès de la connaissance, Les Plessis-Robinson, 305–25.

Rodowick, D. N. (1997), 'The Memory of Resistance', in I. Buchanan, *A Deleuzian Century?*, special issue of *The South Atlantic Quarterly* (96:3), 417–39.

Rose, G. (1984), 'The New Bergsonism: Deleuze', in G. Rose, *Dialectic of Nihilism*, Oxford, Basil Blackwell, 87–108.

Smith, D. W. (1996), 'Deleuze's Theory of Sensation: Overcoming the Kantian Duality', in P. Patton (ed.), *Deleuze: A Critical Reader*, Oxford, Basil Blackwell, 29–57.

Williams, J. (1997), ' Deleuze on J. M. W. Turner: Catastrophism in Philosophy?', in K. Ansell Pearson (ed.), *Deleuze and Philosophy: The Difference Engineer*, London and New York, Routledge, 233–47.

Zourabichvili, F. (1996), 'Six Notes on the Percept', in P. Patton (ed.), *Deleuze: A Critical Reader*, Oxford, Basil Blackwell, 188-217.

Works by Guattari

(1992/1995), *Chaosmose*, Paris, Galilée; *Chaosmosis: an ethico-aesthetic paradigm*, trans. P. Bains and J. Pefanis, Sydney, Power Publications.

(1996) *The Guattari Reader*, ed. G. Genosko, Oxford, Basil Blackwell.

Other works in philosophy, theory, science, and the philosophy of science

Adorno, T. W. (1966/1973), *Negative Dialektik*, Frankfurt am Main, Suhrkamp; *Negative Dialectics*, trans. E. B. Ashton, London, Routledge & Kegan Paul.

Adorno, T. W. and Horkheimer, M. (1979), *Dialectic of Enlightenment*, trans. J. Cumming, London, Verso.

Aldridge, S. (1996), *The Thread of Life: The Story of Genes and Genetic Engineering*, Cambridge, Cambridge University Press.

Alliez, E. (1995), *De l'impossibilité de la phénoménologie. Sur la philosophie française contemporaine*, Paris, Librairie Philosophique J. Vrin.

Ansell Pearson, K. (1997), *Viroid Life: Perspectives on Nietzsche and the Transhuman Condition*, London, Routledge.

——(1997b), 'Life Becoming Body: On the "Meaning" of Post Human Evolution', *Cultural Values* (1: 2), 219–41.

——(1998, forthcoming), 'Creative Evolution/Involution: Exposing the Transcendental Illusion of Organismic Life', in J. Mullarkey (ed.), *The New Bergson*, Manchester, Manchester University Press.

——(1999, forthcoming), 'Spectropoiesis and Rhizomatics: Learning to Live with Death and Demons', in G. Banham and C. Blake (eds), *Evil Spirits: Nihilism and the Fate of Modernity*, Manchester, Manchester University Press.

Appel, T. A. (1987), *The Cuvier–Geoffroy Debate: French Biology in the Decades before Darwin*, Oxford, Oxford University Press.

Aquinas, T. (1988), *The Philosophy of Thomas Aquinas: Introductory Readings*, ed. C. Martin, London, Routledge.

Aristotle (1985), *Metaphysics* (Books VII–X), trans. M. Furth, Indianapolis, Hackett.

——(1986), *De Anima (On the Soul)*, trans. H. Lawson-Tancred, Harmondsworth, Middx., Penguin.

Atkinson, P. (1997), 'The Morphology of Time', *Antithesis* (8: 2), 217–39.

Balibar, E. (1997), *Spinoza: From Individuality to Transindividuality*, Delft, Eburon (Mededelingen vanwege het Spinozahuis 71), 3–36.

Barnett, J. S. (1988), *Biology and Freedom: An Essay on the Implications of Human Ethology*, Cambridge, Cambridge University Press.

Barrow, J. D. and Tipler, F. (1996), *The Anthropic Cosmological Principle*, Oxford, Oxford University Press.

Bataille, G. (1955), *Prehistoric Painting. Lascaux or the Birth of Art*, trans. A. Wainhouse, Lausanne, Skira and London, Macmillan.

——(1987), *Eroticism*, trans. M. Dalwood, London, Marion Boyars.

——(1988), *The Accursed Share* (Vol. 1), trans. R. Hurley, New York, Zone Books.

——(1992), *The Theory of Religion*, trans. R. Hurley, New York, Zone Books.

——(1993), *The Accursed Share* (Vols 2 and 3), trans. R. Hurley, New York, Zone Books.

Bateson, G. (1973), *Steps to an Ecology of the Mind*, St Albans, Paladin.

——(1978), *Mind and Nature*, London, Fontana Collins.

Battersby, C. (1998), *The Phenomenal Woman: Feminist Metaphysics and the Patterns of Identity*, Oxford, Polity Press.

Baudrillard, J. (1993), *Symbolic Exchange and Death*, trans. I. H. Grant, London, Sage.

——(1994), *Simulacra and Simulation*, trans. S. F. Glaser, Ann Arbor, University of Michigan Press.

Beardsworth, R. (1996), 'Nietzsche, Freud, and the Complexity of the Human: Towards a Philosophy of Failed Digestion', *Tekhnema: A Journal of Philosophy and Technology* 3 (Spring), 113–41.

——(1998), 'Thinking Technicity', *Cultural Values* (2: 1), 70–87.

Behe, M. J. (1996), *Darwin's Black Box: The Biochemical Challenge to Evolution*, London and New York, The Free Press.

Benjamin, A. (1993), *The Plural Event. Descartes, Hegel, Heidegger*, London, Routledge.

Benjamin, W. (1979), *Illuminations*, trans. H. Zohn, London, Fontana/Collins.

——(1996), *Selected Writings. Volume 1: 1913–1926*, ed. M. Bullock and M. W. Jennings, Cambridge, MA, and London, The Belknap Press of Harvard University Press.

Boothby, R. (1991), *Death and Desire*, London, Routledge.

Burkhardt, Jr, R. W. (1995), *The Spirit of System: Lamarck and Evolutionary Biology*, Cambridge, MA., Harvard University Press.

Cairns-Smith, A. G. (1996), *Evolving the Mind: On the Nature of Matter and the Origin of Consciousness*, Cambridge, Cambridge University Press.

Canguilhem, G. (1992), 'Machine and Organism', in J. Crary and S. Kwinter, *Incorporations*, New York, Zone Books, 44–70.

——(1994), *A Vital Rationalist: Selected Writings*, trans. A. Goldhammer, New York, Zone Books.

——(1997), 'On *Histoire de la folie* as an Event', in A. I. Davidson (ed.), *Foucault and his Interlocutors*, Chicago, University of Chicago Press, 28–32.

Čapek, M. (ed.) (1976), *The Concepts of Space and Time*, Dordrecht, D. Reidel.

Cilliers, P. (1997), *Complexity and Postmodernism: Understanding Complex Systems*, London, Routledge.

Clarke, A. (1997), *Being There. Putting Brain, Body, and World Together Again*, Cambridge, MA, MIT Press.

Compton, J. J. (1988), 'Phenomenology and the Philosophy of Nature', *Man and World* (21), 65–89.

Coveney, P. and Highfield, R. (1995), *Frontiers of Complexity*, London, Faber & Faber.

Csanyi, V. (1989), *Evolutionary Systems and Society: A General Theory of Life, Mind, and Culture*, Durham and London, Duke University Press.

——(1996), 'Organization, Function, and Creativity in Biological and Social Systems', in E. L. Khalil and K. E. Boulding (eds), *Evolution, Order, and Complexity*, London and New York, Routledge, 146–81.

Csanyi, V. and Kampis, G. (1985), 'Autogenesis: The Evolution of Replicative Systems', *Journal of Theoretical Biology* (114), 303–21.

Dale, C. (1997), 'Falling from the Power to Die', *Antithesis*, special issue on 'Time and Memory', (8: 2), 139–54.

Darwin, C. (1985), *The Origin of Species*, Harmondsworth, Middx., Penguin.

——(1987), *Notebooks 1836–1844: Geology, Transmutation of Species, Metaphysical Enquiries*, transcribed and ed. by P. H. Barrett *et al.*, Cambridge, Cambridge University Press.

——(1993), *The Portable Darwin*, ed. D. M. Porter and P. W. Graham, Harmondsworth, Middx., Penguin.

Davidson, D. (1980), 'The Individuation of Events' (1969), in *Actions and Events*, Oxford, Clarendon Press.

Davies, P. (1995), *About Time*, Harmondsworth, Middx., Penguin.

Dawkins, R. (1983), *The Extended Phenotype*, Oxford, Oxford University Press.

——(1989), *The Selfish Gene*, Oxford, Oxford University Press.

——(1990), 'Parasites, Desiderata Lists and the Paradox of the Organism', *The Evolutionary Biology of Parasitism*, Symposia of the British Society for Parasitology, Vol. 27, ed. A. E. Keymer and A. F. Read, 63–73.

——(1991), *The Blind Watchmaker*, Harmondsworth, Middx., Penguin.

——(1995), *River out of Eden*, London, Weidenfeld & Nicolson.

de Chardin, T. (1965), *The Phenomenon of Man*, London, Collins.

Deely, J. (1994), *New Beginnings: Early Modern Philosophy and Postmodern Thought*, Toronto, University of Toronto Press.

Dennett, D. C. (1995), *Darwin's Dangerous Idea: Evolution and the Meanings of Life*, London, Allen Lane.

Depew, D. J. and Weber, B. H. (1988), 'Consequences of Nonequilibrium Thermodynamics for the Darwinian Tradition', in B. H. Weber, D. J. Depew, and J. D. Smith, *Entropy, Information, and Evolution: New Perspectives on Physical and Biological Evolution*, Cambridge, MA, MIT Press, 317–54.

——(1996), *Darwinism Evolving: Systems Dynamics and the Genealogy of Natural Selection*, Cambridge, MA, MIT Press.

Derrida, J. (1978), *Writing and Difference*, trans. A. Bass, London, Routledge & Kegan Paul.

——(1979), 'Living On', trans. J. Hulbert, in H. Bloom *et al.*, *Deconstruction and Criticism*, New York, Continuum, 75–176.

——(1981), *Dissemination*, trans. B. Johnson, London, Athlone Press.

——(1986), *Glas*, trans. J. P. Leavey Jr and R. Rand, London and Lincoln, NB, University of Nebraska Press.

——(1987), *Positions*, trans. A. Bass, London, Athlone Press.

——(1993), *Aporias*, trans. T. Dutoit, Stanford, CA, Stanford University Press.

——(1994), *Spectres of Marx*, trans. P. Kamuf, London, Routledge.

——(1994b), 'Nietzsche and the Machine' (Interview with Richard Beardsworth), *Journal of Nietzsche Studies* 7 (Spring), 7–67.

——(1995), *The Gift of Death*, trans. D. Wills, Chicago, University of Chicago Press.

——(1997), *Politics of Friendship*, trans. G. Collins, London, Verso.

Descartes, R. (1975), *A Discourse on Method; Meditations and Principles*, trans. J. Veitch, London, Dent.

Deutsch, D. (1997), *The Fabric of Reality*, London, Allen Lane.

Dobzhansky, T. and Boesiger, E. (1983), *Human Culture. A Moment in Evolution*, New York, Columbia University Press.

Duns Scotus, J. (1987), *Philosophical Writings*, trans. A. Wolter, Indianapolis, Hackett.

Edelman, G. (1994), *Bright Air, Brilliant Fire: On the Matter of the Mind*, Harmondsworth, Middx., Penguin.

Eigen, M. (1992), *Steps Towards Life: A Perspective on Evolution*, Oxford, Oxford University Press.

Ellenberger, H. F. (1970), *The Discovery of the Unconscious*, Chapter 7, 'Sigmund Freud and Psychological Analysis', New York, Basic Books, 418–571.

Ereshefsky, M. (ed.) (1992), *The Units of Evolution: Essays on the Nature of Species*, Cambridge, MA, MIT Press.

Fichte, J. G. (1962/1848), *Die Anweisung zum Seligen Leben*, Stuttgart, Verlag Freies Geistesleben; *The Way Towards the Blessed Life*, trans. W. Smith, London, John Chapman.

——(1987), *The Vocation of Man*, trans. P. Preuss, Indianapolis, Hackett.

——(1994), *Introductions to the Wissenschaftslehre and Other Writings*, trans. and ed. D. Breazeale, Indianapolis, Hackett.

Foley, R. (1997), *Humans Before Humanity*, Oxford, Basil Blackwell.

Foucault, M. (1976), *The Birth of the Clinic*, trans. A. M. Sheridan, London, Routledge.

——(1977), 'Theatrum Philosophicum' in *Language, Counter-Memory, and Practice*, trans. D. F. Bouchard and S. Simon, Oxford, Basil Blackwell, 165–99.

——(1986), *The Order of Things: An Archaeology of the Human Sciences*, London, Routledge.

——(1992), *The Archaeology of Knowledge*, trans. A. M. Sheridan Smith, London, Routledge.

——(1997), *Ethics: the Essential Works 1*, ed. P. Rabinow, London, Allen Lane.

Fox, R. F. (1988), *Energy and the Evolution of Life*, New York, W. H. Freeman.

Freud, S. (1987), 'Thoughts for the Times on War and Death' (1915) and 'Civilization and Its Discontents' (1922), in Freud, *Civilization, Society, and Religion*, (Vol. 12 of the Penguin Freud Library), Harmondsworth, Middx., Penguin, 77–91; 243–341.

——(1990), 'Totem and Taboo' (1913), in Freud, *The Origins of Religion* (Vol. 13 of the Penguin Freud Library), Harmondsworth, Middx., Penguin, 43–225.

——(1990b), *Case Histories I* (Vol. 8 of the Penguin Freud Library), Harmondsworth, Middx., Penguin.

——(1991), 'Beyond the Pleasure Principle' (1920), in Freud, *On Metapsychology* (Vol. 11 of the Penguin Freud library), Harmondsworth, Middx., Penguin, 269–339.

——(1991b), 'Anxiety and Instinctual Life', in *New Introductory Lectures on Psychoanalysis* (1933 [1932]) (Vol. 2 of the Penguin Freud Library), Harmondsworth, Middx., Penguin, 113–45.

——(1991c), *Case Histories II* (Vol. 9 of the Penguin Freud Library), Harmondsworth, Middx., Penguin.

Frolov, I. T. (1991), *Philosophy and History of Genetics: the Inquiry and the Debates*, London, Macdonald & Co.

Georgescu-Roegen, N. (1971), *The Entropy Law and the Economic Press*, Cambridge, MA, Harvard University Press.

Glass, B. (1959), 'The Germination of the Idea of Biological Species', in B. Glass *et al.*, *Forerunners of Darwin 1745–1859*, Baltimore, Johns Hopkins University Press.

Goodwin, B. (1994), 'Organisms and Minds: The Dialectics of the Animal–Human Interface in Biology' in T. Ingold (ed.), *What is an Animal?*, London, Routledge, 100–10.

——(1995), *How the Leopard Changed its Spots*, London, Phoenix.

Goodwin, B. and Saunders, P. (eds) (1988), *Theoretical Biology: Epigenetic and Evolutionary Order from Complex Systems*, Edinburgh, Edinburgh University Press.

Gould, S. J. (1977), *Ontogeny and Phylogeny*, Cambridge, MA, Harvard University Press.

——(1980), 'Is a New and General Theory of Evolution Emerging?', *Paleobiology* (6: 1), 119–30.

——(1996), *Life's Grandeur*, London, Jonathan Cape.

Grosz, E. (1995), *Space, Time, and Perversion*, London, Routledge.

Hampshire, S. (1981), *Spinoza*, Harmondsworth, Middx., Penguin.

Harman, P. M. (1982), *Energy, Force, and Matter: The Conceptual Development of Nineteenth-Century Physics*, Cambridge, Cambridge University Press.

Hegel, G. W. F. (1970/1980), *Phaenomeologie des Geistes*, Frankfurt, Suhrkamp; *Phenomenology of Spirit*, trans. A. V. Miller, Oxford, Clarendon Press.

——(1970), *Philosophy of Nature* (Vol. 3), trans. M. J. Petry, London, Allen & Unwin.

——(1995), *Logic*, trans. A. V. Miller, New York, Humanities Press.

——(1995b), *Lectures on the History of Philosophy: Medieval and Modern Philosophy*, trans. E. S. Haldane and F. H. Simson, Lincoln, University of Nebraska Press.

Heidegger, M. (1991), *The Principle of Reason*, trans. R. Lilly, Bloomington, Indiana University Press.

——(1995), *The Fundamental Concepts of Metaphysics: World, Finitude, Solitude*, trans. W. McNeill and N. Walker, Bloomington, Indiana University Press.

Hjelmslev, L. (1969), *Prologemena to a Theory of Language*, trans. F. J. Whitfield, Madison, University of Wisconsin Press.

Ho, M. W. and Saunders, P. T. (1984), *Beyond Neo-Darwinism: An Introduction to the New Evolutionary Paradigm*, London, Academic Press.

Hume, D. (1965), 'On Suicide', in Hume *Of the Standard of Taste and Other Essays*, Indianapolis, Bobbs-Merrill, 151–61.

——(1985), *A Treatise of Human Nature*, Harmondsworth, Middx., Penguin.

——(1993), *An Enquiry Concerning Human Understanding*, Indianapolis, Hackett.

Ingold, T. (ed.) (1994), *What is an Animal?*, London, Routledge.

Jacob, F. (1974), *The Logic of Living Systems*, trans. B. E. Spillman, London, Allen Lane.

Jameson, F. (1990), *Late Marxism. Adorno, or, The Persistence of the Dialectic*, London, Verso.

Jantsch, E. (1980), *The Self-Organizing Universe*, Oxford, Pergamon Press.

Jantsch, E. and Waddington, C. H. (eds.) (1977), *Evolution and Consciousness: Human Systems in Transition*, Reading, MA, Addison-Wesley.

Johnson, C. (1993), *System and Writing in the Philosophy of Jacques Derrida*, Cambridge, Cambridge University Press.

Johnson, L. (1988), 'The Thermodynamic Origin of Ecosystems: A Tale of Broken Symmetry', in B. H. Weber, D. J. Depew and J. D. Smith, *Entropy, Information, and Evolution: New Perspectives on Physical and Biological Evolution*, Cambridge, MA, MIT Press, 75–105.

Jonas, H. (1973), 'Spinoza and the Theory of the Organism', in M. Grene (ed.), *Spinoza: A Collection of Critical Essays*, New York, Anchor Books.

Kampis, G. (1991), *Self-Modifying Systems in Biology and Cognitive Science*, Oxford, Pergamon Press.

Kant, I. (1974/1978), *Kritik der reinen Vernunft*, Frankfurt, Suhrkamp; *Critique of Pure Reason*, trans. N. Kemp Smith, London, Macmillan.

——(1974/1982), *Kritik der Urteilskraft*, Frankfurt, Suhrkamp; *Critique of Judgement*, trans. J. C. Meredith, Oxford, Oxford University Press.

——(1991), 'What is Orientation in Thinking?', in *Kant: Political Writings*, ed. H. Reiss, Cambridge, Cambridge University Press, 237–50.

Kass, L. R. (1978), 'Teleology and Darwin's *The Origin of Species*: Beyond Chance and Necessity', in S. F. Spicker (ed.), *Organism, Medicine, and Metaphysics: Essays in Honour of Hans Jonas on his 75th Birthday*, Dordrecht, D. Reidel, 97–120.

Kauffman, S. A. (1993), *The Origins of Order: Self-Organization and Selection in Evolution*, Oxford, Oxford University Press.

——(1995), *At Home in the Universe: The Search for Laws of Complexity*, London, Viking.

Kennedy, J. S. (1992), *The New 'Anthropomorphism'*, Cambridge, Cambridge University Press.

Klossowski, P. (1969), *Nietzsche et le cercle vicieux*, Mercure de France.

Lacan, J. (1992), *The Ethics of Psychoanalysis*, trans. D. Porter, London, Routledge.

Lamarck, J. B. (1914), *Zoological Philosophy: An Exposition with Regard to the Natural History of Animals*, London, Macmillan.

——(1964), *Hydrogeology*, trans. A. V. Carozzi, Urbana, University of Illinois Press.

Leibniz, G. W. (1973), *Philosophical Writings*, ed. G. H. R. Parkinson, London, Dent.

——(1991), *Leibniz's Monadology* (An Edition for Students), trans. and ed. N. Rescher, London, Routledge.

Lenoir, T. (1982), *The Strategy of Life: Teleology and Mechanics in Nineteenth Century Biology*, Dordrecht, D. Reidel.

Lombard, L. B. (1998), 'Ontologies of Events', in S. Laurence and C. Macdonald (eds), *Contemporary Readings in the Foundations of Metaphysics*, Oxford, Basil Blackwell, 277–94.

Lyotard, J. F. (1989), 'The Sign of History', *The Lyotard Reader*, ed. A. Benjamin, Oxford, Basil Blackwell, 393–412.

——(1991), *The Inhuman: Reflections on Time*, trans. G. Bennington and R. Bowlby, Stanford, CA, Stanford University Press.

——(1993a), *Political Writings*, trans. B. Readings and K. P. Geiman, London, UCL Press.

——(1993b), *Moralités postmodernes*, Paris, Galilée.

——(1993c), *Libidinal Economy*, trans. I. Hamilton Grant, London, Athlone Press (notably 'The Desire Named Marx', 95–155).

——(1994), 'Nietzsche and the Inhuman', interview with R. Beardsworth, *Journal of Nietzsche Studies* (7), 67–131.

McFarland, D. (1996), *Animal Behaviour*, Harlow, Essex, Addison Wesley Longman.

Macherey, P. (1979), *Hegel ou Spinoza*, Paris, François Maspero.

Marcuse, H. (1968), *Negations: Essays in Critical Theory*, trans. J. J. Shapiro, Boston, MA, Beacon Press.

——(1972), *One Dimensional Man*, London, Abacus.

——(1987), *Eros and Civilisation: A Philosophical Inquiry into Freud*, London, Ark.

Margulis, L. (1981), *Symbiosis in Cell Evolution*, San Francisco, W. H. Freeman.

Margulis, L. and Fester, R. (1991), *Symbiosis as a Source of Evolutionary Innovation: Speciation and Morphogenesis*, Cambridge, MA, MIT Press.

Margulis, L. and Sagan, D. (1986), *Microcosmos. Four Billion years of Evolution from our Microbial Ancestors*, New York, Summit Books.

——(1995), *What Is Life?*, London, Weidenfeld & Nicolson.

Marx, K. (1982/1975), 'Okonomisch-philosophische Manusckripte' (1st and 2nd edns), in *Marx/Engels Gesamtausgabe* (Erste Abteilung, Band 2), 187–438; 'Economic and Philosophical Manuscripts of 1844' in *Marx: Early Writings*, trans. R. Livingstone and G. Benton, Harmondsworth, Middx., Penguin, 279–401.

——(1973), *Grundrisse*, trans. M. Nicolaus, Harmondsworth, Middx., Penguin.

——(1990), *Capital: A Critique of Political Economy* (Vol. 1), trans. B. Fowkes, Harmondsworth, Middx., Penguin.

Maturana, H. and Varela, F. (1980), *Autopoiesis and Cognition: The Realization of the Living*, Dordrecht, D. Reidel.

——(1992), *The Tree of Knowledge: The Biological Roots of Human Understanding*, Boston and London, Shambhala.

Maynard Smith, J. (1993), *The Theory of Evolution*, Cambridge, Cambridge University Press.

Mayr, E. (1991), *One Long Argument: Charles Darwin and the Genesis of Modern Evolutionary Thought*, Harmondsworth, Middx., Penguin.

——(1994), 'Typological versus Population Thinking', in E. Sober (ed.), *Conceptual Issues in Evolutionary Biology*, Cambridge, MA, MIT Press, 157–60.

Merleau-Ponty, M. (1962), *Phenomenology of Perception*, trans. C. Smith, London, Routledge.

——(1963), *The Structure of Behaviour* (*La Structure du Comportment*), trans. A. L. Fisher, New York, Beacon Press.

——(1964), *The Primacy of Perception*, trans. J. M. Edie, Evanston, IL, Northwestern University Press.

——(1964b), *Sense and Non-sense*, trans. H. L. Dreyfus and P. A. Dreyfus, Evanston, IL, Northwestern University Press.

——(1964c), *Signs*, trans. R. C. McCleary, Evanston, IL, Northwestern University Press.

——(1978), *The Visible and the Invisible (followed by Working Notes)*, trans. A. Lingis, Evanston, IL, Northwestern University Press.

——(1988), *In Praise of Philosophy and Other Essays*, trans. J. Wild *et. al.*, Evanston, IL, Northwestern University Press.

——(1994), *La Nature: Notes Cours du College de France*, Paris, Éditions du Seuil.

Millett, N. (1997), 'The Trick of Singularity', *Theory, Culture and Society* (14: 2), 51–67.

Monod, J. (1971), *Chance and Necessity: An Essay on the Natural Philosophy of Modern Biology*, New York, Alfred A. Knopf.

Morton, P. (1984), *The Vital Science: Biology and the Literary Imagination 1860–1900*, London, Allen & Unwin.

Müller-Sievers, H. (1997), *Self-Generation: Biology, Philosophy and Literature around 1800*, Cambridge, Cambridge University Press.

Mumford, L. (1967), *The Myth of the Machine: Technics and Human Development*, London, Secker & Warburg.

Nietzsche, F. (1966), *Beyond Good and Evil*, trans. W. Kaufmann, New York, Random House.

——(1968), *The Will to Power*, trans. R. J. Hollingdale and W. Kaufmann, New York, Random House.

——(1969), *Thus Spoke Zarathustra*, trans. R. J. Hollingdale, Harmondsworth, Middx., Penguin.

——(1974), *The Gay Science*, trans. W. Kaufmann, New York, Random House.

——(1979), *Ecce Homo*, trans. R. J. Hollingdale, Harmondsworth, Middx., Penguin.

——(1979b), *Twilight of the Idols*, trans. R. J. Hollingdale, Harmondsworth, Middx., Penguin.

——(1986), *Human, All Too Human*, trans. R. J. Hollingdale, Cambridge, Cambridge University Press.

——(1987), *Nietzsche Sämtliche Werke: Kritische Studienausgabe* (in 15 vols), ed. G. Colli and M. Montinari, Berlin and New York, Walter de Gruyter.

——(1994), *On the Genealogy of Morality*, trans. C. Diethe, Cambridge, Cambridge University Press.

Osborn, H. F. (1918), *The Origin and Evolution of Life. On the Theory of Action, Reaction and Interaction of Energy*, New York, Charles Scribner's Sons.

Plato (1961), *The Collected Dialogues*, ed. E. Hamilton and H. Cairns, New Jersey, Princeton University Press.

Pollack, R. (1994), *Signs of Life: The Language and Meanings of DNA*, Harmondsworth, Middx., Viking Penguin.

Prigogine, I. and Stengers, I. (1985), *Order out of Chaos*, London, Flamingo.

Ray, C. (1991), *Time, Space, and Philosophy*, London, Routledge.

Rayner, A. D. M. (1997), *Degrees of Freedom: Living in Dynamic Boundaries*, London, Imperial College Press.

Reanney, D. (1995), *The Death of Forever: A New Future for Human Consciousness*, London, Souvenir Press.

Roque, A. J. (1985), 'Self-Organization: Kant's Concept of Teleology and Modern Chemistry', *Review of Metaphysics* 39 (Summer), 107–35.

Rose, G. (1992), *The Broken Middle*, Oxford, Basil Blackwell.

——(1996), *Mourning Becomes the Law. Philosophy and Representation*, Cambridge, Cambridge University Press.

Rosen, R. (1991), *Life Itself. A Comprehensive Inquiry into the Nature, Origin, and Fabrication of Life*, New York, Columbia University Press.

Rosset, C. (1993), *Joyful Cruelty. Toward a Philosophy of the Real*, ed. and trans. D. F. Bell, Oxford, Oxford University Press.

Ruse, M. (ed.) (1989), *What the Philosophy of Biology Is: Essays Dedicated to David Hull*, Dordrecht, Kluwer.

Russell, B. (1992), *The Philosophy of Leibniz*, London, Routledge.

Ruyer, R. (1946), *Éléments de Psycho-Biologie*, Paris, Presses Universitaires de France.

——(1988), 'There is no Subconscious: Embryogenesis and Memory', trans. R. Scott Walker, *Diogenes* 142 (Summer), 24–46.

Sagan, D. (1992), 'Metametazoa: Biology and Multiplicity', in J. Crary and S. Kwinter (eds), *Incorporations*, New York, Zone Books, 362–85.

Sapp, J. (1994), *Evolution by Association*, Oxford, Oxford University Press.

Sartre, J.-P. (1972), *The Transcendence of the Ego: An Existentialist Theory of Consciousness*, trans. F. Williams and R. Kirkpatrick, New York, Octagon Books.

——(1972b), 'The Singular Universal', in J. Thompson (ed.), *Kierkegaard: A Collection of Critical Essays*, New York, Anchor Books, 230–66.

——(1989), *Being and Nothingness*, trans. H. E. Barnes, London, Routledge.

Savage-Rumbaugh, S. and Lewin, R. (1994), *Kanzi. The Ape at the Brink of the Human Mind*, New York, Doubleday.

Schmidt, A. (1970), *The Concept of Nature in Marx*, London, New Left Books.

Schrodinger, E. (1992), *What is Life?*, Cambridge, Cambridge University Press.

Serres, M. (1975), *Feux et signaux de brume: Zola*, Paris, Grasset.

——(1977), *La naissance de la physique dans le texte de Lucrece*, Paris, Les Éditions de Minuit.

——(1983), 'Language and Space: From Oedipus to Zola', in Serres, *Hermes. Literature, Science, Philosophy*, ed. J. V. Harari and D. F. Bell, Baltimore, Johns Hopkins University Press, 39–54.

Simondon, G. (1995/1992), *L'individu et sa genèse physico-biologique*, Grenoble, Jerome Millon (originally published 1964); trans. 'Genesis of the Individual' (Introduction only), in J. Crary and S. Kwinter, *Incorporations*, New York, Zone Books, 296–320.

Singer, C. (1950), *A History of Biology. A General Introduction to the Study of Living Things*, London, H. K. Lewis & Co.

Smolin, L. (1997), *The Life of the Cosmos*, London, Weidenfeld & Nicolson.

Sober, E. (1994), 'Evolution, Population Thinking, and Essentialism', in E. Sober (ed.), *Conceptual Issues in Evolutionary Biology*, Cambridge, MA, MIT Press, 161–91.

Spinoza, B. (1955), *The Ethics* (with *Correspondence*), trans. R. H. M. Elwes, New York, Dover Press.

——(1992), *Ethics/Treatise on the Emendation of the Intellect/Selected Letters*, trans. S. Shirley, Indianapolis, Hackett.

Stengers, I. (1997), *Power and Invention: Situating Science*, trans. P. Bains, Minneapolis, University of Minnesota Press.

Stonier, T. (1992), *Beyond Information: The Natural History of Intelligence*, London, Springer Verlag.

Taylor Parker, S. and Baars, B. (1990), 'How Scientific Usages Reflect Implicit Theories: Adaptation, Development, Instinct, Learning, Cognition, and Intelligence', in S. Taylor Parker and K. R. Gibson (eds), *'Language' and Intelligence in Monkeys and Apes*, Cambridge, Cambridge University Press, 65–97.

Thorpe, W. H. (1979), *The Origins and Rise of Ethology*, London, Heinemann.

Tinbergen, N. (1969), *The Study of Instinct*, Oxford, Clarendon Press.

Tomlin, E. W. F. (1955), *Living and Knowing*, London, Faber & Faber.

Toulmin, S. (1982), *The Return to Cosmology: Postmodern Science and the Theology of Nature*, Berkeley, University of California Press.

Van Gelder, T. and Port, R. S. (1996), 'It's About Time: An Overview of the Dynamical Approach to Cognition', in H. Geirsson and M. Losonsky (eds), *Readings in Language and Mind*, Oxford, Basil Blackwell, 326–53.

Varela, F. *et al.* (1995), *The Embodied Mind: Cognitive Science and the Human Experience*, Cambridge, MA, MIT Press.

Vernadsky, V. (1998) (originally published 1926), *The Biosphere*, trans. D. B. Langmuir, New York, Copernicus.

von Baer, K. (1973), 'The Controversy over Darwinism', in D. L. Hull, *Darwin and his Critics*, Cambridge, MA, Harvard University Press, 416–27.

von Bertalanffy, L. (1968), *General System Theory*, London, Allen Lane.

von Goethe, J. W. (1988), *Scientific Studies* (Vol. 12 of *The Collected Works*), ed. and trans. D. Miller, Princeton, NJ, Princeton University Press.

von Schelling, F. W. J. (1994), *On the History of Modern Philosophy*, trans. A. Bowie, Cambridge, Cambridge University Press.

——(1995), *Ideas for a Philosophy of Nature*, trans. E. E. Harris and P. Heath, Cambridge, Cambridge University Press.

von Uexküll, J. (1926), *Theoretical Biology*, trans. D. L. Mackinnon, London, Kegan Paul.

——(1934/1992), 'A Stroll Through the Worlds of Animals and Men: A Picture Book of Invisible Worlds', *Semiotica* (89: 4), 319–91.

von Uexküll, Thure (1992), 'The sign theory of Jakob von Uexküll', *Semiotica* (89: 4), 279–315.

Watson, J. (1970), *The Double Helix: A Personal Account of the Discovery of the structure of DNA*, Harmondsworth, Middx., Penguin.

Weismann, A. (1882), *Studies in the Theory of Descent* (2 vols), trans. R. Medola, with a prefatory note by Charles Darwin, London, Sampson Low, Marston, Searle, & Rivington.

——(1892/1893), *Das Keimplasma. Eine Theorie der Vererbung*, Jena, Gustav Fischer; *The Germ-Plasm. A Theory of Heredity*, trans. W. Newton Parker and H. Ronnfeldt, London, Walter Scott.

——(1909), 'The Selection Theory', in A. C. Seward, *Darwin and Modern Science: Essays in Commemoration of the Centenary of the Birth of Charles Darwin and of the Fiftieth Anniversary of the Publication of 'The Origin of Species'*, Cambridge, Cambridge University Press, 18–66.

Wesson, R. (1991), *Beyond Natural Selection*, Cambridge, MA, MIT Press.

Whitehead, A. N. (1947), *Adventures of Ideas*, Cambridge, Cambridge University Press.

——(1978), *Process and Reality: An Essay in Cosmology*, (from the Gifford Lectures of 1927–8), Corrected Edition, ed. D. R. Griffin and D. W. Sherburne, London, The Free Press.

Wicken, J. S. (1987), *Evolution, Thermodynamics, and Information: Extending the Darwinian Paradigm*, Oxford, Oxford University Press.

——(1988), 'Thermodynamics, Evolution, and Emergence: Ingredients for a New Synthesis', in B. H. Weber, D. J. Depew, and J. D. Smith, *Entropy, Information, and Evolution: New Perspectives on Physical and Biological Evolution,* Cambridge, MA, MIT Press, 139–73.

Woolhouse, R. S. (1993), *Descartes, Spinoza, Leibniz. The Concept of Substance in Seventeenth-Century Metaphysics,* London, Routledge.

Worringer, W. (1927), *Form in Gothic,* authorized translation by H. Read, London, G. P. Putnam's Sons Ltd.

Wright, R. (1994), *The Moral Animal: The New Science of Evolutionary Psychology,* London, Little, Brown, & Co.

Yovel, Y. (1992), *Spinoza and Other Heretics: The Adventures of Immanence,* Princeton, NJ, Princeton University Press.

Zizek, S. (1989), *The Sublime Object of Ideology,* London, Verso.

——(1991), *For They Know Not What They Do: Enjoyment as a Political Factor,* London, Verso.

——(1997), *The Plague of Fantasies,* London, Verso.

Works in literature

Artaud, A. (1965), *Anthology,* ed. J. Hirschman, San Francisco, City Light Books.

——(1993), *The Theatre and Its Double,* trans. V. Corti, London, Calder.

Butler, S. (1914), 'Darwin Among the Machines', in Butler, *A First Year in Canterbury Settlement,* London, A. C. Fifield.

——(1985), *Erewhon,* Harmondsworth, Middx., Penguin.

Crane, S. (1994), *The Red Badge of Courage,* Middx., Penguin.

Dickinson, E. (1969), *A Choice of Emily Dickinson's Verse,* selected with an Introduction by T. Hughes, London, Faber & Faber.

Fitzgerald, F. Scott (1965), *The Crack-Up with other Pieces and Stories,* Harmondsworth, Middx., Penguin.

Hardy, T. (1981), *Tess of the d'Urbervilles: A Pure Woman,* London, Macmillan.

Kafka, F. (1961), *Metamorphosis and Other Stories,* trans. W. and E. Muir, Harmondsworth, Middx., Penguin.

Kundera, M. (1991), *Immortality,* trans. P. Kussi, London, Faber & Faber.

Lawrence, D. H. (1961), *Fantasia of the Unconscious and Psychoanalysis of the Unconscious,* London, Heinemann.

——(1964), *Studies in Classic American Literature,* London, Heinemann.

——(1985), *Study of Thomas Hardy and Other Essays,* ed. B. Steele, Cambridge, Cambridge University Press.

——(1986), *Mornings in Mexico,* Harmondsworth, Middx., Penguin.

——(1988), *Reflections on the Death of a Porcupine and other Essays,* ed. M. Herbert, Cambridge, Cambridge University Press.

——(1993), *Complete Poems,* Harmondsworth, Middx., Penguin.

——(1994), *Sons and Lovers,* Harmondsworth, Middx., Penguin.

——(1995), *Apocalypse,* ed. M. Kalnins, Harmondsworth, Middx., Penguin (based on the Cambridge University Press edition).

——(1995b), *The Rainbow,* ed. M. Kinkead-Weekes, Harmondsworth, Middx., Penguin (based on the Cambridge University Press edition).

——(1995c), *Women in Love,* ed. D. Farmer, L. Vasey and J. Worthen, Harmondsworth, Middx., Penguin (based on the Cambridge University Press edition).

——(1997), *Kangaroo*, ed. B. Steele, Harmondsworth, Middx., Penguin (based on the Cambridge University Press edition).

Zola, E. (1954), *Germinal*, Harmondsworth, Middx., Penguin.

——(1962), *Thérèse Raquin*, Harmondsworth, Middx., Penguin.

——(1995), *Doctor Pascal*, trans. E. A. Vizetelly, Stroud, Alan Sutton Publishing.

——(1996), *La Bête Humaine*, trans. R. Pearson, Oxford, Oxford University Press.

INDEX

261

Printed in the United States
by Baker & Taylor Publisher Services